# Statistical Physics
# An Entropic Approach

# Statistical Physics

## An Entropic Approach

IAN FORD

*Department of Physics and Astronomy,*
*University College London, UK*

WILEY

A John Wiley & Sons, Ltd., Publication

This edition first published 2013


*Registered office*
John Wiley & Sons Ltd, The Atrium, Southern Gate, Chichester, West Sussex, PO19 8SQ, United Kingdom

For details of our global editorial offices, for customer services and for information about how to apply for permission to reuse the copyright material in this book please see our website at www.wiley.com.



*Library of Congress Cataloging-in-Publication Data*
Ford, Ian, 1962-
Statistical physics : an entropic approach / Ian Ford.
pages cm
Includes index.
ISBN 978-1-119-97531-1 (hardback) – ISBN 978-1-119-97530-4 (paper) 1. Statistical physics–Textbooks. I. Title.
QC311.5.F67 2013
536′.7015195–dc23

2012051177

A catalogue record for this book is available from the British Library.

HB ISBN: 9781119975311
PB ISBN: 9781119975304

Set in 10/12pt Times by Laserwords Private Limited, Chennai, India

With love to Helen, Alex and Jonathan; to my father Derek, and in memory of my mother, Phyllis.

# Contents

# Preface

I think I have been more confused about the nature of entropy than almost anything else I've encountered in physics. I remember I was initially mystified by the analysis of static forces, and again by the concept of the Green's function: but entropy still causes me to ask myself: do I really understand this? And I don't think I'm alone.

For me, the solution to this unease was to teach statistical physics and to fix firmly in my mind what message I was to deliver. There were several possibilities. Was I to adhere to the information theoretic point of view that so appealed to me as an undergraduate, or was I to focus instead on the central role of dynamics, whether deterministic or stochastic? Which of the entropies of Boltzmann or Gibbs should I present as more fundamental? But these are fairly refined matters, and the message had to address deeper issues. Students would inevitably ask 'what is entropy?', and I realised that I needed to have a simple answer, and that the word 'disorder' was not going to do.

This book takes a look at statistical thermodynamics with the question 'what is entropy?' very much to the fore. I want to show that up to a point, entropy is actually rather ordinary. It is a property of matter, if a little less familiar than energy, pressure and density, but connected to them all through the relationships of classical thermodynamics. We can measure it with relatively simple equipment such as a thermometer and a source of heat.

Having established this, the change in the entropy of participants in thermodynamic processes can be discussed, and then we encounter the not-so-ordinary concept of the generation of entropy 'out of nothing'. So we then develop statistical mechanics to try to find a microscopic view of what this quantity might represent, and to explain the classical laws of thermodynamics. Along the way, we build a powerful understanding of the properties of condensed matter, the traditional realm of application of these statistical ideas. But still, what is entropy?

The answer is uncertainty: entropy is uncertainty commodified. At least this is the interpretation that makes best sense to me. We do not or cannot measure all the details of the present state of the world and so when processes occur, we are not quite sure what will happen, even if we believe that we understand the physics at the microscopic level. Our certainty about the future is less than our certainty about the present, unless we are dealing with very special systems. This is a matter of intuition and needs to be accommodated in any incomplete model of evolution at the macroscopic scale. The increased uncertainty is quantified as an increase in the total entropy of the world, and that is what entropy is. The most remarkable thing is that we can measure uncertainty with a thermometer.

But maybe it is not as straightforward as that? Entropy and the second law of thermodynamics have been subjects of lengthy discussion over the years. The fact that a supposedly basic law of Nature has received repeated attention and fomented disagreements for decades, while other laws have been happily absorbed without dissent, can indicate several things. The most positive conclusion is that the issue is multifaceted, making it really important and interesting, and well worth the effort of trying to understand it. A less encouraging conclusion is that perhaps people are discussing quite different matters, and this has led to confusion. The word entropy has been applied to many technical and nontechnical concepts, and we have to be careful what we are saying. The property has been discussed in quite abstract and philosophical terms, as well as in terms of the hard thermodynamics of the laboratory. The often-quoted advice of von Neumann to Shannon, to name his proposed information measure 'entropy' on the grounds that nobody quite knew what entropy was, illustrates the situation perfectly. A great deal has been written about the matter, including some that I have not found helpful, and this has done nothing to dispel my feeling of unease.

Anyway, it is my sincere hope that the interpretation presented here will not be viewed as unhelpful. I want to provide a treatment that appeals to intuition without leaving too many loose ends in the mathematics, employing detailed examples to reinforce the somewhat dry concepts. The book comprises a treatment of classical thermodynamics, with the focus particularly on the role played by entropy, and the development of equilibrium statistical thermodynamics, all suitable for a second year undergraduate course. Later on, I provide a discussion of nonequilibrium statistical physics, in a manner intended to secure the idea of entropy as a measure of uncertainty. The dynamics of probability, and its application to Brownian motion, are included as lines of development. Towards the very end, I discuss fluctuation relations, which seem to me to provide insight into the behaviour of thermodynamic systems away from equilibrium, and into the very process of entropy generation, since they establish a link with dynamics.

Nevertheless, the book is quite definitely intended for undergraduates. I assume familiarity with elementary ideas of thermal behaviour from an introductory course on the properties of matter, as well as exposure to suitable mathematics and the principles of quantum mechanics. Some material will be challenging at this level: hence an entropy hazard warning sign will appear in a few places! It is a *short* book, and obviously has associated deficiencies in the level of detail, particularly in the coverage of experimental support for some of the models. It is the focus on the nature of entropy that I hope will set it apart from the many other introductory books available on the subject of statistical physics, some of which I refer to in Further Reading. Otherwise, the reader might question the need for yet another treatment! On the other hand, I wrote this book to alleviate the personal unease I felt towards the concept of entropy, and to reach a position that I felt could be taught and defended; whether anyone else can find value in the undertaking is, of course, a huge bonus.

Caution: Entropy

I would like to express my gratitude to colleagues and students at UCL and elsewhere who have stimulated my thoughts on these topics or have offered encouragement and advice, in particular Richard Spinney, Brian Cowan, Rainer Klages, Rosemary Harris, Paul Tangney and Veronika Brázdová. I thank Roy Axell for introducing me to entropy all those years ago and I am grateful to the people at Wiley for this opportunity.

Ian Ford
UCL, December 2012

Instructors can access PowerPoint files of the illustrations presented within this text, for teaching, at http://booksupport.wiley.com.

# 1 Disorder or Uncertainty?

This book is not a novel, and I think it is acceptable to give away the plot at the very outset. Entropy is a thermal property of matter, and when real (as opposed to idealised) macroscopic changes take place in the world, the total amount of entropy goes up. This is the celebrated second law of thermodynamics, so celebrated, in fact, that saying 'the second law' alone is often enough to convey which field it relates to. It is due to the efforts of Ludwig Boltzmann (1844–1906) and Josiah Willard Gibbs (1839–1903) that we now connect thermodynamic entropy with statistical ideas; with the uncertainty that prevails in the microscopic state of the world if we have only limited information about it. The growth of entropy when constraints on a system are removed, to initiate change, is a consequence of an increase in this uncertainty: the number of possibilities for the microscopic state goes up, and so does the entropy.

It is often said that the rise in entropy is related to the natural tendency for disorder to increase, and while this can sometimes help to develop intuition, it can be misleading. The atoms of a crystalline solid held within a thermally insulated box have evidently chosen to arrange themselves as a regular lattice. They might instead have arranged themselves as a liquid with the same total energy, but at a lower temperature since some of the kinetic energy would need to be converted into potential energy in order to melt the solid. But they did not. Nature sometimes has a preference for spatially ordered instead of disordered systems: if we set up the system in the molten state, the material would spontaneously freeze.

A better interpretation is that the spatially ordered arrangement of atoms in the solid has a larger number of underlying microstates than the cooler, but spatially disordered fluid. The disorder in atomic velocities is larger at the higher temperature (and even here I would rather say the *uncertainty* in velocities is larger) and this gives a greater overall uncertainty surrounding the actual microstate of the system, when in equilibrium, if the atoms are arranged as a solid. The selection rule imposed by Nature for the choice of macrostate is to maximise the uncertainty.

An uncertain situation might convey the idea of disorder or untidiness, but we need to take care when we build analogies between entropy and untidy situations. My desk is very disordered, but this does not mean that it has more entropy than it would have if I were

*Statistical Physics: An Entropic Approach*, First Edition. Ian Ford.

to tidy it. A disordered desk and a tidy desk are just two particular arrangements of the system. But if I defined the term 'untidy' to encompass a certain set of arrangements of items on my desk, while another, much smaller, set of arrangements is classed as 'tidy', then I could start to make statistical statements about the likely condition (tidy/untidy) of my desk in the future, as long as I had a model of how the arrangement of items changed from day to day, as a result of my usual activities. I could define 'tidy' such that the fraction of desk area showing through the jumble is greater than 75%, say. Then a tidy desktop (few configurations, lots of desk showing) would most likely develop into an untidy desktop (many configurations, less desk showing) as the days (or even minutes!) passed. An untidy desk would probably remain untidy, though its evolution into a tidy desk is not beyond all expectation.

But this is as far as ideas concerning the loss of order and gain of untidiness should be taken. A key point is that we could start the process with everything scattered randomly over the desk. This is not a tidy or an ordered initial condition. It is, on the other hand, a *definite* initial condition, with no uncertainty attached to it. If entropy is uncertainty, then a definite initial state has the same (zero) amount of entropy whether it is tidy or untidy, ordered or disordered. It is the certainty in configuration that is lost if we fail to follow the details of the desktop dynamics as time progresses, not the tidiness or the order. The rise in this uncertainty is equivalent to the increase in entropy.

As an extension to this reasoning, the initial condition might be that the system is in one of a certain number of configurations, perhaps similar to one another, but perhaps completely different: an arbitrary collection of my favourite desktop arrangements. Such a slightly indefinite initial state would evolve into a more indefinite state: a low but nonzero entropy situation evolves into one with a higher entropy. This is a more sophisticated description of the evolution of a complex system than a picture of order turning into disorder. This is the meaning of the second law.

Really, discussions of desks or even rooms becoming untidy should include shutting the door to the room (and maybe putting up an entropy hazard warning sign!). We leave the occupant to rearrange things according to his or her wishes. The configuration of the room changes with time and, from the other side of the door, we do not know exactly how it proceeds. All we can do is occasionally ask the occupant for some information that does not specify the exact arrangement, but instead is more generic, such as how much desk is showing. Our knowledge about the state of affairs inside the room is steadily impaired, and eventually goes to a minimum, based on what we can discover remotely.

This is how we interrogate a macroscopic system, allowing us to close in on the meaning of thermodynamic entropy. The macroscopic equilibrium state of a gas is described by a measurable density and temperature, but this is insufficient to specify its exact microscopic state, which would be a list of the locations and velocities of all the atoms, at least from a classical physics perspective. This is an occasion when admitting 'I do not know what is going on' is extremely profound. Thermodynamic entropy is a measure of this uncertainty: it is proportional to the logarithm of the number of microscopic configurations compatible with the available measurements or information. We can categorise those configurations into different classes, such as 'gas concentrated in a corner' or 'gas spread out uniformly in the container', and then estimate the likelihood that the system might be found in each class, as long as probabilities for each microscopic configuration are provided. We choose these probabilities on the basis of what we might know about the dynamics or by sophisticated 'best judgement'.

For an isolated system in equilibrium, equal probabilities for all configurations are often assumed, which is perhaps an oversimplification, but it implies that the system is most likely to be found in the macroscopic class that possesses the greatest number of configurations. If the system were disturbed by the release of some constraint (say a change in confining volume), it would eventually find a new equilibrium, and again take the class with the most microscopic states. In equilibrium, the macroscopic state with the greatest uncertainty is chosen. In this way, an arrow of macroscopic change (or of time, loosely) emerges and it is characterised by entropy increase.

It is sometimes said that the universe is falling to bits, or that everything is going wrong, but this a profoundly pessimistic view of the events that we attempt to describe with the second law. The statement that disorder is always on the increase carries the same gloomy view about the future. But does the interpretation that uncertainty is increasing offer anything more positive?

The evolution of the universe is a consequence of the rules of interaction between the component particles and fields, many of which we have determined in detail in the laboratory. These rules recognise no such thing as pessimism or decline. The universe is simply following a dynamical trajectory. But one of the core features of the dynamics is that transfers take place between participants in a way that seems to favour the sharing out of energy or space between them. The attributes of the universe are being mixed up in a manner that is hard to follow and our failure to grasp and retain the detail of all this is what is meant by the growth of uncertainty. However, we could interpret this failure as a reflection of the richness of the dynamics of the world and all its possibilities. We could perhaps view the second law more positively as a statement about the extraordinary complexity and promise that the universe can offer as it evolves.

The growth of entropy is our rationalisation of this complexity. We can explain the direction of macroscopic change, including events taking place in a test tube as well as processes occurring in the wider cosmos, on the basis of a simply stated and implemented rule of Nature. We can do this without having to delve too deeply into the microscopic laws: it seems that in certain important ways they all have a similar effect. The second law is a reflection of an underlying imperative to mix, share and explore, such that certain macroscopic events happen frequently, because they are nearly inevitable under such circumstances, while others occur more rarely.

So if we wish to ascribe a motivation to the workings of the universe, instead of arguing that the natural direction of change is towards disorder and destruction, we might regard the dynamics as essentially egalitarian and, as an indirect consequence, potentially benevolent. Particles of a gas with more than their fair share of energy naturally tend to pass some to their slower neighbours. Energy will flow, but this does not mean that the exceptional cannot arise. The toolbox of physical processes available to the world is so well stocked that the flow can be partly intercepted and put to use in building and maintaining complex structures. Nature will find opportunities to feed off energy flows in extraordinary ways: mixing and sharing seem to have the capacity to build as well as to dissipate, at least until the mixing is completed. These are themes that are worth developing.

# 2

# Classical Thermodynamics

Our main focus is statistical thermodynamics, but it is important to consider first the discipline of classical thermodynamics since it provides a framework and back-story to the main developments in the book. In this chapter, we describe the basic rules with special consideration given to the role of entropy, and in the next, we enlarge on some of the applications. The discussion of statistical thermodynamics starts in Chapter 4.

## 2.1 The Classical Laws of Thermodynamics

Thermodynamics emerged from the empirical science of calorimetry, the measurement of the generation and transfer of thermal energy, or heat, and from the development of technology to extract mechanical energy, or work, from a heat source. It was then extended to include consideration of the properties of matter and transformations between phases such as solids, liquids and gases. It is a theory of the macroscopic transfer of heat and mass, events that are known as thermodynamic processes. Strictly the focus of the theory is on systems that are in thermal equilibrium, the situation reached when all the processes have ceased. It is summed up in the four classical laws of thermodynamics, which are statements of empirical observation:

**Zeroth law** If two systems are in thermal equilibrium with a third system, then they are in equilibrium with each other; in fact there is a single system property (called temperature) that serves to indicate whether systems are in thermal equilibrium;

**First law** There is a system property called energy that is conserved, but can take several different forms that interconvert;

**Second law** There is a system property called entropy that, if the system is isolated from its environment, either increases or (in principle) remains constant during thermodynamic processes;

**Third law** The entropy of a system is a universal constant (set to zero) at the absolute zero of temperature.

*Statistical Physics: An Entropic Approach*, First Edition. Ian Ford.

Entropy appears in two of these laws, and is a central concept in thermodynamics. It has acquired a reputation for being hard to understand, and for this reason, entropy will be the focus of the discussion of classical thermodynamics in this chapter. Energy is a much more familiar concept: we buy it, we 'use' it and we read about it on food packaging, but it is possible to develop some intuition for entropy as well.

In the early development of classical thermodynamics, there was little fundamental understanding of what entropy actually represented. This situation was transformed when Boltzmann and Gibbs (and others) invented statistical mechanics towards the end of the nineteenth century, although there are still controversies to this day. To repeat the claim made in the previous chapter, it can be understood to represent *uncertainty*, and, in a limited sense, *disorder* – a lack of information about the detail of a system. Its evolution has been associated with the winding down of the universe after the initial impetus of the Big Bang. Philosophers suggest that it plays a role in our perception of the directionality of time. A startling set of notions to emerge from the simple science of calorimetry and the technology of the steam engine!

We shall see in later chapters what the fundamental insight of statistical mechanics was, and understand why it is written, in mathematical notation, on Boltzmann's gravestone. However, we can get a general feel for entropy by studying a simple example, before extending to more general systems. An example can also provide us with a grounding in the sometimes confusing concepts of heat, work and energy. We shall consider the ideal, monatomic, classical gas or ideal gas for short.

## 2.2 Macroscopic State Variables and Thermodynamic Processes

The statement of the first law of thermodynamics conveys to us something of the nature of thermodynamics and the phenomenology of classical thermodynamic processes. It concerns the conservation and interconversion of *energy*, an example of a parameter, variable or function of state (a quantity that specifies the macroscopic condition of a physical system when it is in equilibrium). Other examples include pressure, temperature and volume: all measurable and familiar in macroscopic physics. We shall call them *state variables*. They describe the equilibrium condition of a system without reference to any previous history.

By equilibrium, we mean that there is no time dependence in the condition of a system, which includes the absence of fluxes of energy or matter through the system. A thermodynamic process can be a transfer of energy or matter into or through a system, or some internal change such as freezing, often brought about by a change in one of the constraints imposed on it, that has the ultimate effect of altering one or more of the state variables of the system.

There are two types of macroscopic state variable. There are those that are proportional to the amount of material in the system, such as energy, that we called *extensive*, and those such as temperature that do not change if we replicate a system to make a larger one: these we call *intensive*. Further examples are given in Figure 2.1, which also sketches the 'world-view' that we take in thermodynamics. According to this view, we focus our attention on the behaviour of a system, which could be a flask of helium, a lump of

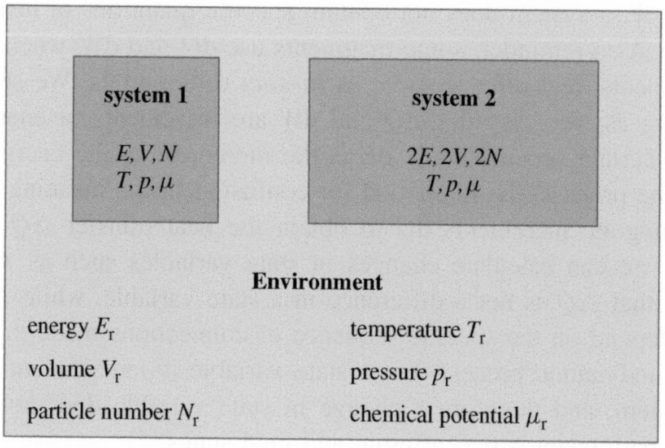


**Figure 2.1** The world-view according to thermodynamics. An environment is characterised by the macroscopic properties labelled by a suffix r. The one that might be unfamiliar is chemical potential, which we discuss later. Systems coupled to this environment are characterised by similar properties, shown here without a suffix. System 2 is simply two copies of system 1 joined together. Intensive state variables do not change under such replication, but extensive variables double. Furthermore, when in equilibrium, it is the intensive state variables of the system that normally equal those of the environment, for reasons that we shall come to later.

steel or a bottle of milk, and regard everything else as the environment, characterised by just a few parameters and an ability to exchange various quantities with the system. The environment is often assumed to be very large in extent compared with the system of interest.

In thermodynamics, attention is often given to the internal energy, defined to be the total energy of a system minus any bulk translational or rotational kinetic energy, and minus the potential energy due to any externally imposed fields, such as gravitational energy. It therefore comprises just the kinetic and potential energy of internal motion or interactions. We shall find it more convenient, however, to work with the sum of the internal energy and any externally imposed potential energy. We shall use the symbol $E$ to represent this energy.

Energy conservation is a rather fundamental principle in physics, and the energy of a system may therefore be changed only by transfers from the environment brought about by heat flow, for example, or by distorting it using a mechanical force and thereby performing work. It helps perhaps to regard these as transfers of kinetic and potential energy, respectively. Work is an energy transfer brought about by the application of an external force of some kind. It corresponds to a transfer of potential energy from the environment, such as the fall of a weight under gravity to move the piston that compresses a gas. Heat transfer is an energy change brought about by passing molecular kinetic energy into a system, through collisions at an interface, for example. Then the first law states that the state variable $E$ can receive incremental contributions from the environment in the form of heat $dQ$ and work $dW$. We then write the first law of thermodynamics in the form $dE = dQ + dW$.

Since $dQ$ and $dW$ represent incremental changes in energy of the system associated with different transfer processes, they do not represent increments in (purported) state variables $Q$ and $W$: a system does not contain specific quantities of heat or work, only a certain energy. As a reminder, some treatments use $đQ$ and $đW$ when specifying heat and work increments, and refer to them as *inexact* differentials. We shall not use this notation. As long as we grasp that $\mathrm{d}Q$ and $\mathrm{d}W$ are increments of energy that specify the *course* of a certain process while $\mathrm{d}E$ is the increment in the energy state variable *resulting* from the process, the likelihood for confusion in the meaning is minimal. We can certainly integrate increments $\mathrm{d}Q$ to obtain the heat transfer $\Delta Q = \int \mathrm{d}Q$ over a process, just as we can calculate changes in state variables such as $\Delta E = \int \mathrm{d}E$, but we always note that $\Delta Q$ is not a difference in a state variable, while $\Delta E$ is. The heat transfer might depend on the specific sequence of connections made to sources of heat during the thermodynamic process, but a state variable is independent of the previous history of a system, and therefore a change in state variable does not depend on the thermodynamic path taken between initial and final states.

It is worth pointing out that the conservation of energy embodied by the first law holds whether the initial and final states are in or out of equilibrium. However, most state variables in thermodynamics describe systems that are in equilibrium. For example, the state variable temperature, which is mentioned in the zeroth law of thermodynamics as an indicator of whether two systems are in thermal equilibrium, most definitely is an equilibrium property. Of course, we frequently apply the concept of temperature to a system when it is heating up or cooling down, and therefore out of equilibrium, but this view only really holds if the system is only mildly perturbed away from equilibrium, which means that heat flows should not be too large. In the same way, the state variables pressure and entropy are properly ascribed only to equilibrium states, but in certain circumstances, the concepts can be stretched to apply to nonequilibrium situations, which we return to briefly in Section 2.14 and again in Chapter 15.

## 2.3 Properties of the Ideal Classical Gas

We shall frequently use the monatomic ideal classical gas to illustrate aspects of thermodynamics. An ideal gas consists of particles that do not interact with each other, but only with the walls of the container in which they are confined. The equation of state of the ideal classical gas is

$$pV = NkT, \tag{2.1}$$

where $p$ is the pressure, $V$ is the volume, $N$ is the number of particles and $T$ is the temperature. This is also known as the ideal gas law. The pressure, volume and temperature of a gas characterise its equilibrium state, and satisfy the equation of state, irrespective of whether the state was established by compressing, expanding, heating or cooling a previous state. The remaining symbol in (2.1) is Boltzmann's constant $k$, which is numerically equal to $1.38 \times 10^{-23}$ $\mathrm{JK}^{-1}$. This equation involves the concepts of pressure and temperature; so even though they might be very familiar to us, we should consider the nature of these state variables and what they mean empirically.

Pressure is readily interpreted by picturing a gas as a collection of many particles, with a range of velocities such that they collide with each other and with the walls of the container. Pressure is the average normal force per unit area exerted on the wall

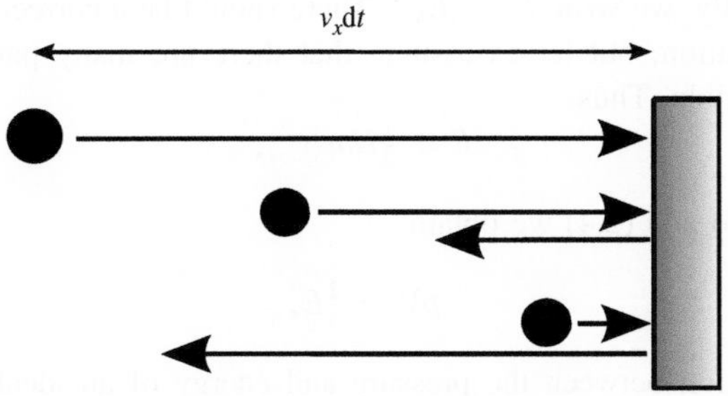


**Figure 2.2** All gas particles with a positive velocity $v_x$, and located within a distance $v_x$d$t$ of a wall, collide with it in a period of time d$t$, giving rise to momentum transfer and hence pressure.

and can be calculated in the following way. Imagine that a particle of mass $m$ and velocity in the $x$-direction $v_x$ collides with a wall and is reflected perfectly, as illustrated in Figure 2.2. The change in particle momentum in the $x$-direction is $-2mv_x$; so the momentum transferred to the wall is $2mv_x$. This is not the only particle that hits the wall in time d$t$. All particles with positive velocity $v_x$ and situated less than a distance $v_x \mathrm{d}t$ from the wall will make a collision; so the momentum change in the time interval is $2mv_x n(v_x)\mathrm{d}v_x \mathcal{A} v_x \mathrm{d}t$ where $n(v_x)\mathrm{d}v_x$ is the number of particles per unit volume with velocity between $v_x$ and $v_x + \mathrm{d}v_x$ and $\mathcal{A}$ is the wall area. The force on the wall is the rate of transfer of momentum to it, according to Newton's second law; so we divide this expression by d$t$ and sum over all positive velocity cohorts, giving a force

$$2m\mathcal{A}\int_0^\infty n(v_x)v_x^2\mathrm{d}v_x = \frac{2m\mathcal{A}n_+\int_0^\infty n(v_x)v_x^2\mathrm{d}v_x}{\int_0^\infty n(v_x)\mathrm{d}v_x} = 2m\mathcal{A}n_+\langle v_x^2\rangle, \qquad (2.2)$$

where $n_+ = \int_0^\infty n(v_x)\mathrm{d}v_x$ is the total number density of particles with positive $v_x$ (i.e. travelling towards the wall). In equilibrium, this is just $n/2$, where $n = N/V$ is the total particle number density.

The brackets in the final expression indicate an average defined with respect to the weighting $n(v_x)\mathrm{d}v_x/n_+$, which is essentially a probability that a particle should have an $x$-component of velocity in the region of width $\mathrm{d}v_x$ around $\mathrm{d}v_x$. Dividing by the wall area gives the pressure in the form $p = mn\langle v_x^2\rangle$. Then we note that in an equilibrium state, the mean square velocity components are time independent and statistically equivalent, as there is no mean flow, which implies that $\langle v_x^2\rangle = \langle v_y^2\rangle = \langle v_z^2\rangle = (1/3)\langle v^2\rangle$ where $v$ is the magnitude of the velocity, in which case we obtain

$$pV = \tfrac{1}{3}Nm\langle v^2\rangle. \qquad (2.3)$$

Next, we relate the energy of an monatomic ideal gas in equilibrium to the same statistical property $\langle v^2\rangle$ of the particle velocity distribution. The energy is entirely kinetic, as there are no interactions between the particles, making the potential energy zero. The mean energy of one atom is $\langle E_1\rangle = (1/2)m\langle v^2\rangle$ where $m$ is the atomic mass. The mean energy of $N$ atoms, assuming them to be statistically independent, is $\langle E_N\rangle =$

$(1/2)Nm\langle v^2\rangle$. Finally, we write $E = \langle E_N\rangle$. There should be a correction to exclude bulk translation and rotation, but let us assume that there are many particles and that the correction is negligible. Thus

$$E = \tfrac{1}{2}Nm\langle v^2\rangle, \tag{2.4}$$

and combining (2.3) and (2.4) we obtain

$$pV = \tfrac{2}{3}E, \tag{2.5}$$

which is a connection between the pressure and energy of an ideal gas confined to a container of volume $V$.

Together with (2.1), this implies that

$$E = \tfrac{3}{2}NkT, \tag{2.6}$$

for the monatomic ideal classical gas, which suggests that the state variable temperature appearing in the ideal gas law is a measure of the mean kinetic energy per particle of a system in thermal equilibrium. Temperature is also supposed to be an indicator of thermal equilibrium between systems and we can see how that operates in this example. Boyle's law, an empirical property of a rarefied gas, simply states that the product $pV$ does not change for an ideal gas after a compression or expansion, as long as the initial and final states are in thermal equilibrium with a given environment. It makes perfect sense that the ideal gas law should equate this product to the expression $NkT$ such that ideal gases with the same value of $T$ are in thermal equilibrium, or *isothermal*.

In this way, Boyle's law provides us with an empirical temperature scale. An ideal gas can act as a thermometer through the value of $pV$ it acquires when placed in thermal equilibrium with different environments. In order that the reading on the thermometer should not depend on how much ideal gas we use, we should make the temperature scale a function of the value of $pV/N$, which would be an intensive state variable. For convenience, the quantity $pV/Nk$ for an ideal gas in thermal equilibrium with an environment consisting of pure water at its triple point of equilibrium between the solid, liquid and gas phases is used to define a reference temperature of 273.16 K. If the ideal gas is brought into thermal contact with an environment that is not isothermal with the water triple point mixture, then the equilibrium value of $pV/Nk$ for the ideal gas will differ from 273.16 K and will serve to denote the temperature of the environment. We could have used the combination $pV/N$ as an empirical temperature measured in joules, but the benefit of retaining $k$ is that we can celebrate Boltzmann's contribution to gas physics by having a constant named after him.

## 2.4 Thermodynamic Processing of the Ideal Gas

Now we consider some simple thermodynamic processes involving the compression of an ideal gas, possibly with heat transfer between the system and an environment at a constant temperature. Arbitrary rates of heat transfer or compression will in general disturb the thermal equilibrium between a system and its environment, and strictly temperature can only be determined once equilibrium has been restored. But we can imagine that the process is conducted very slowly, such that thermal equilibrium between system and environment can be approximately maintained throughout, and the compression of the system then proceeds through a sequence of isothermal equilibrium states. Alternatively, the gas could be thermally isolated from the environment and slowly compressed through a sequence of equilibrium states each with a well-defined temperature. This idealisation of a thermodynamic process is often invoked, and is called a *quasistatic* process. Otherwise, a process is said to be *nonquasistatic*.

Let us consider the quasistatic performance of mechanical work that brings about the compression of a gas in a cylinder. The work done on the gas is simply the applied force $f$ times the distance moved by the piston $dx$. It is a transfer of potential energy from the environment. If the compression is slow, the gas is always well approximated by an equilibrium state, and we can assume that it exerts a uniform pressure $p$ against the piston head, of area $\mathcal{A}$, and that the force $p\mathcal{A}$ equals the applied external force $f$. We write $f\mathrm{d}x = -(f/\mathcal{A})(-\mathcal{A}\mathrm{d}x) = -p\mathrm{d}V$, where $\mathrm{d}V$ is the change in system volume (here negative) brought about by the compression.

If the piston were moved nonquasistatically, various complications would ensue: the pressure of the gas might not be uniform, the applied and resistive forces might not balance, and shock waves, convective motion or sound might be generated. All this makes such a process hard to analyse.

For a quasistatic compression, in contrast, and in the absence of heat transfer, we can simply state that $\mathrm{d}E = \mathrm{d}W = -p\mathrm{d}V$. We then proceed using (2.5):

$$\mathrm{d}E = \tfrac{3}{2}\mathrm{d}(pV) = \tfrac{3}{2}(p\mathrm{d}V + V\mathrm{d}p) = -p\mathrm{d}V, \tag{2.7}$$

so

$$\frac{5}{2}\frac{\mathrm{d}V}{V} = -\frac{3}{2}\frac{\mathrm{d}p}{p}, \tag{2.8}$$

such that $(5/2)\ln V = -(3/2)\ln p + \text{constant}$, and hence

$$pV^{\frac{5}{3}} = \text{constant}, \tag{2.9}$$

which describes the quasistatic, *adiabatic* (meaning thermally isolated) compression of a monatomic ideal classical gas on a $p - V$ diagram. It may be contrasted with the condition $pV = \text{constant}$ for quasistatic isothermal compression, associated with (2.1), where the temperature is held constant by allowing heat transfer between the system and the environment. By inserting the equation of state (2.1), the adiabatic compression can be represented on a $T - V$ diagram as $TV^{2/3} = \text{constant}$, indicating that the temperature rises during the compression. But is anything held constant? We shall see.

Consider next the change in energy due to a process of heat input, with the volume held constant so that no external work is done on the system, and once again with the number of particles in the system fixed. We now write the first law as $\mathrm{d}E = \mathrm{d}Q$. When heat is injected into a system, we expect its temperature to change since it is being disturbed from a previously isothermal state. It is of interest to calculate a heat capacity, defined as the amount of heat required to raise the temperature of the system by a specified amount. For the monatomic ideal classical gas, we have

$$\frac{\mathrm{d}Q}{\mathrm{d}T} = \frac{\mathrm{d}E}{\mathrm{d}T} = \frac{\mathrm{d}}{\mathrm{d}T}\left(\frac{3}{2}NkT\right) = \frac{3}{2}Nk, \tag{2.10}$$

and this is referred to as the heat capacity at constant volume, denoted $C_V$ and defined in general by

$$C_V = \left(\frac{\partial E}{\partial T}\right)_{N,V}. \tag{2.11}$$

If we relax the condition of constant volume and allow mechanical work $\mathrm{d}W$ to take place during the process, we write instead

$$\mathrm{d}Q = \mathrm{d}E - \mathrm{d}W. \tag{2.12}$$

Note that the convention employed throughout is that $\mathrm{d}Q$ and $\mathrm{d}W$ denote heat and work energy *given to* the system. Assuming the process is quasistatic, $\mathrm{d}W = -p\mathrm{d}V$ and the system temperature remains spatially uniform. If we divide $\mathrm{d}Q$ by the change in temperature that accompanies the process we get

$$\left(\frac{\mathrm{d}Q}{\mathrm{d}T}\right)_q = \frac{\mathrm{d}E}{\mathrm{d}T} + p\frac{\mathrm{d}V}{\mathrm{d}T}. \tag{2.13}$$

The label $q$ is there to emphasise that the heat input is made quasistatically. We do not add a quasistatic label to the derivatives of energy and volume with respect to temperature because all the quantities involved are state variables that characterise equilibrium states. By definition, the change in a state variable does not depend on the rate at which we conduct the process (as long as we let the final state come to equilibrium, of course). The incremental changes in energy, volume and temperature brought about by the process are independent of the history. In contrast, the delivery of heat is process specific, and only for a quasistatic process can the ratio $\mathrm{d}Q/\mathrm{d}T$ be related to the particular form in (2.13).

If we imagine that the delivery of heat to the system is brought about by thermal contact with an environment with a quasistatically increasing temperature $T_{\mathrm{r}}(t)$, then the temperature of the system $T$ will take on the same time dependence. The pressure evolves according to the equation of state, and remains spatially uniform, such that there are no convection currents, thermal or pressure gradients induced during the process, and the energy then changes according to (2.6), fully specifying the right hand side of (2.13). We conclude that if quasistatic work is performed while the temperature is raised, then the amount of heat drawn from the environment will be affected: in short, the heat capacity of the system will differ from the constant volume case.

So heat capacity is a generic term, and it depends on the conditions that are imposed. It is usual to consider a particular case where external work is performed at constant pressure. With $p$ held constant during the process, the last term in (2.13) becomes $\mathrm{d}(pV)/\mathrm{d}T = \mathrm{d}(NkT)/\mathrm{d}T = Nk$, and the heat capacity, at constant pressure, of the monatomic ideal classical gas is then

$$C_p = \tfrac{5}{2}Nk. \tag{2.14}$$

Note that $C_p/C_V = 5/3$. This ratio is denoted $\gamma$ for more general systems that we shall consider in Section 3.5. Also note that this ratio is the same as the exponent in (2.9) describing adiabatic compression. It turns out that this is no coincidence.

## 2.5 Entropy of the Ideal Gas

We now consider the thermodynamic processing of a monatomic ideal classical gas that leads to the concept of thermodynamic entropy. We transfer heat quasistatically from an environment into a system and consider the rather innocuous looking quantity $(dQ/T)_q$ that characterises an incremental stage in such a process. It is the heat transfer to a system modulated by its temperature, and the latter is equal to the temperature of the environment, since the change is quasistatic. The environmental temperature is assumed to change slowly and produce a consequent slow evolution in the properties of the system. Using the first law, we can write

$$\left(\frac{\mathrm{d}Q}{T}\right)_q = \frac{\mathrm{d}E + p\mathrm{d}V}{T}, \tag{2.15}$$

and therefore, using the energy–pressure relation,

$$\left(\frac{\mathrm{d}Q}{T}\right)_q = \frac{3}{2}\frac{\mathrm{d}(pV)}{T} + p\frac{\mathrm{d}V}{T} = \frac{5}{2}p\frac{\mathrm{d}V}{T} + \frac{3}{2}V\frac{\mathrm{d}p}{T} = \frac{5}{2}Nk\frac{\mathrm{d}V}{V} + \frac{3}{2}Nk\frac{\mathrm{d}p}{p}. \tag{2.16}$$

If we sum such incremental changes over a complete quasistatic heat transfer process from equilibrium state $a$ to equilibrium state $b$, we get

$$\int_a^b \left(\frac{\mathrm{d}Q}{T}\right)_q = \left[\frac{5}{2}Nk\ln V + \frac{3}{2}Nk\ln p\right]_a^b = \left[\frac{3}{2}Nk\ln\left(pV^{\frac{5}{3}}\right)\right]_a^b. \tag{2.17}$$

The left hand side is a measurable quantity: the heat transferred to the system during a quasistatic process modulated by the changing inverse system temperature as the process takes place. The right hand side can be expressed in terms of the initial and final pressure and volume of the gas $(p_a, V_a)$ and $(p_b, V_b)$, noting that $N$ remains the same, and we write

$$\int_a^b \left(\frac{\mathrm{d}Q}{T}\right)_q = Nk\ln\left(\frac{p_b^{\frac{3}{2}}V_b^{\frac{5}{2}}}{p_a^{\frac{3}{2}}V_a^{\frac{5}{2}}}\right) = S(p_b, V_b, N) - S(p_a, V_a, N) = \int_a^b \mathrm{d}S, \tag{2.18}$$

where we have defined a property of the monatomic ideal classical gas called *entropy*, a name coined in 1865 by Rudolf Clausius (1822–1888), that takes the form

$$S(p, V, N) = Nk\ln\left(\frac{p^{\frac{3}{2}}V^{\frac{5}{2}}}{C(N)}\right). \tag{2.19}$$

The entropy $S$ of a monatomic ideal classical gas is a state variable, since it is a function of state variables pressure and volume, and we shall see that it turns out to have some rather special properties. The quantity $C$ in the denominator is included to make the argument of the logarithm dimensionless. It can, in principle, depend on system properties that do not change as a result of the process, and the only one in this case is $N$, the number of particles, and so we write $C(N)$.

Note that in order to raise the temperature of twice the amount of gas, we would need twice the amount of heat. The change in entropy of the gas resulting from the process of quasistatic heat transfer $\mathrm{d}S = (\mathrm{d}Q/T)_q$ is therefore also proportional to the amount of material; so entropy is extensive, like the volume, energy, or the number of particles itself, while pressure and temperature are intensive. Thus we require

$$S(p, 2V, 2N) = 2S(p, V, N) \quad \Rightarrow 2Nk \ln\left(\frac{p^{\frac{3}{2}}(2V)^{\frac{5}{2}}}{C(2N)}\right) = 2Nk \ln\left(\frac{p^{\frac{3}{2}}V^{\frac{5}{2}}}{C(N)}\right), \tag{2.20}$$

and we deduce that $C$ scales in a particular way with system size: we write $C = \hat{c}N^{\frac{5}{2}}$ where $\hat{c}$ is now independent of the state variables $p$, $T$, $N$ and $V$.

There are alternative forms for the entropy of the ideal gas, obtained by inserting the relation between pressure and energy:

$$S(E, V, N) = Nk \ln\left(\frac{V(\frac{2}{3}E)^{\frac{3}{2}}}{\hat{c}\,N^{\frac{5}{2}}}\right), \tag{2.21}$$

or by inserting the equation of state

$$S(T, V, N) = Nk \ln\left(\frac{(kT)^{\frac{3}{2}}}{\hat{c}N/V}\right). \tag{2.22}$$

The entropy per particle of the gas clearly increases as the temperature increases or as the density $n = N/V$ decreases, and we start to acquire some intuition about the behaviour of this new state variable.

So entropy is a property of the gas. It is not some vague, hard-to-understand concept: it can be written as a perfectly well-defined function of state variables. We have focussed on the ideal gas example in order to demonstrate this explicitly. For more complicated systems, such as a liquid or a solid, it is not always so easy to derive a mathematical expression for entropy but nevertheless it can be calculated, in principle, through the *defining* relationship for entropy differences:

$$\int_a^b \left(\frac{\mathrm{d}Q}{T}\right)_q = S(b) - S(a). \tag{2.23}$$

We call the left hand side the *Clausius integral* and this is the Clausius expression for entropy change. It is easy to measure entropy (or rather *differences* in entropy) experimentally using this relationship, essentially using a thermometer. In addition, notice something special about this definition. We have shown explicitly for an ideal gas that $S$ is a function of state variables and is therefore a state variable itself. However, we see that it is related to a sum of increments of a process variable, namely the heat transfer. A system does not possess a quantity of heat $Q$, as we have emphasised earlier: it only receives incremental contributions $\mathrm{d}Q$ to its energy $E$ during a thermodynamic process. Summing the $\mathrm{d}Q$ over the quasistatic process from state $a$ to state $b$ will in general produce a $\Delta Q$ that depends on the thermodynamic path, specifically the history of the compression, expansion and coupling to heat sources that is taken between them. But by summing the $\mathrm{d}Q/T$, we obtain something that is not specific to the path, meaning that it is a difference in a state variable $S$. Dividing the inexact differential $\mathrm{d}Q$ by the state variable $T$ produces an increment, known as an exact differential, in the state variable $S$. At the moment, we must regard the extension of all this to materials other than ideal gases as a conjecture, but a proof using the machinery of Carnot cycles will be given in Section 2.9.

Finally, we use (2.17) to note that $pV^{5/3}$ = constant describing the quasistatic adiabatic compression of the monatomic ideal gas is nothing more than the condition $S$ = constant. Quasistatic adiabatic compression of a system, characterised of course by $\mathrm{d}Q = 0$ at every incremental stage of the process, is *isentropic* (constant $S$) by analogy with an isothermal compression at constant $T$.

We have established the entropy of an ideal gas explicitly, but what are those special properties referred to earlier? Now it is time to find out.

## 2.6 Entropy Change in Free Expansion of an Ideal Gas

The core message of the second law of thermodynamics is that the entropy of an *isolated* system cannot decrease in spontaneous thermodynamic processes: those brought about, usually, by the release of a constraint. In almost all circumstances it increases, but in some special cases it can remain the same. Let us illustrate this with examples.

We first consider what is called a free, or Joule expansion of an ideal gas. Initially the gas is held inside a container of volume $V_0$ at pressure $p_0$ and temperature $T_0$. The container is situated inside a larger volume $V_1$ that is otherwise empty and thermally isolated from the environment. The container bursts and an expansion into $V_1$ takes place, involving gas flow, shock waves, pressure gradients, sound generation and so on and the process is nonquasistatic. The process is illustrated in Figure 2.3. When everything has settled down into a new equilibrium state, the final entropy can be identified as:

$$S(T_1, N, V_1) = Nk \ln\left(\frac{(kT_1)^{\frac{3}{2}}}{\hat{c}N/V_1}\right), \tag{2.24}$$

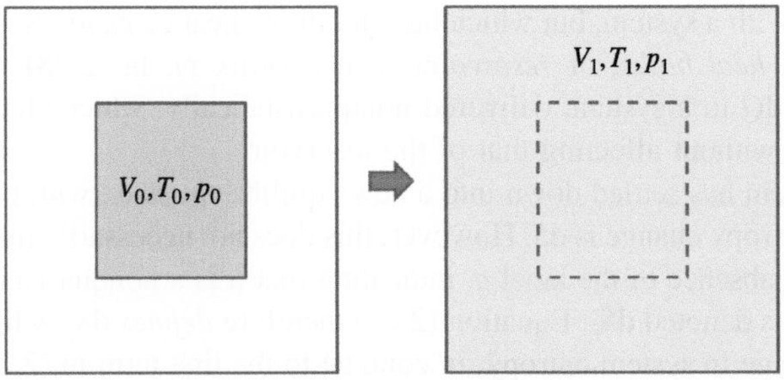


Figure 2.3 A container of volume $V_0$ holds a gas in equilibrium at temperature $T_0$ and pressure $p_0$. It then ruptures, giving rise to the free expansion of the gas into the larger, thermally isolated volume $V_1$, and once equilibrium has been restored, the temperature and pressure are $T_1$ and $p_1$, respectively.

while the initial entropy was

$$S(T_0, N, V_0) = Nk \ln \left( \frac{(kT_0)^{\frac{3}{2}}}{\hat{c}N/V_0} \right). \tag{2.25}$$

Since the gas does no work in expanding against a vacuum, and no heat is supplied from the environment since the outer container is thermally isolated, there is no change in the system energy. This implies that the final temperature is the same as the initial temperature, because the energy is given by $E = (3/2)NkT$ and so the entropy change is

$$\Delta S = S(T_0, N, V_1) - S(T_0, N, V_0) = Nk \ln \left( \frac{V_1}{V_0} \right). \tag{2.26}$$

Note that the unknown constant $\hat{c}$ does not appear in a difference of ideal gas entropies.

Clearly, the system entropy has increased as a consequence of the free expansion since $V_1 > V_0$, which is our first example of the second law in action. But this is a new aspect of entropy: we conclude that it does not change solely as a result of quasistatic heat transfers, as implied by (2.23). We related the change in entropy of a system to an incremental quasistatic heat input by

$$\mathrm{d}S = \left( \frac{\mathrm{d}Q}{T} \right)_q, \tag{2.27}$$

but this has to be modified for a nonquasistatic process such as free expansion because in such cases there is entropy change but no heat transfer. We consider instead the expression

$$\mathrm{d}S = \frac{\mathrm{d}Q}{T_\mathrm{r}} + \mathrm{d}S_i, \tag{2.28}$$

where $T_\mathrm{r}$ is the temperature of the heat source and the corresponding version for a finite change $\Delta S = \int \mathrm{d}Q/T_\mathrm{r} + \Delta S_i$.

As suggested in Figure 2.1, in thermodynamics we often invoke an environment that exchanges heat with a system, but which has an infinite heat capacity. Such environments are often called *heat baths*, or *reservoirs* (hence suffix r). In (2.28) we consider the transfer of heat $\mathrm{d}Q$ to a system, delivered nonquasistatically, which changes the system temperature but without affecting that of the reservoir.

Once the system has settled down into a new equilibrium state, with time-independent properties, its entropy change is $\mathrm{d}S$. However, this does not necessarily match the quantity $\mathrm{d}Q/T_\mathrm{r}$ (note the absence of the label $q$, indicating that it is a nonquasistatic process) and the discrepancy is denoted $\mathrm{d}S_i$. Equation (2.28) therefore *defines* $\mathrm{d}S_i$, which we shall call the *internal* change in system entropy, in contrast to the first term in (2.28) that involves heat flow from the *external* reservoir. In the case of free expansion, there was no heat transfer and the change in system entropy was entirely internal; so the corresponding expression would be $\Delta S = \Delta S_i$, such that $\Delta S_i = Nk \ln(V_1/V_0)$. The 'natural' direction of change of the system brought about by the rupture of the container is accompanied by a positive $\Delta S_i$. We need to investigate further the properties of this contribution.

## 2.7 Entropy Change due to Nonquasistatic Heat Transfer

Consider a process of heat exchange between a reservoir at temperature $T_r$ and a monatomic ideal classical gas initially at temperature $T_0$. The system variables $N$ and $V$ are fixed. No work is done on the system and the heat transfer is nonquasistatic. After equilibrium is reestablished, we assume that the system has acquired the temperature $T_r$, according to the zeroth law, and the change in system entropy is

$$\Delta S = S(T_r, N, V) - S(T_0, N, V) = Nk \ln\left(\frac{(kT_r)^{\frac{3}{2}}}{\hat{c}N/V}\right) - Nk \ln\left(\frac{(kT_0)^{\frac{3}{2}}}{\hat{c}N/V}\right) = \frac{3}{2}Nk \ln\frac{T_r}{T_0}. \tag{2.29}$$

We could be heating or cooling the system, so this entropy change could be positive or negative, depending on whether $T_r$ is greater than or less than $T_0$. A positive or negative change in system entropy is not in conflict with the second law as the system is not *isolated*. Of more significance is the sign of

$$\Delta S_i = \Delta S - \int \frac{dQ}{T_r} = \Delta S - \frac{\Delta E}{T_r} = \frac{3}{2}Nk\left(\ln\frac{T_r}{T_0} - \frac{T_r - T_0}{T_r}\right), \tag{2.30}$$

where we have employed (2.29), the first law $\Delta Q = \Delta E$ describing the heat transfer and system energy change during the nonquasistatic process, and $E = (3/2)NkT$. It is crucial to notice that the contribution $\Delta S_i$ to the system entropy change is positive whether $T_0$ is greater than or less than $T_r$, for both cooling or heating, as illustrated in Figure 2.4.

$\Delta S_i$ is the *internal* change in entropy and is a consequence of the nonquasistatic nature of the heat exchange between the environment and the ideal gas. It is never negative for this process. It is also referred to as the *dissipative* contribution to entropy change, or simply as entropy production. It is the central player in the second law of thermodynamics: it is the source that causes the total entropy to increase in a spontaneous nonquasistatic process. The entropy of an ideal gas can be increased or decreased, for example, by heating or cooling, but the internal entropy change seems never to be negative.

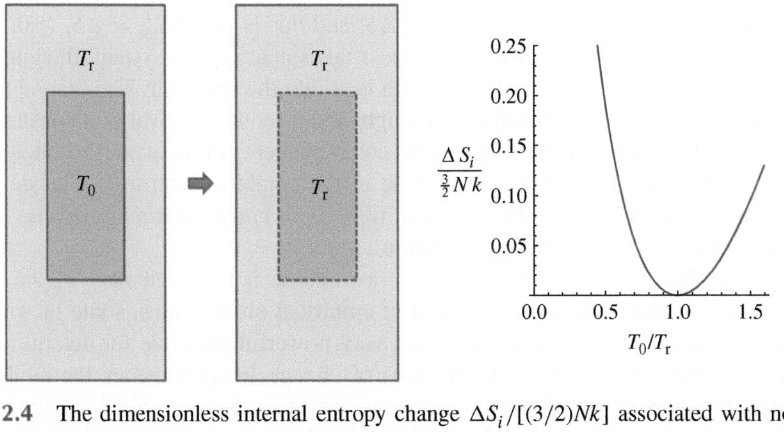


**Figure 2.4** The dimensionless internal entropy change $\Delta S_i/[(3/2)Nk]$ associated with nonquasistatic heat transfer between a reservoir at temperature $T_r$ and an ideal gas system initially at temperature $T_0$, assuming that they eventually become isothermal.

But how are we to interpret $\Delta S_i$? This is not straightforward. The most we can say at this point is that it is associated with heat flows and the transient departure from equilibrium brought about by differences in temperature between the system and environment. Only when such differences and flows become infinitesimal, and the rate of the corresponding process quasistatic, can entropy production be eliminated. If the temperature difference $T_r - T_0 = \delta T$ is small, we can insert the approximation $\ln(T_r/T_0) = -\ln(1 - \delta T/T_r) \approx \delta T/T_r + (1/2)(\delta T/T_r)^2$ in (2.30) to demonstrate that the corresponding internal entropy change is $\delta S_i \propto (\delta T)^2$: second order in the initial temperature difference. This is enough to justify the claim that $\Delta S_i = 0$ for a quasistatic process, during which the temperature of the environment is changed extremely slowly, allowing only tiny temperature mismatches between the environment and system.

We might point out that heat bath temperatures are meant to remain constant; so to be more explicit, we should represent a general quasistatic heat transfer process as a sequence of thermal equilibrations of the system with a set of heat baths at various slightly different temperatures, each of which brings about a small change $\delta T$ in system temperature and a small contribution $\delta S_i$ to internal entropy change. The sum of the $\delta T$ makes a finite overall temperature change $\Delta T \approx \sum \delta T$ but the overall internal entropy change is $\Delta S_i \approx \sum \delta S_i \propto \sum (\delta T)^2$, and this vanishes since a sum of second order infinitesimal contributions is negligible. Such gentle coupling of the system to slightly warmer or colder heat baths never produces significant heat flows and so the process does not lead to internal entropy change.

Consider now the entropy change of the reservoir that exchanges heat with the system in Figure 2.4. The reservoir is supposed to be large enough that the exchange of heat does not affect its temperature. It remains in equilibrium and hence suffers no internal entropy change. This is a defining feature of a reservoir: of course it is an idealisation of a real source of heat, but a very useful one. Its change in entropy in the heat transfer process is simply $\Delta S_r = -\Delta Q/T_r$, the negative sign indicating that the transfer of heat to the reservoir is equal and opposite to the transfer of heat $\Delta Q$ to the system. Since the change in entropy of the system is $\Delta S = \Delta S_i + \Delta Q/T_r$, we see that the exchange of heat between the reservoir and the system simply transfers entropy from one to the other. The change in entropy of the 'universe', here taken to be the combination of the ideal gas system and the reservoir, is $\Delta S_{tot} = \Delta S + \Delta S_r$ and this is just $\Delta S_{tot} = \Delta S_i \geq 0$. All the entropy production associated with the process takes place in the system, though once it is generated, it can be transported along with heat into the reservoir. The second law can be cast as 'changes in equilibrium state brought about by the removal of a constraint and the subsequent spontaneous heat flow between components of an overall isolated system will always be accompanied by an increase in the combined entropy'. Clausius put it more succinctly in 1865: *Die Entropie der Welt strebt einem Maximum zu:* the entropy of the universe evolves towards a maximum.

The claim that $\Delta S_i$ is never negative is axiomatic: it is a statement of the second law. It turns out that it rationalises a host of empirical observations, some of which we have already seen with the ideal gas. It is a very powerful principle for determining the direction (or more properly the destination) of change in spontaneous thermodynamic processes; that heat should flow to equalise temperatures, for example. We cannot prove it from the other laws of classical thermodynamics. Next we demonstrate how it emerged historically from empirical studies of heat transfer and engine design.

## 2.8 Cyclic Thermodynamic Processes, the Clausius Inequality and Carnot's Theorem

We consider taking the ideal gas around a cyclic heating and cooling process driven by a time-dependent reservoir temperature $T_{\text{r}}(t)$, or more properly a sequence of reservoirs with slightly different temperatures. We assume the initial and final situations are at equilibrium but the process need not be quasistatic. Cyclic means that the conditions at the end of the process are the same as those at the start. The system state variables are returned to their initial values, and this includes the entropy. We integrate (2.28) to give

$$\oint \frac{\mathrm{d}Q}{T_{\text{r}}(t)} = S_{\text{final}} - S_{\text{initial}} - \Delta S_i, \tag{2.31}$$

where the time dependence of the reservoir temperature has been made explicit, and where the integration sign implies a cyclic process. Since $S_{\text{final}} = S_{\text{initial}}$ for a cycle and $\Delta S_i \geq 0$, this means that

$$\oint \frac{\mathrm{d}Q}{T_{\text{r}}(t)} \leq 0, \tag{2.32}$$

which is known as the *Clausius inequality*. It is a practical statement of the second law in terms of thermal driving and heat transfers. It is an equality for quasistatic process conditions, for which $\Delta S_i = 0$, and where we can replace the reservoir temperature by the system temperature, and write $\oint (\mathrm{d}Q/T)_q = 0$.

The Carnot cycle is a famous example of a quasistatic cyclic process, named after Sadi Carnot (1796–1832). In its simplest form, it is a sequence of quasistatic expansions and compressions of an ideal gas, as illustrated in Figure 2.5. The first stage in the cycle is an isothermal expansion in contact with a hot reservoir. This moves the system along a path known as an *isotherm* from the top left to the top right of the cycle shown. The second stage is an adiabatic expansion from top right to bottom right, taking the system

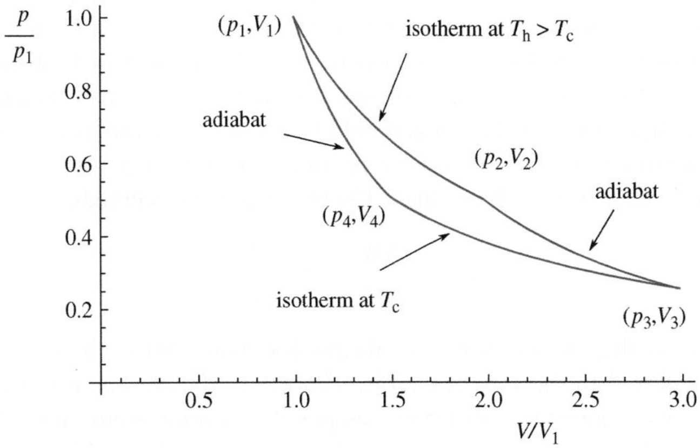


**Figure 2.5** Sketch of the quasistatic isothermal and adiabatic expansions and compressions that make up a Carnot cycle operating on an ideal gas.

along a path known as an *adiabat* or isentrope. The third is an isothermal compression in contact with a cold reservoir, and the fourth is an adiabatic compression that returns the system to its original state.

The cycle is driven by a time-dependent environmental pressure, synchronised with the thermal coupling and decoupling between the system and the two reservoirs. Since it is quasistatic, there is no entropy generation during the process, and furthermore the change in system entropy for one complete cycle should be zero. This can easily be demonstrated using the entropy function (2.22) and the adiabatic condition $TV^{2/3} = \text{constant}$. The entropy of the gas changes only during the isothermal expansion and compression stages of the cycle; so the change over a cycle is

$$\begin{aligned}\Delta S &= Nk\left[\ln\left(\frac{(kT_{\rm h})^{\frac{3}{2}}}{\hat{c}N/V_2}\right) - \ln\left(\frac{(kT_{\rm h})^{\frac{3}{2}}}{\hat{c}N/V_1}\right)\right] + Nk\left[\ln\left(\frac{(kT_{\rm c})^{\frac{3}{2}}}{\hat{c}N/V_4}\right) - \ln\left(\frac{(kT_{\rm c})^{\frac{3}{2}}}{\hat{c}N/V_3}\right)\right] \\ &= Nk\ln\left(\frac{V_2}{V_1}\frac{V_4}{V_3}\right),\end{aligned} \tag{2.33}$$

in the notation used in the diagram, and since $T_{\rm h}V_2^{2/3} = T_{\rm c}V_3^{2/3}$ and $T_{\rm h}V_1^{2/3} = T_{\rm c}V_4^{2/3}$, we have $V_2/V_1 = V_3/V_4$ and $\Delta S$ vanishes.

The cycle is designed to convey heat from the hot to the cold reservoir, with the conversion of some of this flow into mechanical work, obtained by employing the system expansions and compressions to move a load, for example. The hot reservoir passes heat $\Delta Q_{\rm h}$ into the system during the isothermal expansion from $V_1$ to $V_2$. Since $\Delta E = 0$ for the system during this expansion, this heat is converted into work performed on the environment, which is equal to $\int_{V_1}^{V_2} p\mathrm{d}V = NkT_{\rm h}\ln(V_2/V_1)$. Similarly, the cold reservoir receives heat $\Delta Q_{\rm c}$ equal to the work done on the system during the isothermal compression, which is $-\int_{V_3}^{V_4} p\mathrm{d}V = -NkT_{\rm c}\ln(V_4/V_3) = NkT_{\rm c}\ln(V_2/V_1)$. The work done on the environment per cycle is $\Delta W = \Delta Q_{\rm h} - \Delta Q_{\rm c} = Nk(T_{\rm h} - T_{\rm c})\ln(V_2/V_1)$. We have an idealised device that converts heat into work, a type of *heat engine* known as a Carnot engine. Furthermore, it can operate in both directions: the input of mechanical work can pump heat from the cold to the hot reservoir, like a refrigerator.

The sequence can be used to investigate the efficiency $\eta_C$ of a Carnot engine. This is the amount of work produced as a proportion of the amount of heat taken from the hot reservoir, or $\Delta W/\Delta Q_{\rm h}$. Carnot demonstrated in 1824 that this efficiency depended solely on the temperatures of the hot and cold heat reservoirs: this is known as Carnot's theorem. Interestingly, he obtained these results by using what was called the caloric theory of heat, which is now discredited. Using our present methods we can write

$$\eta_C = \frac{\Delta W}{\Delta Q_{\rm h}} = 1 - \frac{T_{\rm c}}{T_{\rm h}}. \tag{2.34}$$

The point here is that the efficiency is always less than 100%: there is always waste heat $\Delta Q_{\rm c}$ transferred to the cold reservoir. The most efficient engines need to operate between as great a temperature difference as possible. Carnot went on to show that any engine based on the quasistatic expansion and compression of an arbitrary substance had the same efficiency as the equivalent ideal gas Carnot engine or engines. In actual fact, most heat engines in Carnot's time were dreadfully inefficient, and the upper limit was nowhere near attainable. But the implications of his analysis were far reaching. In the next section, we show that a consideration of cyclic processes reveals that any *real* engine operating nonquasistatically has a lower efficiency than if it operated quasistatically, and that this reduction in efficiency is associated with the generation of entropy. For this reason, Carnot is regarded as the father of the second law.

## 2.9 Generality of the Clausius Expression for Entropy Change

First, we consider a Carnot cycle operating on a material, or *working substance*, for which we do not know an equation of state, or the heat capacities, or indeed whether an entropy state variable may be defined at all. Nevertheless, we can take it through a cycle of compressions and expansions. Let us actually consider a reversed cycle: an isothermal *compression* at $T_h$, adiabatic expansion, isothermal *expansion* at $T_c$ and adiabatic compression back to the starting point. All this is done quasistatically. Such a reverse cycle acts as a refrigerator or a heat pump. An amount of heat $\Delta Q_c'$ is taken from the cold reservoir and $\Delta Q_h'$ is delivered to the hot reservoir, per cycle, and this requires an input of work $\Delta W' = \Delta Q_h' - \Delta Q_c'$. The flows of energy are illustrated in Figure 2.6.

Now imagine that we use an ideal gas Carnot engine to drive a Carnot heat pump employing the arbitrary working substance. We design the cycles for our engine and pump to make sure that the output work per cycle of one matches the input work per cycle of the other: $\Delta W = \Delta W'$. From Figure 2.6 we can deduce that

$$\Delta Q_h - \Delta Q_c = \Delta Q_h' - \Delta Q_c'. \tag{2.35}$$

Taken together, these two systems quasistatically take heat $\Delta Q_h - \Delta Q_h'$ from the hot reservoir, per cycle, and deliver an equal amount $\Delta Q_c - \Delta Q_c'$ to the cold reservoir. The operation of the machine is reversible, in the sense that we could have the ideal gas Carnot cycle operating in a forward direction (as an engine) and the arbitrary

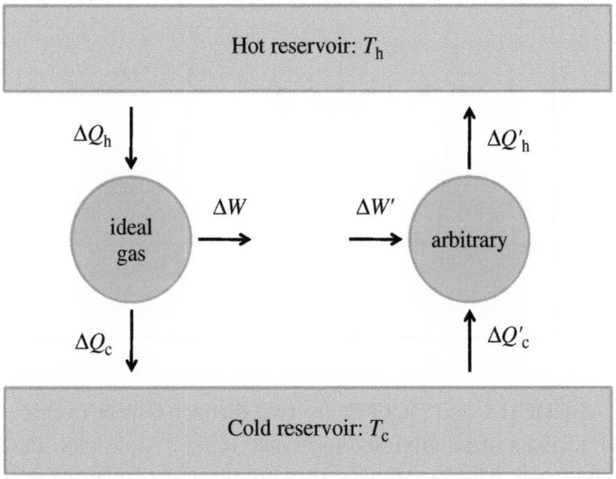


**Figure 2.6** Two Carnot engines operating between a hot and cold reservoir, one acting forwards on an ideal gas, and the other in reverse on an arbitrary substance.

substance Carnot cycle operating backwards (as a heat pump), or vice versa. Now, Carnot regarded it as inadmissible that a machine could be designed, even in principle, that could pump heat from a cold to a hot reservoir without some external input of work. But if $\Delta Q_{\rm h} \neq \Delta Q'_{\rm h}$, then in one of the directions of operation this is precisely what we have got. The only conclusion is that $\Delta Q_{\rm h} = \Delta Q'_{\rm h}$ and the two cycles cancel each other out.

Now let us consider the Clausius integral for the Carnot cycle taken by the ideal gas:

$$\oint \left(\frac{\mathrm{d}Q}{T}\right)_q = 0 = \frac{\Delta Q_{\rm h}}{T_{\rm h}} - \frac{\Delta Q_{\rm c}}{T_{\rm c}}, \tag{2.36}$$

and the arbitrary working substance:

$$\oint \left(\frac{\mathrm{d}Q}{T}\right)_q = -\frac{\Delta Q'_{\rm h}}{T_{\rm h}} + \frac{\Delta Q'_{\rm c}}{T_{\rm c}} = -\frac{\Delta Q_{\rm h}}{T_{\rm h}} + \frac{\Delta Q_{\rm c}}{T_{\rm c}} = 0, \tag{2.37}$$

using (2.36). Therefore we can write

$$\left(\frac{\mathrm{d}Q}{T}\right)_q = \mathrm{d}S, \tag{2.38}$$

for an arbitrary substance undergoing a portion of a Carnot cycle, where $\mathrm{d}S$ is an increment in a state variable $S$. If the arbitrary substance were an ideal gas, this would be the entropy quantity explored in earlier sections. Entropy is therefore a *general* property of matter defined through the Clausius integral. We deduced this in spite of not knowing the shape of the isotherms and adiabats on the $p - V$ plot for the arbitrary substance. The final step is to notice that any quasistatic cyclic process on the $p - V$ plot can be represented as a modified sequence of overlapping Carnot cycles, as illustrated in Figure 2.7; so the path taken can be quite general, allowing us to evaluate entropy differences between any two equilibrium states of the arbitrary substance.

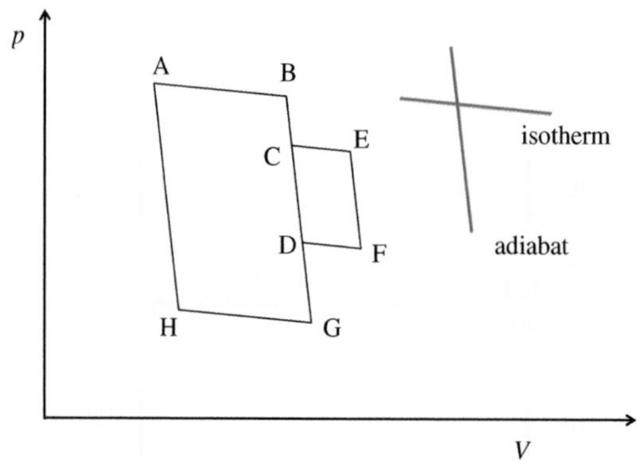


**Figure 2.7** Paths ABCDGHA and DCEFD are two distinct Carnot cycles comprising sections along isotherms and adiabats of an arbitrary substance in the $p - V$ plot. Path ABCEFDGHA is not a Carnot cycle: four heat baths are involved (there are four isotherms in the path). However, it is equivalent to the path ABCD(DCEFD)GHA: a Carnot cycle within another Carnot cycle. Clearly, any cyclic path can be broken down into elementary Carnot cycles.

It follows that if any substance receives heat from a reservoir nonquasistatically, it contributes to the change in its entropy through

$$dS = \frac{dQ}{T_r} + dS_i, \tag{2.39}$$

where we include the internal entropy production. Furthermore, for a cyclic process we find that the Clausius inequality (2.32) emerges just as before. Nonquasistatic heat transfer to a system from a reservoir creates entropy, to a degree defined in (2.39). All substances possess entropy and the capacity to generate it when thermally processed, as specified by the second law, just as we found explicitly was the case for the ideal gas.

Finally, we assess the efficiency of an ideal gas heat engine that follows the path shown in Figure 2.5 but this time with the formerly isothermal segments taken at a nonquasistatic rate. We wait at the end of each for thermal equilibrium to be restored before embarking on the quasistatic adiabatic segments. We again deduce that the heat extracted from the hot reservoir $\Delta Q_h$ is equal to the work done on the environment ($\Delta E = 0$ still for the isothermal processing of an ideal gas), but now according to (2.39) this is equal to $(\Delta S_{12} - \Delta S_i^h)T_h$ where $\Delta S_{12} = S(T_h, V_2, N) - S(T_h, V_1, N) = Nk \ln(V_2/V_1)$ from (2.22), and $\Delta S_i^h$ is the entropy generated internally during the expansion of the gas in contact with the hot reservoir. Similarly, the heat transferred to the cold reservoir $\Delta Q_c$ during the nonquasistatic compression is equal to the work done on the system by the environment and is equal to $-(\Delta S_{34} - \Delta S_i^c)T_c$, where $\Delta S_{34} = S(T_c, V_4, N) - S(T_c, V_3, N) = Nk \ln(V_4/V_3)$ and $\Delta S_i^c$ denotes the entropy generated internally during the compression of the gas in contact with the cold reservoir. As before, the efficiency is the total work per cycle divided by the heat extracted from the hot reservoir, so

$$\eta = \frac{\Delta W}{\Delta Q_h} = \frac{\Delta Q_h - \Delta Q_c}{\Delta Q_h} = 1 + \frac{\left[Nk \ln\left(\frac{V_4}{V_3}\right) - \Delta S_i^c\right] T_c}{\left[Nk \ln\left(\frac{V_2}{V_1}\right) - \Delta S_i^h\right] T_h}$$

$$= 1 - \frac{\left[Nk \ln\left(\frac{V_2}{V_1}\right) + \Delta S_i^c\right] T_c}{\left[Nk \ln\left(\frac{V_2}{V_1}\right) - \Delta S_i^h\right] T_h} = 1 - \frac{T_c}{T_h} - \frac{(\Delta S_i^c + \Delta S_i^h)T_c}{\Delta Q_h} \leq \eta_C, \tag{2.40}$$

from (2.34), and we see that the entropy production causes the efficiency of the engine to fall below the Carnot efficiency. Turning this around, the failure to extract as much work from a heat flow as might theoretically be possible, or indeed any inefficiency of energy transformation, is due to entropy generation, giving the latter an additional intuitive meaning as a measure of the wastage of heat in the operation of heat engines.

## 2.10 Entropy Change due to Nonquasistatic Work

Our next example of entropy production concerns the nonquasistatic performance of work on an ideal gas without any heat transfer. Since $\Delta Q$ is zero, we expect to find that the overall change in entropy is entirely the $\Delta S_i$ arising from internal generation, as was the case with free expansion. Consider a monatomic ideal classical gas in a thermally insulated cylinder with a mismatch between its pressure and that of the environment. What happens if we release the piston?

Equation (2.9) tells us that $pV^{5/3}$ = constant characterises a quasistatic expansion or compression of the gas in the absence of heat exchange. This means that following the release of the piston at a point where the system has volume $V_0$ and pressure $p_0$, the piston will oscillate on a $p - V$ diagram about the external pressure $p_r$, along an adiabat as long as we assume the motion is quasistatic. If so, it would behave as same as an undamped spring and the system entropy would remain constant.

However, the motion is not quasistatic. Intuitively we know that a real gas would not oscillate for ever: eventually the motion will cease when the bulk kinetic energy of the oscillation is converted into heat. We view this as a consequence of the finite viscosity of the gas. The final pressure when the piston comes to rest would be expected to be equal to the reservoir pressure $p_r$. Since the cylinder is insulated, the heat generated by viscosity is not passed to the environment and the final equilibrium temperature would therefore be greater than the temperature of the system when at pressure $p_r$ on the adiabat during undamped oscillation. The raised temperature at the equilibrium pressure $p_r$, together with the equation of state $pV = NkT$, means that the final volume $V_f$ of the gas will lie to the right of the adiabat. Hence the entropy of the final equilibrium state, proportional to $\ln p_r V_f^{5/3}$, is greater than the initial entropy, proportional to $\ln p_0 V_0^{5/3}$. The development on a $p - V$ plot corresponding to this history is illustrated in Figure 2.8. Entropy has been generated during the spontaneous equilibration in pressure between system and environment.

We introduced viscous damping as a familiar process whereby motion in a mechanical system is naturally brought to a halt. It is often known as dissipation: the kinetic energy is dissipated as heat. The production of entropy that we have just deduced is somehow connected with the conversion of energy from what mechanical engineers would regard as a high quality coherent form (piston motion) into a low quality incoherent form (motion of the atoms). Since the system receives no heat transfer from the environment,

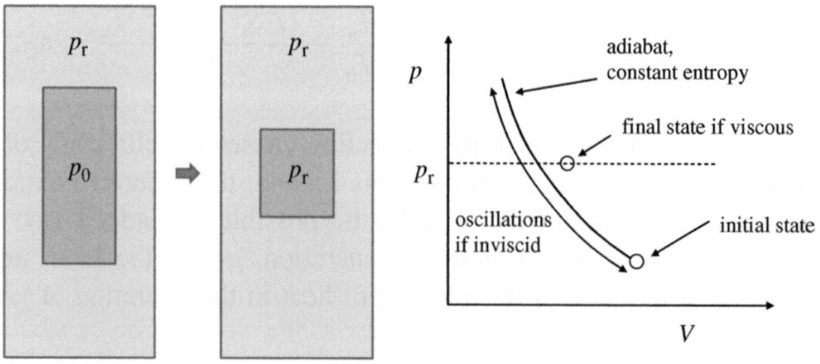


**Figure 2.8** Illustration on a $p - V$ plot of the increase in entropy of a monatomic ideal classical gas associated with the removal of an initial mismatch between system pressure $p$ and reservoir pressure $p_r$. The kinetic energy possessed by the oscillation (that would persist for a truly ideal gas) is eventually converted to heat, and hence to a volume increase at the final pressure $p_r$, if the gas is viscous. The final state therefore lies above the adiabat passing through the initial state. The final value of $pV^{5/3}$ is greater than the initial value, and hence the entropy of the system has increased.

the entropy increase is from internal generation $\Delta S_i$ alone. It is clearly appropriate to describe internal entropy production as a dissipative entropy change. In addition, the phenomenon of viscosity, or of friction in general, is related to the generation of entropy associated with flows driven by pressure gradients: another intuitive link, this time between entropy production and the wastage of mechanical energy.

## 2.11 Fundamental Relation of Thermodynamics

It is useful at this point to establish an important general connection between the entropy of a system and various other state variables. The result is often called the *fundamental relation of thermodynamics*. We have seen that the entropy change can be related to a quasistatic heat transfer to a system: $\mathrm{d}S = (\mathrm{d}Q/T)_q$ and that the work done in a quasistatic mechanical process of volume change may be written $(\mathrm{d}W)_q = -p\mathrm{d}V$. The argument we gave for the latter holds for any substance, not just an ideal gas, although there might be additional terms. Therefore, for a quasistatic process of heat transfer and volume change the increment in energy, according to the first law, can be written as

$$\mathrm{d}E = T\mathrm{d}S - p\mathrm{d}V. \tag{2.41}$$

If the energy change $\mathrm{d}E$ were brought about by a nonquasistatic process, then $\mathrm{d}Q$ for the process would not be equal to $T\mathrm{d}S$ and $\mathrm{d}W$ would not be equal to $-p\mathrm{d}V$. Nevertheless, once equilibrium has been restored, (2.41) would still specify $\mathrm{d}E$ in terms of the incremental changes in equilibrium state variables $\mathrm{d}S$ and $\mathrm{d}V$. It has to, because changes in the system variables can be brought about by a variety of thermodynamic processes, heating and squeezing in different sequences and at different rates, but once equilibrium is restored, the details of the process are irrelevant as far as changes in $E$, $T$, $S$, $p$ and $V$ are concerned. Therefore, if (2.41) is true for a quasistatic process, it is true for all processes.

The fundamental relation provides far reaching connections between state variables. It is often written as

$$\mathrm{d}S = \frac{1}{T}\mathrm{d}E + \frac{p}{T}\mathrm{d}V, \tag{2.42}$$

and since we have an expression (2.21) for $S(E,V,N)$ for the monatomic ideal classical gas, this relationship can easily be checked. Since

$$S(E,V,N) = Nk \ln\left(\frac{V\left(\frac{2}{3}E\right)^{\frac{3}{2}}}{\hat{c}\, N^{\frac{5}{2}}}\right), \tag{2.43}$$

we have $\mathrm{d}S = (3/2)Nk\mathrm{d}E/E + Nk\mathrm{d}V/V$ plus a term proportional to $\mathrm{d}N$ that we shall return to shortly: as $E = (3/2)NkT$ and $pV = NkT$ this clearly matches (2.42).

Quite generally, we know we can write an entropy increment, when considered as a function of $E$ and $V$, in the form

$$\mathrm{d}S(E,V) = \left(\frac{\partial S}{\partial E}\right)_V \mathrm{d}E + \left(\frac{\partial S}{\partial V}\right)_E \mathrm{d}V, \tag{2.44}$$

which then implies that

$$\frac{1}{T} = \left(\frac{\partial S}{\partial E}\right)_V, \tag{2.45}$$

and

$$\frac{p}{T} = \left(\frac{\partial S}{\partial V}\right)_E. \tag{2.46}$$

We just showed that these are satisfied by the entropy function for the ideal gas, but they are quite general relationships between state variables. These results provide us with a means to *define* temperature and pressure if we happen to have at our disposal a mathematical expression for the entropy.

But we neglected to explore the change in entropy brought about by a change in number of particles $N$. We expect to be able to write

$$\mathrm{d}S(E,V,N) = \left(\frac{\partial S}{\partial E}\right)_{V,N}\mathrm{d}E + \left(\frac{\partial S}{\partial V}\right)_{E,N}\mathrm{d}V + \left(\frac{\partial S}{\partial N}\right)_{E,V}\mathrm{d}N, \tag{2.47}$$

and to this end we define the so-called chemical potential $\mu$:

$$\frac{\mu}{T} = -\left(\frac{\partial S}{\partial N}\right)_{E,V}, \tag{2.48}$$

such that the fundamental relation may be extended to

$$\mathrm{d}E = T\mathrm{d}S - p\mathrm{d}V + \mu\mathrm{d}N. \tag{2.49}$$

Chemical potential might be unfamiliar, but it is just another state variable, an equilibrium property of matter. It turns out to have a rather special role to play, but first, to develop some intuition, let us derive an expression for the chemical potential of the ideal gas. We have

$$\left(\frac{\partial S}{\partial N}\right)_{E,V} = \left(\frac{\partial}{\partial N}\right)_{E,V} Nk\ln\left(\frac{V(\frac{2}{3}E)^{\frac{3}{2}}}{\hat{c}\,N^{\frac{5}{2}}}\right) = k\ln\left(\frac{V(\frac{2}{3}E)^{\frac{3}{2}}}{\hat{c}\,N^{\frac{5}{2}}}\right) - \frac{5}{2}k, \tag{2.50}$$

so

$$\mu = kT\ln\left(\frac{N}{V}\frac{\hat{c}\,\mathrm{e}^{\frac{5}{2}}}{(kT)^{\frac{3}{2}}}\right). \tag{2.51}$$

where e is the base of natural logarithms. We can make this expression more compact by introducing a temperature dependent length $\lambda_{\mathrm{th}}(T)$ defined by

$$\lambda_{\mathrm{th}}^3 = \frac{\hat{c}\,\mathrm{e}^{\frac{5}{2}}}{(kT)^{\frac{3}{2}}}, \tag{2.52}$$

such that we get, for an ideal gas,

$$\mu(N,V,T) = kT\ln\left(\frac{N}{V}\lambda_{\mathrm{th}}^3\right), \tag{2.53}$$

where we explicitly note that the chemical potential is a function of state variables $N$, $V$ and $T$. The quantity $\lambda_{th}$ is known as the thermal de Broglie wavelength, for reasons that will become apparent in Chapter 9.

At constant particle density $N/V$, the temperature dependence of the chemical potential of the gas is given by $(\partial\mu/\partial T)_{N,V} = \mu/T - 3k/2$. In Chapter 9, we shall find that a criterion for classical, as opposed to quantum, behaviour is that the density of the gas should be much less than $\lambda_{th}^{-3}$, and from (2.53) this clearly means that $\mu/T < 0$. Therefore, the chemical potential of the classical ideal gas decreases with temperature at constant density. On the other hand, the chemical potential in (2.53) clearly increases with density at constant temperature. This dependence on density and temperature is sketched in Figure 2.9. Notice that we cannot determine the absolute value of the chemical potential as we do not (yet) know the value of the constant $\hat{c}$. Also note that we do not sketch the chemical potential for temperatures approaching absolute zero since it is there that we might expect quantum behaviour.

Notice that the entropy of the ideal gas may be written in the form

$$S = Nk\left(\frac{5}{2} - \frac{\mu}{kT}\right) = Nk\left[\frac{5}{2} - \ln\left(n\lambda_{th}^3\right)\right], \tag{2.54}$$

where $n = N/V$ is the particle density. The dependence of the chemical potential on temperature and gas density ties in with that of the entropy: as the density increases at constant temperature, for example by isothermal compression, so does its chemical potential, while its entropy decreases.

Now we enlarge on the special role that chemical potential plays in thermodynamics. It is an indicator of equilibrium between isothermal systems that are able to exchange particles. In this respect it is analogous to the temperature, which according to the zeroth law is an indicator of equilibrium between systems able to exchange energy in the form of heat. Systems that can exchange heat and particles evolve so as to equalise their temperatures and chemical potentials. This behaviour is a consequence of the second law, as will be demonstrated in Section 3.2. We have already employed the assertion of the equalisation of temperatures after heat exchange in Section 2.7, without any special comment because it is so familiar. Particle exchange is less so, but it does make intuitive

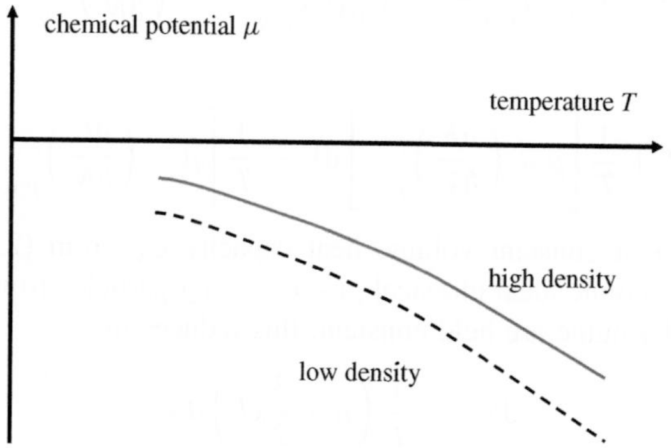


**Figure 2.9** Sketch of the chemical potential of a classical ideal gas as a function of temperature and particle density. Note that it is a negative quantity in the classical regime.

sense to claim that a flow of particles is driven by a difference in densities, and that the flow stops when density differences are eliminated. An example would be the diffusion of a trace gas into a large volume from a small canister once it is opened. Another example is the osmotic exchange of a solvent across a semipermeable membrane separating solutions with different solute concentrations. In view of the relationship between chemical potential and particle density in (2.53) for an ideal gas, differences in density are synonymous with differences in $\mu$, if the systems in question are isothermal, and the elimination of density differences is equivalent to the equalisation of chemical potentials. Taking the analogy with heat flow further, we might suspect that the nonquasistatic flow of particles between systems driven by a chemical potential difference will lead to entropy production. Let us examine this more closely.

## 2.12 Entropy Change due to Nonquasistatic Particle Transfer

In the following example we show that nonquasistatic particle flows generate entropy. We consider a system able to exchange particles and energy with an environment, or reservoir, with chemical potential $\mu_r$ and temperature $T_r$. Such a reservoir can be called a *particle bath* as well as a heat bath. It is supposed to be so large that its chemical potential is not changed if it supplies particles to a system, just as a heat bath can supply heat without changing its temperature. Neither will it suffer any internal generation of entropy during a process. We suppose that the gas remains isothermal with the reservoir during exchange, namely that its temperature $T$ is equal to $T_r$ throughout, and that the system volume is fixed.

As a preliminary, we recast the extended fundamental relation as

$$dS = \frac{1}{T}dE + \frac{p}{T}dV - \frac{\mu}{T}dN, \tag{2.55}$$

but now it is our intention to identify the dependence of the entropy change on an increment in the *temperature* instead of energy, in addition to increments in volume and particle number. We insert

$$dE = \left(\frac{\partial E}{\partial T}\right)_{V,N} dT + \left(\frac{\partial E}{\partial V}\right)_{T,N} dV + \left(\frac{\partial E}{\partial N}\right)_{T,V} dN, \tag{2.56}$$

and obtain

$$dS = \frac{C_V}{T}dT + \frac{1}{T}\left[p + \left(\frac{\partial E}{\partial V}\right)_{T,N}\right]dV - \frac{1}{T}\left[\mu - \left(\frac{\partial E}{\partial N}\right)_{T,V}\right]dN, \tag{2.57}$$

using the definition of constant volume heat capacity $C_V$ from (2.11). For a system consisting of a monatomic ideal classical gas, receiving particles from a reservoir while its temperature and volume are held constant, this reduces to

$$dS = -\frac{1}{T}\left(\mu - \frac{3}{2}kT\right)dN. \tag{2.58}$$

An integral of such increments would be an analogue of the Clausius definition of entropy change (2.23), but this time corresponding to a quasistatic particle transfer, driven by a time-dependent chemical potential $\mu_r(t)$ of the reservoir, with the rate of change slow enough such that the chemical potential of the system $\mu$ remains equal to $\mu_r$ throughout.

If we wish to consider nonquasistatic particle exchange between system and reservoir, (2.58) will need to be revised along the lines of (2.28). We write

$$dS = -\frac{1}{T}\left(\mu_r - \frac{3}{2}kT_r\right)dN + dS_i \tag{2.59}$$

as a *definition* of the internal entropy change $dS_i$ associated with a nonquasistatic process of isothermal particle transfer to a system from a particle bath, with the system starting and ending in equilibrium. We use the reservoir chemical potential as a reference point in the construction of (2.59), just as we used $T_r$ in the construction of (2.28).

We consider a heat and particle bath at fixed temperature $T_r$ and chemical potential $\mu_r$. The ideal gas system has an initial temperature equal to $T_r$ and an initial chemical potential $\mu_0 \neq \mu_r$ that depends on temperature and the initial particle density $N_0/V$, according to (2.53). The system is coupled to the heat and particle bath and we assert that after a while a new equilibrium is established with the system chemical potential becoming equal to $\mu_r$. This is illustrated in Figure 2.10. The change in system entropy is then an extension to (2.59):

$$\Delta S = -\frac{1}{T_r}\left(\mu_r - \frac{3}{2}kT_r\right)\Delta N + \Delta S_i, \tag{2.60}$$

where $\Delta N = N_1 - N_0$ is the change in number of particles in the system and $\Delta S_i$ is the entropy generated internally. From (2.54) the change in system entropy is

$$\Delta S = N_1 k\left(\frac{5}{2} - \frac{\mu_r}{kT_r}\right) - N_0 k\left(\frac{5}{2} - \frac{\mu_0}{kT_r}\right) = \frac{5}{2}k(N_1 - N_0) - \frac{N_1\mu_r}{T_r} + \frac{N_0\mu_0}{T_r}, \tag{2.61}$$

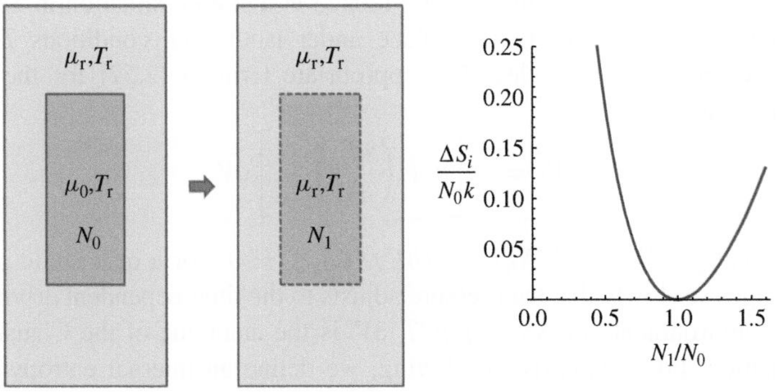


**Figure 2.10** Entropy generation $\Delta S_i$ brought about by nonquasistatic particle exchange between an ideal gas system and a particle bath.

and we rearrange (2.60) and use (2.53) to get

$$\begin{aligned}\Delta S_i &= \left(\frac{5}{2}-\frac{3}{2}\right)k(N_1-N_0)-\frac{N_1\mu_r}{T_r}+\frac{N_0\mu_0}{T_r}+\frac{\mu_r}{T_r}(N_1-N_0)\\ &= k(N_1-N_0)+\frac{N_0}{T_r}(\mu_0-\mu_r)\\ &= k(N_1-N_0)+kN_0\ln\left(\frac{N_0}{N_1}\right). \qquad (2.62)\end{aligned}$$

The entropy of the reservoir changes by $\Delta S_r = -[\mu_r - (3/2)kT_r]\Delta N_r/T_r$ arising from the change $\Delta N_r = -\Delta N$ in its number of particles, and this is equal and opposite to the first term in (2.60) for the system entropy change $\Delta S$. By definition, the reservoir does not suffer any internal generation of entropy. The change in the combined entropy of system and environment is therefore $\Delta S_i$, and from (2.62), which is sketched as a function of $N_1/N_0$ in Figure 2.10, we conclude that irrespective of whether particles are flowing into or out of the ideal gas system, the internal entropy change $\Delta S_i$ is never negative, in line with our consideration of nonquasistatic heat flow in Section 2.7 and Figure 2.4. The behaviour of $\Delta S_i$ as a function of initial mismatch in relevant properties is very similar in the two cases.

Once again, if the mismatch in chemical potential is small, such that the change in particle content of the system during the process $\delta N = N_1 - N_0$ is small, then expansion of the logarithm in (2.62) allows us to show that $\delta S_i \propto (\delta N)^2$, and for a sequence of equilibrations with a set of particle baths each with a slightly different chemical potential, and each leading to a particle transfer $\delta N$, the overall entropy change is $\Delta S_i \sim \sum (\delta N)^2$, a sum of second order small quantities. In the limit of quasistatic particle exchange, the overall entropy production goes to zero.

## 2.13 Entropy Change due to Nonquasistatic Volume Exchange

Finally, we consider entropy generation associated with the equalisation of pressure between a system and its environment. We looked at a similar situation in Section 2.10. For simplicity, we allow this to take place under isothermal conditions $T = T_r$, and without the exchange of particles. The appropriate term in (2.57) for the change in system entropy is

$$dS = \frac{1}{T}\left[p + \left(\frac{\partial E}{\partial V}\right)_{T,N}\right]dV, \qquad (2.63)$$

and for an ideal gas $E = (3/2)NkT$, so $(\partial E/\partial V)_{T,N} = 0$. For a quasistatic process, the volume changes so slowly that the pressure adjusts to the time-dependent driving pressure $p_r(t)$ of the environment; so $\Delta S = \int (p/T)dV$ is the analogue of the Clausius integral for this situation. For nonquasistatic driving, we define an internal entropy production through

$$dS = \frac{p_r}{T_r}dV + dS_i, \qquad (2.64)$$

for an incremental process, and $\Delta S = (p_r/T_r)\Delta V + \Delta S_i$ for a finite change. Bearing in mind that $T$ and $N$ are constant, we can use (2.22) to show that $\Delta S = Nk \ln(V_1/V_0)$ for a nonquasistatic change in gas volume from $V_0$ to $V_1$, so

$$\Delta S_i = Nk \ln\left(\frac{V_1}{V_0}\right) - \frac{p_r}{T_r}(V_1 - V_0) = \frac{p_r V_1}{T_r}\left[\ln\left(\frac{V_1}{V_0}\right) - \frac{(V_1 - V_0)}{V_1}\right], \tag{2.65}$$

since the final system pressure $p_r$ is equal to $NkT_r/V_1$. This constitutes the total entropy production of ideal gas and environment for the conditions of the nonquasistatic process, and as it takes a form similar to (2.30) and (2.62) it is never negative, irrespective of the ratio of volumes $V_0/V_1$, as we now have come to expect.

## 2.14 General Thermodynamic Driving

We have seen that thermodynamic processes driven by separate mismatches in temperature, pressure and chemical potential between an ideal gas and its environment, leading to outcomes that correspond to our intuitive expectations, are accompanied by an increase in the total entropy. For a thermodynamic process that involves incremental changes in all three driving parameters, the dissipative or internal contribution $dS_i$ to the change in the entropy of the system may be written as

$$dS_i = dS - \frac{dE}{T_r} - \frac{p_r}{T_r}dV + \frac{\mu_r}{T_r}dN, \tag{2.66}$$

where $T_r$, $p_r$ and $\mu_r$ are properties of the environment, and might be functions of time. Notice that we have employed the expression involving the increment $dE$ rather than the form involving $dT$ in (2.56), for convenience of notation. We assert that $dS_i$ in (2.66) is never negative, not only for an ideal gas, but for any choice of thermodynamic system and environment. This is a statement of the second law.

This can be made apparent in the following way. An environment characterised by time-dependent parameters $T_r$, $p_r$ and $\mu_r$ drives a system such that its state variables $E$, $V$ and $N$ also acquire a time dependence. If the system is only mildly perturbed from equilibrium during such a nonquasistatic driving process, it is reasonable to characterise it using spatially uniform time-dependent state variables $T$, $p$ and $\mu$ as well. Intuition suggests that the latter typically lag behind the environmental variables that drive them, as sketched in Figure 2.11. We shall return to this viewpoint in Section 15.1. The system would also be characterised by a time-dependent entropy $S(t)$ related in the standard way to the time-dependent state variables. Combining (2.66) with (2.49) we can write

$$\frac{dS_i}{dt} = \left(\frac{1}{T(t)} - \frac{1}{T_r(t)}\right)\frac{dE}{dt} + \left(\frac{p(t)}{T(t)} - \frac{p_r(t)}{T_r(t)}\right)\frac{dV}{dt} - \left(\frac{\mu(t)}{T(t)} - \frac{\mu_r(t)}{T_r(t)}\right)\frac{dN}{dt}, \tag{2.67}$$

and the second law asserts that the sum of the three terms on the right hand side will be positive. Notice that the contributions take the form of a mismatch in intensive state variables of system and environment, such as $1/T$, multiplied by an increment in an extensive state variable of the system, such as $dE$.

Intuitively this representation makes sense. For example, if $T < T_r$ we know from experience that heat flows into the system, in which case $dE > 0$ (for constant $N$ and $V$)

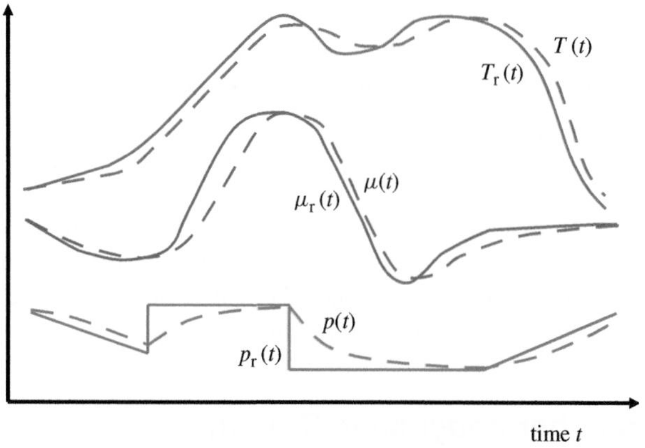


**Figure 2.11** Sketch of the way intensive system properties are driven by, but lag behind, time-dependent reservoir properties, as a consequence of nonquasistatic energy, volume and particle exchanges. Such a picture holds only if the rates of change are not too fast, such that thermodynamic properties such as temperature are approximately valid out of equilibrium.

and the first term on the right hand side of the above equation is positive. Conversely, if $T > T_r$ at some point during the process, we know that heat will flow out of the system, such that $dE < 0$ and the term is still positive. Similarly if the system pressure (over $T$) is greater than/less than the reservoir pressure (over $T_r$), we know from experience that the natural change $dV$ in system volume is positive/negative. Thus positivity of the second contribution to the right hand side of (2.67) seems to emerge. Finally, if the system chemical potential (over $T$) is less than or greater than the ratio $\mu_r/T_r$, which suggests a difference in particle density between system and reservoir, we expect $dN$ to be positive/negative.

The second law makes the claim that in all empirical situations the sum of the contributions to $dS_i$ is never negative. A powerful rephrasing is that energy, volume and particles flow as they do, when constraints are changed, *because the total entropy must increase*. The second law hence seems to offer a rationale for the natural evolution of any thermodynamic process.

## 2.15 Reversible and Irreversible Processes

We continue to consider the evolution of a system driven by a time-dependent environment. The only way for the total entropy *not* to increase is for the system variables to match the driving environmental variables exactly. There should be no time lag between them. This intuitively requires an extremely slow rate of change: a quasistatic process. The terms on the right hand side in (2.67) then turn out to be second order in the small mismatch between the system and its environment, as we saw earlier, and therefore negligible even when summed over the duration of the process.

An idealised process of this kind has a very special character. Imagine going forward through some quasistatic process driven by a time evolution in environmental variables.

If the evolution of the environmental variables were then reversed, the system variables would follow them, without any time lag, and once again no overall entropy would be produced. Eventually, we could recover the initial state of both system and environment, but only if we perform the forward and backward processes very slowly and hence with no overall increase in entropy.

This sort of entropy conserving thermodynamic process is called *reversible*. We regard it as synonymous with the word quasistatic, although in principle there might be dynamical systems that undergo discontinuous changes in properties even when driven quasistatically, and this might affect the way in which equilibrium between system and environment can be maintained throughout the process. A reversible process is an idealisation of a real process: no process can actually develop so slowly that no entropy is generated, but it is an idealisation that is very useful in thermodynamics. In contrast, all real processes are characterised by entropy production. The entropy generated going forward cannot be effaced by a reversal of the external driving forces in an attempt to restore the initial state: in recognition of this they are called *irreversible* processes.

The reversibility of a Carnot engine or pump was invoked in Section 2.9 to support the assertion that the efficiency of such systems does not depend on the working substance, although the word was used in a general sense. We see now that thermodynamic reversibility has a technical sense associated with zero entropy production. If an engine were driven at a finite rate it would not perform a Carnot cycle and it would not be reversible. It would generate entropy going forward, and if we were to drive it backwards as a pump, it would also generate entropy. It would be an *irreversible* cyclic process and (2.40) demonstrates that it would be less efficient than a reversible cycle.

## 2.16 Statements of the Second Law

This chapter began with a listing of the four laws of thermodynamics, but most of the discussion has concerned the second law. This is because it involves the topic of entropy, about which more can be said than any other concept in thermodynamics. It is a law that concerns nonquasistatic thermodynamic processes. Traditionally, presentations of statistical physics have focussed largely on the equilibrium properties of matter, and how they change in a quasistatic process. From such a perspective, the second law seems strangely peripheral: here it has taken centre stage.

The second law states that all spontaneous thermodynamic processes, initiated by the removal of a constraint, increase the entropy of the universe. In fact this is a slightly incomplete statement: as entropy is a property of a system in equilibrium, the idea to be conveyed is that if the universe, after the constraint is removed, could ever be brought back into equilibrium, it would have an entropy that is larger than it was before the process started. Quasistatic processes are engineered to leave the entropy unchanged by taking an infinite amount of time to complete: these are thermodynamically reversible, but clearly not realistic. The second law forbids certain directions, or better put *destination*s, of a spontaneous process; of all possible end-points, Nature excludes those where the total entropy of all the relevant components is lower than at the start. The second law may therefore be expressed as a number of statements of the form ‘this does not happen’. They normally correspond to familiar experience.

- **Clausius statement:** Heat cannot flow spontaneously from a system at a low temperature to a system at a higher temperature;
- **Kelvin statement:** It is impossible in a cyclic process to convert heat completely into mechanical work (i.e. no engine can have an efficiency of 100%).

If the Kelvin statement is violated, it can be shown by use of a Carnot engine that the Clausius statement is violated too.

Other statements in the same spirit include:

- Matter cannot flow spontaneously from a system with a low chemical potential to a system at the same temperature and a higher chemical potential;
- A system at a low pressure will not expand at the expense of a system at a high pressure;
- A gas will not spontaneously contract to occupy a volume smaller than its container;
- Energy is reduced in 'quality' by natural processes (bulk potential energy is high quality energy, heat is low quality). This is equivalent to saying 'energy is dissipated'.

We have shown that these phenomena all appear to be manifestations of the second law. But we have not *proved* that the second law should hold: this is beyond the remit of classical thermodynamics and why it is one of the axiomatic laws from which thermodynamic behaviour can be deduced. However, a proof of sorts will be presented using ideas of random dynamics in Chapter 17.

A corollary to these ideas, to be developed further in Chapter 3, is that after a process is initiated, Nature chooses one particular destination over all others, namely the available end-point of the process with the *greatest* total entropy. The argument for this goes as follows. Imagine some other possible final state of the system, with lower total entropy. It can be established as an equilibrium state of the system, through the application of a suitable constraint. For example, in a free expansion of a gas from a volume $V_0$ into a volume $V_1$, as in Section 2.6, the evolution could be halted at an intermediate volume $V_{\rm m}$ if an enclosing container with this volume were inserted. The entropy change in such an arrested expansion would be $Nk\ln(V_{\rm m}/V_0)$, lower than the entropy change in the expected final state $Nk\ln(V_1/V_0)$. But the point is that such intermediate states can only be maintained against evolution towards the state with the highest entropy by the existence of the necessary constraints. If the constraints were removed, we would expect to see the system continue to seek a state with higher entropy, in accordance with the second law. The conclusion is that evolution can *only* cease when the entropy is at its highest value consistent with the remaining constraints. Otherwise, we would require constraints that do not exist, such as a container with a volume less than $V_1$.

So when a constraint is removed, the universe evolves spontaneously to a new equilibrium state that *maximises* its overall entropy. The universe, in the words of Clausius, strives to maximise its entropy through the variation of any unconstrained state variables, in other words by rearranging its constituents. The second law seems to be a so-called variational principle, a requirement to maximise or minimise something: it is one of several such principles to be found in physics, a property that makes it rather beautiful, but still somewhat mysterious.

## 2.17 Classical Thermodynamics: the Salient Points

It is worth summarising the main features of classical thermodynamics.

- Thermodynamics is fundamentally a theory of the transfer of heat and matter between macroscopic physical systems, and of the equilibrium states to which such transfers lead, which are defined by time-independent properties and the absence of fluxes of energy or particles within them.
- We often imagine a system to interact with an *environment* that has idealised properties based on its presumed large size. The environment is also referred to as a reservoir, or a heat, volume or particle bath.
- Macroscopic systems are described by state variables. Some are well defined whether the system is in equilibrium or in a state of evolution: examples are energy, volume and the number of particles. Others are properly defined only in equilibrium, such as temperature, pressure, chemical potential and entropy, although to a rough approximation they can be applied to systems that are mildly perturbed from equilibrium.
- Thermodynamics provides relationships between state variables, such as an equilibrium equation of state or a specification of an increment in one variable in terms of increments in others, for example the fundamental relation of thermodynamics.
- The equilibrium state of an isolated system has the maximum value of entropy of all possible macroscopic arrangements of the system, subject to any imposed constraints such as an enclosing volume, or a fixed energy and particle number.
- Removal of constraints typically leads to the evolution of a system and its environment. If such a process takes place at a finite rate, or nonquasistatically, the new equilibrium that is established between them will *always* have a larger total entropy. In a sense, we can consider that the process of evolution is accompanied by a rate of production of entropy.

## Exercises

**2.1** Estimate the entropy generated by a typical physics professor in one day.

**2.2** (a) A piston is used to compress an ideal gas quasistatically from volume $V_i$ to volume $V_f$. If the gas is in thermal contact with a heat bath at temperature $T$, such that the compression is carried out isothermally, calculate the work done on the gas, the change in entropy of the gas and the change in entropy of the heat bath. (b) The compression is repeated but nonquasistatically. Are the three calculated quantities higher, lower, or the same as before? (c) The procedure is repeated but without contact with the heat bath such that the compression is adiabatic and quasistatic. The initial temperature is $T$. Again, what values do the three quantities take? What is the final temperature? (d) Now assume the compression is adiabatic but nonquasistatic. Are the three quantities different from those in (c)? Is the final temperature greater than, less than or the same as the final temperature in (c)?

**2.3** A 1 kg block of steel at temperature 60° C is placed at the bottom of a lake of depth 10 m at temperature 10° C. The specific heat capacity of steel is $420\,\mathrm{JK^{-1}\,kg^{-1}}$ and is approximately temperature independent. Calculate the entropy change of the block, the lake and the universe, after thermal equilibrium has been reached. A crane is used to lift the block extremely slowly from the bottom of the lake to the surface. What is the entropy change of the universe? Then it is dropped back in. Again comment on the entropy change of the universe.

**2.4** Consider two systems each containing an ideal gas of $N$ particles in a volume $V$ but with different temperatures $T_1$ and $T_2$. The volume of each system is kept constant while thermal contact is made between them. After equilibrium has been established, demonstrate that the total entropy of the universe has changed by $\Delta S = 3Nk \ln((T_1 + T_2)/(2(T_1 T_2)^{1/2}))$ and that this is never negative.

# 3

# Applications of Classical Thermodynamics

We discuss some applications of the ideas developed in Chapter 2. We focus on properties of entropy, on the criteria for equilibrium and how they arise from the second law.

## 3.1 Fluid Flow and Throttling Processes

Thermodynamics can be applied to materials in motion, such as fluid flow, as long as we employ local state variables such as pressure and temperature on the understanding that they are only approximate descriptors for systems that are out of equilibrium. It is beyond the scope of this book, but we could develop models for the transport of heat by convection and conduction, or for the evolution of the motion of a fluid caused by gradients in the pressure.

We shall discuss only one fluid flow problem, partly because it is simple, and partly because it involves entropy production, an inevitable feature if we have systems away from equilibrium. It is known variously as plug flow, throttling, or the Joule–Thomson process and it plays a role in the industrial cooling of gases.

We consider a fluid flowing uniformly along a pipe, such that from the point of view of an observer moving at the same speed in the direction of motion, it might be considered to be in equilibrium and to possess a uniform pressure. Real fluids have viscosity and could only be kept in such a state of motion by imposing a pressure gradient, but we ignore this. There is a plug of porous material occupying a section of the pipe that does offer resistance to the flow. In order to force the fluid through the plug, a piston or similar device has to perform work on the fluid upstream of the plug. The flow passes through the plug in a complicated manner, but emerges into the clear section of pipe and eventually settles back into a steady uniform flow, and pushes a downstream piston. There are a number of contradictory features here: the fluid is assumed to be without viscosity, and yet eventually loses the turbulent pattern of flow brought about by the plug. We idealise the behaviour in order to say something about entropy generation.

*Statistical Physics: An Entropic Approach*, First Edition. Ian Ford.

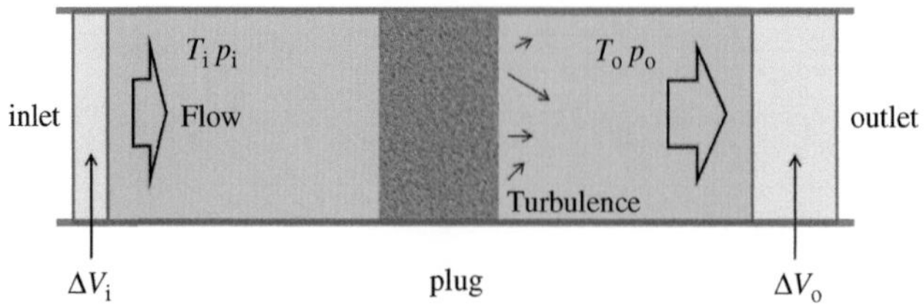


**Figure 3.1** Fluid flow along a pipe and through a resistive plug, giving rise to density, temperature and pressure changes, enthalpy conservation and entropy generation.

The walls of the pipe are thermally insulated, so the balance of energy input and output for a system consisting of a translating tube of fluid passing through the plug consists of work done on the fluid, work performed by it and energy change arising from the difference in temperature between the upstream and downstream flows. Referring to Figure 3.1, we denote the inlet and outlet pressure and temperature as $p_{i,o}$ and $T_{i,o}$, respectively. Assuming that the work processes are quasistatic, we write the balance of overall work done on the fluid as $\Delta W = p_i \Delta V_i - p_o \Delta V_o$, where $\Delta V_i$ and $\Delta V_o$ are the (positive) volumes vacated at the inlet and occupied at the outlet by the tube of fluid in a particular period of time as a result of the flow. The change in energy of the tube is $e(n_o, T_o)\Delta V_o - e(n_i, T_i)\Delta V_i$, where $e$ is the energy per unit volume of the fluid, written as a function of particle density $n$ and temperature. Equating these energy changes, we write

$$p_i \Delta V_i - p_o \Delta V_o = e(n_o, T_o)\Delta V_o - e(n_i, T_i)\Delta V_i, \tag{3.1}$$

or

$$(e(n_i, T_i) + p_i)\Delta V_i = (e(n_o, T_o) + p_o)\Delta V_o. \tag{3.2}$$

This is a conservation law. Essentially, a fluid packet of initial volume $\Delta V_i$ passes down the pipe, changing its volume to $\Delta V_o$ well downstream of the plug, and altering its pressure and temperature too, but maintaining its *enthalpy,* a state variable defined by

$$H = E + pV. \tag{3.3}$$

Typically, the plug produces a pressure drop in the flow ($p_i > p_o$), but, depending on the nature of the gas, it can experience either an increase or a decrease in temperature. We shall not discuss this rich range of behaviour, but instead focus on an ideal gas to investigate the entropy generation. Plug flow is analogous to a steady state version of free expansion, and we expect internal entropy production. We have $E = (3/2)NkT$ and $pV = NkT$; so $H = (5/2)NkT = (5/2)pV$. The drop in pressure therefore gives rise to an increase in volume of a packet of fluid after its passage through the plug. The entropy change of a packet containing $N$ particles is given by

$$\Delta S = Nk \ln\left(\frac{p_o^{\frac{3}{2}} V_o^{\frac{5}{2}}}{p_i^{\frac{3}{2}} V_i^{\frac{5}{2}}}\right) = Nk \ln\left(\frac{p_i}{p_o}\right) > 0, \tag{3.4}$$

using (2.19) together with conservation of enthalpy $\frac{5}{2}p_o V_o = \frac{5}{2}p_i V_i$, and this is entirely due to internal generation, since the heat flow into the fluid tube is zero.

In general, we can show that the second law requires there to be a pressure drop by writing $dH = dE + pdV + Vdp$ and inserting the fundamental relation (2.49) to obtain $dH = TdS + \mu dN + Vdp$ such that

$$dS = \frac{1}{T}dH - \frac{\mu}{T}dN - \frac{V}{T}dp \quad \Rightarrow \quad \left(\frac{\partial S}{\partial p}\right)_{H,N} = -\frac{V}{T} < 0, \tag{3.5}$$

meaning that a drop in pressure of a packet of fluid at constant enthalpy and particle number is associated with an increase in system entropy. In other words, a packet that passes through the plug emerges with the same values of $N$ and $H$, and will have increased its entropy as a consequence of the irreversible processes brought about by the resistance to the flow, and (3.5) tells us that its pressure will have decreased.

## 3.2 Thermodynamic Potentials and Availability

According to the variational statement of the second law of thermodynamics given in Section 2.16, an isolated system moves towards equilibrium by rearranging itself to maximise its entropy. Free expansion of a gas is a good example. This must be modified; however, if the system can exchange energy, volume or particles with its environment. The entropy of the universe is maximised when equilibrium is found, but the change in the entropy of the system can be negative if the system temperature goes down, for example. Is there a different variational principle we can use?

We must apply the second law to both the system and the environment in order to find a principle that indicates how system variables should change. We consider a system in equilibrium characterised by a set of thermodynamic variables, and then place it in contact with an environment to allow them both to evolve towards a new equilibrium state. We require

$$\Delta S + \Delta S_r > 0, \tag{3.6}$$

to a maximal extent, where $\Delta S_r$ is the entropy change of the environment. From the fundamental relation (2.49), we deduce that $T_r \Delta S_r = \Delta E_r + p_r \Delta V_r - \mu_r \Delta N_r$ because the reservoir is at a constant temperature $T_r$, pressure $p_r$ and chemical potential $\mu_r$ throughout the process. By conservation of energy, volume and particle number, the changes in system variables are given by $\Delta E = -\Delta E_r$, $\Delta V = -\Delta V_r$, and $\Delta N = -\Delta N_r$, so

$$T_r \Delta S + T_r \Delta S_r = T_r \Delta S - \Delta E - p_r \Delta V + \mu_r \Delta N > 0. \tag{3.7}$$

This may be summarised in the statement

$$\Delta A = \Delta E + p_r \Delta V - T_r \Delta S - \mu_r \Delta N < 0, \tag{3.8}$$

to a maximal extent, where we define the *availability* $A$ as

$$A = E + p_r V - T_r S - \mu_r N. \tag{3.9}$$

As the initial availability is fixed, the new equilibrium state of the system is characterised by changes that *minimise* the availability of the final state.

Note that this variational principle involves a quantity $A$ that is a function of extensive *system* variables $E$, $V$, $S$ and $N$, and intensive *environmental* variables $p_r$, $T_r$ and $\mu_r$. The intensive system variables $p$, $T$ and $\mu$ do not appear. The minimisation is over extensive system variables, and any internal partitioning of these quantities between different constituents of the system.

However, these variables might be constrained in various ways, depending on the nature of the coupling between system and environment. If the system were isolated, such that $N$, $E$ and $V$ remain constant during all conceivable spontaneous processes, then $A = -T_r S +$ constants and the minimisation of availability would correspond to the maximisation of system entropy $S(E, N, V)$ at constant $N$, $E$ and $V$. This maximisation is over all internal arrangements of the constituents, for example all possible density profiles across the system volume. If the initial profile were nonuniform, we would expect the system to evolve to a uniform profile to maximise its entropy, just as we saw in the case of a free expansion of an ideal gas.

### 3.2.1 Helmholtz Free Energy

Now consider a new situation where the system volume $V$ and particle content $N$ are held constant during the process, but heat transfers are possible such that $E$ might change. The $p_r V$ and $\mu_r N$ terms in the availability are constant and can be ignored; so the availability is $A = E - T_r S +$ constants. Let us see what the minimisation of $A$ would mean for an ideal gas placed in contact with a heat bath. Using (2.21), the reduced availability (i.e. ignoring constants) is

$$A_R(E) = E - T_r S = E - NkT_r \ln\left(\frac{V\left(\frac{2}{3}E\right)^{\frac{3}{2}}}{\hat{c}N^{\frac{5}{2}}}\right). \tag{3.10}$$

The state variable $E$ is unconstrained and its value in the final equilibrium is to be selected variationally. By setting to zero the derivative of this expression with respect to $E$, we find that

$$1 - \frac{3NkT_r}{2E} = 0, \tag{3.11}$$

implying that the final system energy is given by $3NkT_r/2$, and that the system temperature evolves to $T_r$. More generally, we would write

$$\frac{dA_R}{dE} = 1 - T_r\left(\frac{\partial S}{\partial E}\right)_{N,V} = 1 - \frac{T_r}{T} = 0, \tag{3.12}$$

using the relationship (2.45) that gives the same result. We expected this, and indeed assumed that it would happen when considering heat transfer and entropy production in Section 2.7. We now see it as a consequence of the second law.

The form taken by the reduced availability in the final state is $E(T_r, N, V) - T_r S(E(T_r), N, V)$. This combination of state variables has a special name – the *Helmholtz free energy*, and is an example of a so-called *thermodynamic potential*:

$$F(N, V, T) = E - TS. \tag{3.13}$$

The minimised reduced availability for the example of an ideal gas is written, using (2.6) and (2.22), as

$$A_{\rm R}(E(T_{\rm r})) = F_{\rm ig}(N, V, T_{\rm r}) = \frac{3}{2}NkT_{\rm r}\left[1 - \ln\left(\frac{kT_{\rm r}}{(\hat{c}N/V)^{2/3}}\right)\right], \tag{3.14}$$

where the Helmholtz free energy is labelled to be that of an ideal gas (ig).

Note that the Helmholtz free energy has been cast here as a function of the state variables $N$, $V$ and $T$. The energy and the entropy can both be expressed as functions of these variables. These are regarded as the 'natural' variables for $F$, for reasons we shall understand better at the end of this section.

The selection of equilibrium of a system under constraints of fixed $N$, $V$ and $T$ is sometimes portrayed simply as the minimisation of $F = E - TS$, which carries the suggestion that both $S$ and $T$ in this expression are allowed to vary as $E$ is changed. It is clearly more accurate to state that we are to minimise the reduced availability $E - T_{\rm r}S$ over the system energy $E$ as well as any parameters involving the internal arrangement of the constituents of the system. But at some point in this minimisation procedure, the reduced availability can often take the apparent form of a Helmholtz free energy or a sum of such terms.

For example, consider a gas initially constrained to have a nonuniform density profile across a fixed volume. We suppose that in the initial arrangement, $N_1^0$ particles are confined to a subvolume of the system $V_1$, and $N_2^0 = N - N_1^0$ particles reside in the remaining volume $V_2 = V - V_1$, with both parts at arbitrary temperatures, as illustrated in the first situation in Figure 3.2. When the partition between the subvolumes is removed, and when we put the system in contact with a heat bath, we expect it to evolve to equalise the particle densities in the subvolumes, and to bring both subvolumes to the

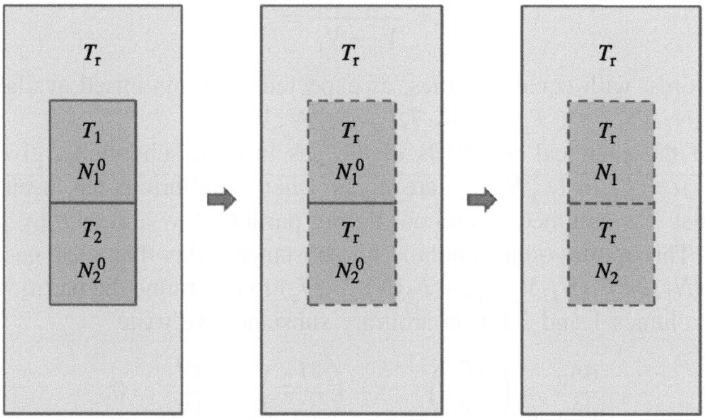


**Figure 3.2** Two stages of availability minimisation to seek a new equilibrium. First, thermal contact is opened between two subvolumes of a system and the environment, bringing both of them to the temperature $T_{\rm r}$. Secondly, particle exchange is opened up between the subvolumes, which leads to an equalisation of their chemical potentials, and for an ideal gas, an equality between particle densities $N_1/V_1$ and $N_2/V_2$. The second stage can be regarded as the minimisation of the Helmholtz free energy of the system over the degree of partition of particles between the subvolumes.

temperature of the heat bath. In order to determine the final equilibrium, we first minimise the availability with respect to the energies $E_1$ and $E_2$ in each subvolume, without changing the particle content in each, which, as we saw above, has the effect of requiring the temperature in each subvolume to equalise with that of the heat bath. This part is represented by the transition from the first to the second situation in Figure 3.2. The reduced availability is then a sum of Helmholtz free energies for the two subvolumes. A further minimisation of the availability is then done with respect to the number of particles $N_1$ in volume $V_1$. Using the free energy just derived for a monatomic ideal classical gas, we write

$$\begin{aligned} A_R(N_1) &= F_{ig}(N_1, V_1, T_r) + F_{ig}(N_2, V_2, T_r) \\ &= \frac{3}{2}N_1 kT_r \left[1 - \ln\left(\frac{kT_r}{\left(\frac{\hat{c}\, N_1}{V_1}\right)^{\frac{2}{3}}}\right)\right] + \frac{3}{2}(N - N_1)kT_r \left[1 - \ln\left(\frac{kT_r}{\left(\frac{\hat{c}\, (N-N_1)}{(V-V_1)}\right)^{\frac{2}{3}}}\right)\right], \end{aligned} \tag{3.15}$$

and we select the equilibrium value of $N_1$ by setting the derivative $\mathrm{d}A_R/\mathrm{d}N_1$ equal to zero:

$$0 = \frac{3}{2}kT_r \left[1 - \ln\left(\frac{kT_r}{\left(\frac{\hat{c}\, N_1}{V_1}\right)^{\frac{2}{3}}}\right) + 2/3\right] - \frac{3}{2}kT_r \left[1 - \ln\left(\frac{kT_r}{\left(\frac{\hat{c}\, (N-N_1)}{(V-V_1)}\right)^{\frac{2}{3}}}\right) + 2/3\right], \tag{3.16}$$

which implies that in the equilibrium state, shown as the third situation in Figure 3.2, the particles are partitioned between the two subvolumes according to

$$\frac{N_1}{V_1} = \frac{N - N_1}{V - V_1} = \frac{N_2}{V_2}, \tag{3.17}$$

or in other words, with equal densities, as expected. The minimised availability is then given by $F_{ig}(N_1, V_1, T_r) + F_{ig}(N_2, V_2, T_r) = F_{ig}(N, V, T_r)$.

Notice that the chemical potentials of the gas in each subvolume, given according to (2.53) by $\mu = kT \ln(\lambda_{th}^3 N/V)$, are equal when equilibrium has been established, a principle that was assumed when considering particle flow and entropy generation in Section 2.10. This applies quite generally for substances other than ideal gases. When we minimise $A_R(N_1) = F_1(N_1, V_1, T_r) + F_2(N_2, V_2, T_r)$ to determine the partition of particles between subvolumes 1 and 2 for an arbitrary substance we write

$$\frac{\mathrm{d}A_R}{\mathrm{d}N_1} = \left(\frac{\partial F_1}{\partial N_1}\right)_{V_1, T_r} + \left(\frac{\partial F_2}{\partial N_2}\right)_{V_2, T_r} \frac{\mathrm{d}N_2}{\mathrm{d}N_1} = 0. \tag{3.18}$$

To proceed further we differentiate (3.13):

$$\mathrm{d}F = \mathrm{d}E - T\mathrm{d}S - S\mathrm{d}T, \tag{3.19}$$

and use the fundamental relation $\mathrm{d}E = T\mathrm{d}S - p\mathrm{d}V + \mu\mathrm{d}N$ to give

$$\mathrm{d}F = -p\mathrm{d}V + \mu\mathrm{d}N - S\mathrm{d}T, \tag{3.20}$$

and then construct the partial derivative

$$\left(\frac{\partial F}{\partial N}\right)_{V,T} = \mu, \tag{3.21}$$

in which case (3.18) reduces to $\mu_1(N_1, V_1, T_r) = \mu_2(N_2, V_2, T_r)$ having inserted $dN_2/dN_1 = -1$ since $N_1 + N_2 = N$, and we conclude that equilibrium requires equality of chemical potentials.

Notice in (3.20) that an increment in $F$ is given in terms of increments in $N$, $V$ and $T$. We could of course relate the latter to increments in other state variables, but the expression would be more complicated. This is one reason why $F$ is considered to be a natural function of variables $N$, $V$ and $T$.

### 3.2.2 Why *Free* Energy?

The terminology of free energy is a little archaic and comes from considering the performance of quasistatic mechanical work on a system constrained throughout the process to be isothermal with respect to an environment at a temperature $T$. The work done is $\Delta W = \Delta E - \Delta Q$, and the quasistatic heat transfer from the environment that arises as a consequence of the process is $\Delta Q = T\Delta S$, so $\Delta W = \Delta(E - TS) = \Delta F$ as $T$ is a constant. For nonquasistatic work, $\Delta Q = T(\Delta S - \Delta S_i)$ in which case $\Delta W = \Delta E - T\Delta S + T\Delta S_i = \Delta(E - TS) + T\Delta S_i = \Delta F + T\Delta S_i$, and therefore

$$\Delta W - \Delta F = T\Delta S_i > 0, \tag{3.22}$$

since $\Delta S_i > 0$. The interpretation is that the work done is partly converted into a change in the free energy of the system and partly wasted through a positive heat transfer to the environment, of magnitude $T\Delta S_i$, that essentially corresponds to friction. As long as the work is done quasistatically, all the mechanical work performed would be stored in the system as a change in the state variable $F$.

Similarly, we could make the system perform work on some external body by way of a thermodynamic process, while keeping the system isothermal with its environment. For example, a compressed gas could be expanded from $V_0$ to $V_1$ so as to push a piston while being maintained at a constant temperature. If the process were quasistatic, the (positive) work done on the environment, $-\Delta W$, would correspond to the negative of $\Delta F = F(N, V_1, T) - F(N, V_0, T)$, the change in the system state variable $F(N, V, T)$ over the process. If the process were nonquasistatic, then the work done on the environment would be given by $-\Delta W = -\Delta F - T\Delta S_i$, which would be lower than the maximum work achievable $(-\Delta F)$ for the corresponding quasistatic process. A process to extract work at a rapid rate from a system is less effective than one carried out more slowly. More of the stored free energy is wasted as frictional heat. The change in the Helmholtz free energy $F$ is therefore a measure of the maximum amount of energy possessed by a system and its environment that can be 'freed' to perform mechanical work during a specified isothermal process. Hence the name has emerged.

### 3.2.3 Contrast between Equilibria

As an example of different forms of the principle of minimisation of thermodynamic availability for different environmental constraints, consider a system containing a pool of water. At a certain instant, a membrane is removed from the surface, allowing the liquid to evaporate into the space above. A new equilibrium is reached when the water vapour pressure reaches the so-called saturated vapour pressure $p_e(T)$, a material property that is an increasing function of temperature, which we shall encounter again in Section 3.9.2.

The internal parameter of the system that is varied in order to minimise availability after the membrane is removed is the degree of partitioning of water molecules between the gas and liquid phases. If the system were thermally isolated while it did this, the principle of maximising system entropy would determine the final equilibrium state. However, such a state would be cooler than the initial liquid, as latent heat is needed to evaporate the liquid, and with the overall energy of the system held constant, this leaves less in the form of molecular kinetic energy, and hence we get a lower final temperature. The final vapour pressure would be the saturated vapour pressure at such a reduced temperature.

If, on the other hand, the system remained in thermal contact with a heat bath throughout the process, heat would flow into the system to keep the temperature the same. The environment would supply the latent heat of evaporation. In the final equilibrium state the vapour pressure would be the saturated vapour pressure at the original pool temperature and it would be higher than the final vapour pressure for the isolated case. This difference in outcome following the removal of the membrane is explained by noting that in the isothermal case, we select the final equilibrium state by (effectively) minimising the Helmholtz free energy of the system rather than by maximising its entropy. Both situations correspond to the minimisation of the availability, but under different constraints.

### 3.2.4 Gibbs Free Energy

Now consider the availability for a condition of constant $N$, but with the system volume no longer constrained. Both the energy and volume of the system are allowed to vary after the release of an initial constraint and the relevant part of $A$ is the combination $A_R(E,V) = E - T_r S(E,V) + p_r V$. We impose

$$0 = \left(\frac{\partial A_R}{\partial V}\right)_E = -T_r\left(\frac{\partial S}{\partial V}\right)_E + p_r = -\frac{T_r}{T}p + p_r, \tag{3.23}$$

and

$$0 = \left(\frac{\partial A_R}{\partial E}\right)_V = 1 - T_r\left(\frac{\partial S}{\partial E}\right)_V = 1 - \frac{T_r}{T}, \tag{3.24}$$

where (2.45) and (2.46) have been used, and the expected equilibrium conditions $p = p_r$ and $T = T_r$ emerge. The equilibrium values of energy and volume are $E(p_r, T_r)$ and $V(p_r, T_r)$. The minimised reduced availability $E(p_r, T_r) - T_r S(E(p_r, T_r), V(p_r, T_r)) + p_r V(p_r, T_r)$ is equal to the so-called *Gibbs free energy* $G$ of the system evaluated under the prevailing conditions of reservoir pressure and temperature, where we define

$$G(N,p,T) = E - TS + pV. \tag{3.25}$$

This is another example of a thermodynamic potential. By analogy with the previous discussion of Helmholtz free energy, if a system that is not open to particle transfers from the environment has freedom to partition itself internally while in contact with a heat and *volume* bath, then the choice of equilibrium state is made by minimising the availability. After minimisation with respect to system energy, which sets the temperature equal to

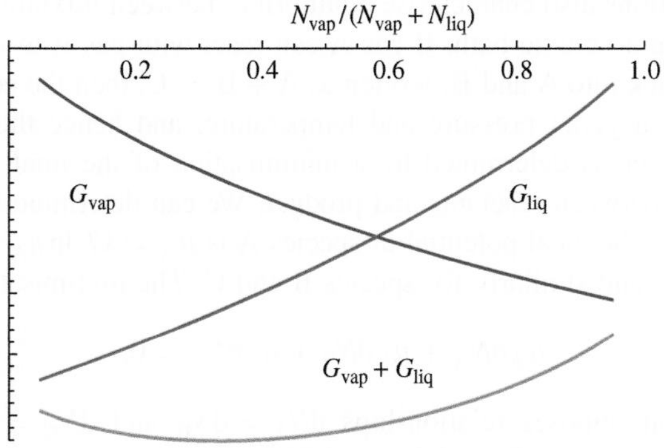


**Figure 3.3** Minimisation of availability for a liquid–vapour system under conditions of constant pressure, volume and particle number. The equilibrium state is such that the particles are partitioned between liquid and vapour phases to minimise the sum of their Gibbs free energies, in this case when $N_{\text{liq}} = 2N_{\text{vap}}$. Note that $G_{\text{vap}}$ decreases as $N_{\text{vap}}$ increases, consistent with (3.28) and $\mu_{\text{vap}} < 0$.

that of the environment, the reduced availability resembles a sum of Gibbs free energies characterising each internal subvolume of a system. The minimum of this function over the partitioning *is* the Gibbs free energy of the final equilibrium state.

For example, consider again a system consisting of a pool of water in equilibrium with its vapour such that both have the same temperature and pressure. We expose the system to an environment at the same temperature $T_r$ but at a different pressure $p_r$ and allow the system to change its volume and any other parameters to seek out a new equilibrium. Clearly both the liquid and vapour need to assume the new pressure $p_r$, but for the vapour to do so at constant temperature, there will need to be some evaporation or condensation of the liquid. The procedure to follow is to minimise $G_{\text{liq}}(N_{\text{liq}}, p_r, T_r) + G_{\text{vap}}(N_{\text{vap}}, p_r, T_r)$ over the partitioning of particles between liquid and vapour, with $N = N_{\text{liq}} + N_{\text{vap}}$ fixed. This is illustrated in Figure 3.3. We write

$$\left(\frac{\partial G_{\text{liq}}}{\partial N_{\text{liq}}}\right)_{p_r,T_r} + \left(\frac{\partial G_{\text{vap}}}{\partial N_{\text{vap}}}\right)_{p_r,T_r} \frac{dN_{\text{vap}}}{dN_{\text{liq}}} = 0, \quad \Rightarrow \quad \left(\frac{\partial G_{\text{liq}}}{\partial N_{\text{liq}}}\right)_{p_r,T_r} = \left(\frac{\partial G_{\text{vap}}}{\partial N_{\text{vap}}}\right)_{p_r,T_r} \tag{3.26}$$

and proceed by differentiating (3.25) and inserting the fundamental relation:

$$dG = dE - TdS - SdT + pdV + Vdp = -SdT + Vdp + \mu dN, \tag{3.27}$$

to show that

$$\left(\frac{\partial G}{\partial N}\right)_{p,T} = \mu. \tag{3.28}$$

We again conclude that when particle exchange is possible between parts of a system, a condition for equilibrium is that each should acquire the same chemical potential.

Incidentally, (3.27) demonstrates that $G$ is a natural function of variables $T$, $p$ and $N$. The convenience of this choice is also evident in the fact that $T$ and $p$ are constrained by the reservoir during the process while $N$ is a constant for the system as a whole.

These considerations also characterise equilibrium between mixtures of reactive gases coupled to a heat and volume bath. If gas A can react with gas B to make gas C, while C can dissociate back into A and B, written as $\mathrm{A}+\mathrm{B} \rightleftharpoons \mathrm{C}$, then the equilibrium volume of the mixture for a given pressure and temperature, and hence the balance between reactants and product, is determined by a minimisation of the total $G$ over the extent of the partitioning between reactants and product. We can determine this balance in the following way. The chemical potential of species A is $\mu_{\mathrm{A}} = kT \ln n_{\mathrm{A}} + \mathrm{const}$, where $n_{\mathrm{A}}$ is the density of A, and similarly for species B and C. The minimisation of $G$ requires

$$\mu_{\mathrm{A}} \mathrm{d}N_{\mathrm{A}} + \mu_{\mathrm{B}} \mathrm{d}N_{\mathrm{B}} + \mu_{\mathrm{C}} \mathrm{d}N_{\mathrm{C}} = 0, \tag{3.29}$$

and as the reaction imposes relationships $\mathrm{d}N_{\mathrm{A}} = \mathrm{d}N_{\mathrm{B}}$ and $\mathrm{d}N_{\mathrm{C}} = -\mathrm{d}N_{\mathrm{A}}$ this corresponds to $\mu_{\mathrm{A}} + \mu_{\mathrm{B}} - \mu_{\mathrm{C}} = 0$, and hence $kT \ln\,(n_{\mathrm{A}} n_{\mathrm{B}}/n_{\mathrm{C}}) = \mathrm{const}$, or $n_{\mathrm{C}} \propto n_{\mathrm{A}} n_{\mathrm{B}}$, a result known as the law of mass action.

### 3.2.5 Grand Potential

We could also consider a system placed in contact with a heat and particle bath such that the release of the initial constraint allows it to change its temperature and particle number at constant volume. The relevant part of the availability is then $A_{\mathrm{R}}(E,N) = E - T_{\mathrm{r}} S(E,N) - \mu_{\mathrm{r}} N$. The minimisation over $E$ at constant $N$ establishes that the system temperature $T$ is equal to that of the environment $T_{\mathrm{r}}$, as before. The condition for minimisation with respect to $N$ is

$$\left(\frac{\partial A_{\mathrm{R}}}{\partial N}\right)_E = T_{\mathrm{r}} \left(\frac{\partial S}{\partial N}\right)_E - \mu_{\mathrm{r}} = \frac{T_{\mathrm{r}}}{T}\mu - \mu_{\mathrm{r}} = 0, \tag{3.30}$$

and hence, as expected, the chemical potential of the system equalises with that of the environment. Having achieved this, any further internal partitioning of the system is determined by minimising a sum of terms of the form

$$\Phi(\mu, V, T) = E - TS - \mu N \tag{3.31}$$

for each subsystem. $\Phi$ is known as the *grand potential* and it is most naturally cast as a function of $\mu$ and $T$, the constraints that are imposed on the system as it evolves, and $V$, the volume of each subsystem that is to be varied to seek the new equilibrium.

For example, consider a system of volume $V$ that can receive water vapour from a pool of water, with both in contact with a heat bath. The system is therefore coupled to an environment that can supply heat and particles, although, in contrast to the case examined in Section 3.2.4, the chemical potential of the source of particles is considered to be fixed. The reduced availability is minimised to yield the condition (3.30), and the equilibrium number of particles in the system is given by

$$N = \frac{V}{\lambda_{\mathrm{th}}^3(T_{\mathrm{r}})} \exp\left(\frac{\mu_{\mathrm{r}}}{kT_{\mathrm{r}}}\right), \tag{3.32}$$

using (2.53). The pressure $p_{\mathrm{e}}(T_{\mathrm{r}}) = NkT_{\mathrm{r}}/V$ is then an expression of the saturated vapour pressure referred to in Section 3.2.3. Recall that the chemical potential of a classical gas, and hence that of the reservoir with which it is in equilibrium, is negative;

so this expression describes a saturated vapour pressure that rises with temperature, as is found experimentally.

The state variables $F$, $G$ and $\Phi$ are collectively known as thermodynamic potentials. They are used to determine the internal partitioning of energy, volume and particles between different subsystems that are coupled to a common environment that is able to supply energy, energy and volume, and energy and particles, respectively, as we have seen. Technically, the enthalpy $H$ encountered in Section 3.1 is also a thermodynamic potential, but not one that finds practical use so easily. Thermodynamic potentials represent the relevant part of the availability, which in turn is a representation of the total entropy of a system and its environment. When a constraint is removed to initiate change, it is a rule of Nature that systems evolve to a new equilibrium state that minimises availability. The second law operates in different guises and a myriad of physical phenomena seem to correspond to consequences of the deceptively simple condition that $\Delta S_i \geq 0$.

## 3.3 Maxwell Relations

James Clerk Maxwell (1831–1879) made seminal contributions to classical thermodynamics, one of which was the derivation of relations that bear his name, and provide connections between derivatives of state variables. They arise from expressions such as

$$dF = -p dV + \mu dN - S dT, \tag{3.33}$$

that relate a small change in a thermodynamic potential to small changes in other state variables. We derived this example in Section 3.2.1.

Now, an increment in the function $F(V, N, T)$ may be written quite generally as

$$dF = \left(\frac{\partial F}{\partial V}\right)_{N,T} dV + \left(\frac{\partial F}{\partial N}\right)_{V,T} dN + \left(\frac{\partial F}{\partial T}\right)_{N,V} dT, \tag{3.34}$$

and therefore we can make identifications

$$-p = \left(\frac{\partial F}{\partial V}\right)_{N,T} \qquad \mu = \left(\frac{\partial F}{\partial N}\right)_{V,T} \qquad -S = \left(\frac{\partial F}{\partial T}\right)_{N,V}, \tag{3.35}$$

one of which is the expression for the chemical potential derived in (3.21). But as the ordering of differentiation in second order partial derivatives is immaterial, such that

$$\left(\frac{\partial}{\partial V}\right)_{N,T}\left(\frac{\partial F}{\partial T}\right)_{N,V} = \left(\frac{\partial}{\partial T}\right)_{N,V}\left(\frac{\partial F}{\partial V}\right)_{N,T}, \tag{3.36}$$

we can deduce that

$$\left(\frac{\partial S}{\partial V}\right)_{N,T} = \left(\frac{\partial p}{\partial T}\right)_{N,V}, \tag{3.37}$$

and this is a Maxwell relation. We shall find this one very useful in Section 3.8.

Another Maxwell relation can be derived starting with

$$\left(\frac{\partial}{\partial N}\right)_{T,V}\left(\frac{\partial F}{\partial T}\right)_{N,V} = \left(\frac{\partial}{\partial T}\right)_{N,V}\left(\frac{\partial F}{\partial N}\right)_{V,T}, \tag{3.38}$$

giving

$$-\left(\frac{\partial S}{\partial N}\right)_{T,V} = \left(\frac{\partial \mu}{\partial T}\right)_{N,V}. \tag{3.39}$$

Alternatively, we could start with the derivative of the Gibbs free energy obtained in (3.27):

$$\mathrm{d}G = -S\,\mathrm{d}T + V\,\mathrm{d}p + \mu \mathrm{d}N. \tag{3.40}$$

Maxwell relations that emerge from (3.40) include

$$\left(\frac{\partial S}{\partial p}\right)_{N,T} = -\left(\frac{\partial V}{\partial T}\right)_{N,p}. \tag{3.41}$$

Maxwell relations are quite general relations, valid for any substance. They are very powerful for establishing connections between different thermodynamic properties, as we shall see.

## 3.4 Nonideal Classical Gas

The monatomic classical ideal gas gave us the means to explore various phenomena involving entropy production in Chapter 2, but it is a very simple system. In order to understand the thermodynamic behaviour of more realistic substances, we need to employ more complex model equations of state. We shall begin this development by considering the virial expansion. The ideal gas equation of state $p = nkT$ is valid when the particle density $n = N/V$ is small, such that particles are typically far apart and the assumption that the particles do not interact with each other is a reasonable approximation, if the interactions are short range. In contrast, the virial expansion consists of expressing the pressure of a nonideal gas as a power series in $n$:

$$\frac{p}{kT} = \sum_{i=1}^{\infty} B_i(T) n^i, \tag{3.42}$$

where $B_1 = 1$, and $B_i$ for $i \geq 2$ is the $i$th virial coefficient, taken to be a function of temperature and fitted to experimental data. The series when truncated at the second virial coefficient $B_2$ is explored further in Section 3.8.

Johannes van der Waals (1837–1923) proposed an equation of state designed to represent the properties of gas and liquid phases of a fluid in a single expression. It takes the form

$$\left[p + a\left(\frac{N}{V}\right)^2\right](V - bN) = NkT, \tag{3.43}$$

where $a$ and $b$ are positive parameters. The argument for the inclusion of the term proportional to the $b$ parameter rests on noting that particles are not points, but possess a finite volume, a consequence of which is that the centres of mass of the particles are only able to move within a volume that is smaller than the container volume $V$. The reduction in available volume is proportional to $N$, and the volume deficit per particle is $b$.

The motivation for the term containing the $a$ parameter is that the particles interact with one another, such that the velocity with which a particle hits the walls of the container is reduced by attractive forces from neighbouring particles to its rear. The gas pressure is related to the impact velocity, as we saw in Section 2.3, and the argument can be adapted to employ a reduced velocity $v_x{}'$ due to deceleration before impact, such that $p = nm\langle v_x'^2\rangle$. Now, the reduction in squared velocity is proportional to the loss of kinetic energy, which is the gain in potential energy per particle on moving from the bulk of the gas to the wall. Writing $v_x^2 - v_x'^2 \propto n$, assuming that the change in potential energy is proportional to the number of particles within a specified neighbourhood of the particle, we conclude that $p = nm\langle v_x^2\rangle - an^2$, where $a$ is a positive constant. Accepting this modification together with the effective change in confining volume yields the van der Waals equation (3.43). We shall revisit this derivation in Chapter 14 when we have acquired further tools from statistical thermodynamics.

## 3.5 Relationship between Heat Capacities

Heat capacities not only tell us how the temperature of a system increases as heat is supplied, but also represent properties of the system entropy. The difference between heat capacities $C_p$ and $C_V$ at constant pressure and constant volume is of particular interest. Starting with the first law for the quasistatic delivery of heat and work to a system with fixed $N$, we write $\mathrm{d}Q = \mathrm{d}E + p\mathrm{d}V = T\mathrm{d}S$ and so

$$C_V = \left(\frac{\mathrm{d}Q}{\mathrm{d}T}\right)_{q;V,N} = \left(\frac{\partial E}{\partial T}\right)_{V,N} = T\left(\frac{\partial S}{\partial T}\right)_{V,N}, \tag{3.44}$$

where the notation of (2.13) is slightly extended, and

$$C_p = \left(\frac{\mathrm{d}Q}{\mathrm{d}T}\right)_{q;p,N} = \left(\frac{\partial E}{\partial T}\right)_{p,N} + p\left(\frac{\partial V}{\partial T}\right)_{p,N} = \left(\frac{\partial H}{\partial T}\right)_{p,N}, \tag{3.45}$$

relating $C_p$ to a derivative of enthalpy. But we can also write

$$C_p = T\left(\frac{\partial S}{\partial T}\right)_{p,N}, \tag{3.46}$$

and then

$$T\mathrm{d}S = T\left(\frac{\partial S}{\partial T}\right)_{V,N}\mathrm{d}T + T\left(\frac{\partial S}{\partial V}\right)_{T,N}\mathrm{d}V = C_V\mathrm{d}T + T\left(\frac{\partial p}{\partial T}\right)_{N,V}\mathrm{d}V, \tag{3.47}$$

using the Maxwell relation (3.37), such that

$$T\left(\frac{\partial S}{\partial T}\right)_{p,N} = C_p = C_V + T\left(\frac{\partial p}{\partial T}\right)_{N,V}\left(\frac{\partial V}{\partial T}\right)_{p,N}, \tag{3.48}$$

showing that the difference in heat capacities is related to properties of the system equation of state $p(N,V,T)$. For an ideal gas, for example, with $p = NkT/V$, we have $(\partial p/\partial T)_{N,V} = Nk/V = p/T$ and $(\partial V/\partial T)_{p,N} = Nk/p$ so we find that $C_p - C_V = Nk$. For most substances $C_p > C_V$, since the pressure and volume increase with temperature under the conditions of the partial derivatives on the right hand side of (3.48).

## 3.6 General Expression for an Adiabat

A relationship between state variables describing a system with constant entropy is called an adiabat, or isentrope, and we can derive the slope of such a line in the $T - V$ diagram. The relationship between constant volume and constant pressure heat capacities can be inserted back into (3.47) to give

$$TdS = C_V dT + (C_p - C_V)\left(\frac{\partial T}{\partial V}\right)_{p,N} dV, \tag{3.49}$$

and a condition for increments of $V$ and $T$ that leave the entropy of a system constant is

$$0 = dT + (\gamma - 1)\left(\frac{\partial T}{\partial V}\right)_{p,N} dV, \tag{3.50}$$

where $\gamma$ is the ratio of heat capacities $C_p/C_V$, such that

$$\left(\frac{\partial T}{\partial V}\right)_{S,N} = -(\gamma - 1)\left(\frac{\partial T}{\partial V}\right)_{p,N} \tag{3.51}$$

specifies adiabats in the $T - V$ diagram.

For an ideal gas, we have $(\partial T/\partial V)_{p,N} = p/Nk = T/V$ and $(\partial T/\partial V)_{S,N} = -(\gamma - 1)T/V$ which for constant $\gamma$ can be integrated to give $TV^{\gamma-1} =$ constant, or equivalently $pV^{\gamma} =$ constant. This is the form taken in (2.9), where the ratio $\gamma = 5/3$ applies. If we define the isobaric coefficient of thermal expansion as $\alpha = V^{-1}(\partial V/\partial T)_{p,N}$, which is evidently the fractional increase in volume of a substance with a fixed number of particles, due to an increase in temperature at constant pressure, then the slope of an adiabat for an arbitrary substance is given by

$$\left(\frac{\partial T}{\partial V}\right)_{S,N} = -\frac{\gamma(V,T) - 1}{V\alpha(V,T)}, \tag{3.52}$$

where we explicitly note the dependence of both $\gamma$ and $\alpha$ on $V$ and $T$.

## 3.7 Determination of Entropy from a Heat Capacity

Since entropy seems to be a central player in thermodynamics, it seems odd that we are not more familiar with instruments that can measure its value, in the manner of a thermometer. We now address how physical data might provide such an instrument, and emphasise that for such an instrument to work, there needs to be a reference system with known entropy.

Since the Clausius integral (2.23) is the most fundamental definition of entropy change, we should start by considering heat transfers and heat capacities. We write, using (3.44):

$$S(N,V,T) - S(N,V,T_0) = \int_{T_0}^{T}\left(\frac{\partial S}{\partial T'}\right)_{V,N} dT' = \int_{T_0}^{T}\frac{C_V(V,T')}{T'}dT', \tag{3.53}$$

or from (3.46)

$$S(N,p,T) - S(N,p,T_0) = \int_{T_0}^{T} \left(\frac{\partial S}{\partial T'}\right)_{p,N} dT' = \int_{T_0}^{T} \frac{C_p(p,T')}{T'} dT', \tag{3.54}$$

and these could lie behind the operation of an 'entropy-ometer' device to measure system entropy indirectly from measurements of temperature and volume or pressure. In order to calibrate it, heat capacities would need to be determined over the range $T_0$ to $T$. The most suitable temperature for the reference system is $T_0 = 0$ K, where the entropy of any system is defined to be zero according to the third law. It is interesting to contrast this with the reference point for the measurement of temperature, the triple point of water, which is rather more accessible to experiment.

## 3.8 Determination of Entropy from an Equation of State

Heat capacity data can provide the system entropy as a function of temperature as discussed in Section 3.7, while the volume or pressure dependence can be extracted from an empirical equation of state $p(N,V,T)$. We write

$$S(N,V,T) - S(N,V_0,T) = \int_{V_0}^{V} \left(\frac{\partial S}{\partial V'}\right)_{N,T} dV' = \int_{V_0}^{V} \left(\frac{\partial p}{\partial T}\right)_{N,V'} dV', \tag{3.55}$$

using the Maxwell relation (3.37). As an example, consider the virial expansion of the pressure of a monatomic nonideal gas, truncated after the second virial coefficient: $pV = NkT(1 + B_2(T)N/V)$, such that $|B_2(T)N/V| \ll 1$. Then

$$\begin{aligned} S(N,V,T) - S(N,V_0,T) &= \int_{V_0}^{V} \left[\frac{Nk}{V'} + \frac{N^2k}{V'^2}\frac{d(TB_2(T))}{dT}\right] dV' \\ &= Nk \ln\left(\frac{V}{V_0}\right) - N^2k\frac{d(TB_2(T))}{dT}\left[\frac{1}{V} - \frac{1}{V_0}\right]. \end{aligned} \tag{3.56}$$

The most appropriate reference case lies at $V_0 \to \infty$, when the properties of the gas approximate to those of an ideal gas and $S(N,V_0,T) \to S_{ig}(N,V_0,T) = Nk \ln\,[V_0(kT)^{3/2}/\hat{c}N]$ according to (2.22), such that

$$S(N,V,T) = Nk \ln\left(\frac{(kT)^{\frac{3}{2}}}{\hat{c}\,N/V}\right) - \frac{N^2k}{V}\frac{d(TB_2(T))}{dT} \tag{3.57}$$

is the first correction to the ideal gas entropy. A similar procedure could be used to obtain the pressure dependence of the entropy.

It is of interest to extract further conclusions from the virial expansion. Using (3.35) a difference in the Helmholtz free energy can be written as

$$\begin{aligned} F(N,V,T) - F(N,V_0,T) &= -\int_{V_0}^{V} p dV' = -NkT \ln\left(\frac{V}{V_0}\right) \\ &\quad + N^2kTB_2(T)\left(\frac{1}{V} - \frac{1}{V_0}\right), \end{aligned} \tag{3.58}$$

and so the change in energy $\Delta E = \Delta F + T\Delta S$ is

$$E(N,V,T) - E(N,V_0,T) = -N^2kT^2\frac{\mathrm{d}B_2(T)}{\mathrm{d}T}\left(\frac{1}{V} - \frac{1}{V_0}\right), \tag{3.59}$$

and we can use the same monatomic ideal gas reference case when $V_0 \to \infty$, namely $E(N,V_0,T) = (3/2)NkT$, to give

$$E(N,V,T) = \frac{3}{2}NkT - \frac{N^2kT^2}{V}\frac{\mathrm{d}B_2(T)}{\mathrm{d}T}. \tag{3.60}$$

The volume dependence indicates that after a free expansion of such a gas, there will be a temperature change, in contrast to the behaviour seen in Section 2.6 for an ideal gas.

The energy expression can be interpreted in the following way. The extra contribution is a potential energy of interactions between the particles, represented by the temperature dependence of the second virial coefficient. The first term is of course the kinetic energy of the system. If we insist that the potential energy is independent of the kinetic energy, or equivalently the temperature, then we require that

$$B_2(T) = b - \frac{a}{kT}, \tag{3.61}$$

where $a$ and $b$ are $T$-independent parameters, such that $\mathrm{d}B_2/\mathrm{d}T = a/kT^2$ and the extra contribution to the energy is independent of $T$. Then the energy and entropy of the gas are $E(N,V,T) = (3/2)NkT - aN^2/V$ and $S(N,V,T) = S_{\mathrm{ig}}(N,V,T) - bN^2k/V$. Furthermore, we can use this model for $B_2(T)$ to write the equation of state in the form

$$p = \frac{NkT}{V}\left(1 + \frac{bN}{V} - \frac{aN}{kTV}\right) \quad \Rightarrow \quad (p + an^2)(1 + bn)^{-1} = nkT, \tag{3.62}$$

where $n = N/V$ is the particle density, which takes a form very reminiscent of the van der Waals equation of state (3.43).

The manipulations of the various expressions in the last few sections illustrate that unexpected connections can be made between measurable material properties. This has been the principal role played by classical thermodynamics since its inception, but the unfortunate aspect is that it involves a considerable amount of calculus, which can be quite unsettling. There have not been many diagrams that help in understanding. Perhaps the effort has yielded results that might appeal only to a constructor of Carnot engines, as it has allowed us to specify the shape of adiabats or equivalently the $T$ and $V$ dependence of entropy, as illustrated in Figure 3.4. Not that there is anything intrinsically wrong with an intense interest in entropy! But in the next section, we encounter phenomena that are much more dramatic and physically important: transformations of phase.

## 3.9 Phase Transitions and Phase Diagrams

A discontinuous phase transition is a fascinating phenomenon whereby a small change in a system parameter, such as temperature, produces a large change in properties of the system. Familiar examples include the melting of a solid to produce a liquid, and the

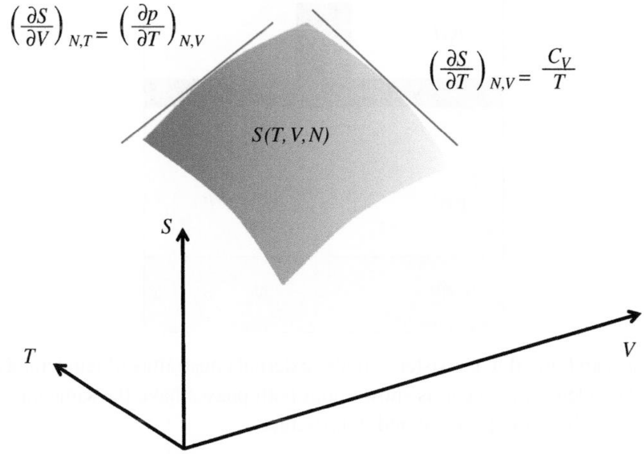


**Figure 3.4** Entropy as a function of $T$ and $V$ for constant $N$, showing how the gradients are related to the heat capacity and the equation of state.

boiling of a liquid to produce a gas, transitions that occur at a particular temperature for a given pressure. The key point is that at the transition temperature, two (or more) *phases*, or distinctly different states of the system, are in coexistence. This means they are in equilibrium with respect to the transfer of particles from one to the other, and this implies that they must have the same chemical potential: we have learnt that this is the indicator of equilibrium between systems able to exchange particles. We need to calculate the chemical potential of different phases and determine the conditions where they become equal. We have already established the chemical potential of the monatomic ideal classical gas in (2.53). We now need to consider other phases, such as liquids and solids.

### 3.9.1 Conditions for Coexistence

Consider a system containing two phases of the same substance that are able to exchange particles with one another, illustrated in Figure 3.5 for gas and liquid phases, in conditions where volume and energy can be exchanged with an environment while the system pressure and temperature are fixed. We showed in Section 3.2.4 that when the system has evolved to an equilibrium, it allocates $N_1$ of its particles to the gas phase and $N_2$ to the liquid phase, in order to minimise the total Gibbs free energy as a function of these variables. We write $G(N,p,T) = G_1(N_1,p,T) + G_2(N_2,p,T)$ and minimise it to give $\mu_1 \mathrm{d}N_1 + \mu_2 \mathrm{d}N_2 = 0$ where $\mu_{1,2}$ are the chemical potentials of the two phases. Since $N_1 + N_2$ is fixed, $\mathrm{d}N_1 = -\mathrm{d}N_2$, and this means that the chemical potentials of the two phases inside the system should be equal.

We found in Section 3.2.4 that the chemical potential of a system is related to its Gibbs free energy in the following way:

$$\mu = \left(\frac{\partial G}{\partial N}\right)_{T,p}. \tag{3.63}$$

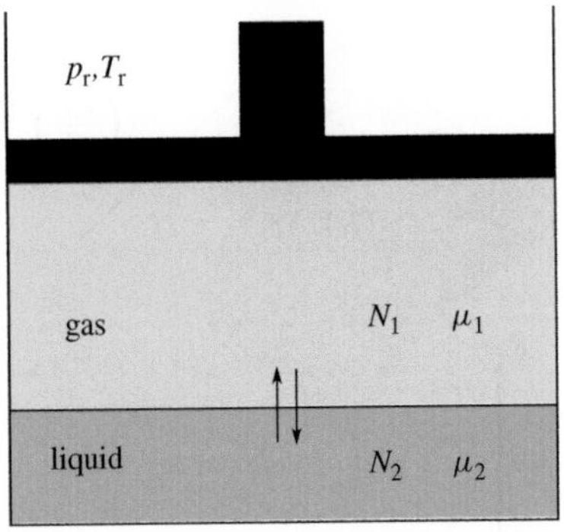


**Figure 3.5** A gas and liquid in coexistence under external constraints of temperature and pressure. Evaporation and condensation occur as shown until both phases have the same chemical potential, $\mu_1 = \mu_2$, as well as the same pressure and temperature.

Note that the state functions $E$, $S$ and $V$ are all extensive, that is, proportional to system size, while $p$ and $T$ are intensive. Thus $G = E - TS + pV$ is also an extensive state variable and we recall that it is a natural function of $N$, $p$ and $T$. The consequence of this is that we should be able to write $G(N,p,T) \propto N$ or $G(N,p,T) = K(p,T)N$, where $K(p,T)$ is a function to be determined. In fact $K(p,T) = (\partial G/\partial N)_{T,p} = \mu$ and hence

$$G = \mu(p,T)N. \tag{3.64}$$

Thus the chemical potential is the Gibbs free energy per particle, and it may be regarded as a natural function of pressure and temperature. Now we explore the consequences of the coexistence condition $\mu_1(p,T) = \mu_2(p,T)$.

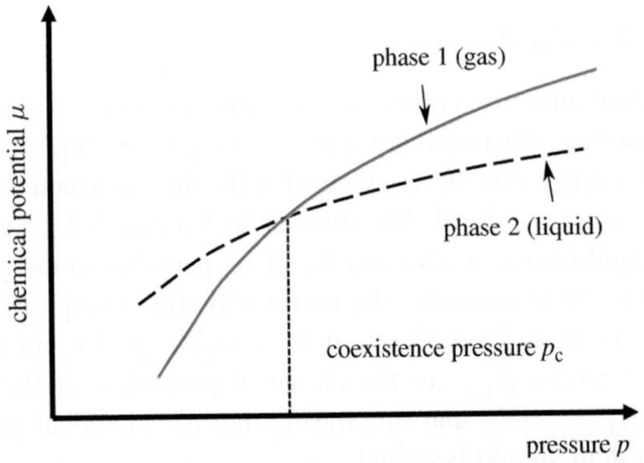


**Figure 3.6** Chemical potentials of two phases (say liquid and gas) as a function of pressure at a given temperature. The phase with the lower chemical potential at a given pressure and temperature is thermodynamically stable, while the other is *metastable*. If the pressure exerted on a system is increased past the coexistence pressure, at constant temperature, a phase transition from phase 1 to phase 2 occurs.

### 3.9.2 Clausius–Clapeyron Equation

The chemical potentials of two phases typically depend on $p$ and $T$ in quite different ways. But at a specified temperature, there should be a pressure at which they become equal, and where phase coexistence is possible. This is illustrated in Figure 3.6. Changing to a different temperature will shift the value of the coexistence pressure. The plot of coexistence pressure against temperature is an example of a *phase diagram*, a description of conditions under which different phases of the system are thermodynamically stable.

We can establish some properties of a boundary between phases on a $p-T$ diagram using (3.27) and (3.64). We write

$$dG = -S\,dT + V\,dp + \mu dN = d(\mu N) = \mu dN + N\,d\mu, \tag{3.65}$$

and so

$$d\mu = -s\,dT + v\,dp \tag{3.66}$$

specifies how the chemical potential of a phase depends on the pressure and temperature. This is known as the Gibbs–Duhem equation, where $s = S/N$ and $v = V/N$ denote the entropy and volume per particle, respectively, also known as the specific entropy and volume. The Gibbs–Duhem equation implies that

$$\left(\frac{\partial \mu}{\partial p}\right)_T = v, \tag{3.67}$$

and it is the different specific volumes of the two phases that give the two curves in Figure 3.6 different (positive) slopes, normally obliging them to intersect somewhere. For example, if phase 1 is a gas, then its specific volume is greater than that of a liquid phase 2, and the gradients of the chemical potentials reflect this. Figure 3.6 also indicates that if the system pressure is lower than the coexistence pressure, then phase 1 is selected, as this has the lower chemical potential and partitioning all the particles into phase 1 minimises the Gibbs free energy. As the pressure crosses the coexistence pressure, a phase transition takes place and the system assumes phase 2, again in order to minimise the Gibbs free energy.

We now consider a temperature $T$ and pressure $p$ where the chemical potentials $\mu_1$ and $\mu_2$ of phase 1 and phase 2 are equal, such that the point $(p, T)$ lies on a boundary in the phase diagram, as illustrated in Figure 3.7. The phase boundary at a given temperature corresponds to the pressure in Figure 3.6 where the two curves of chemical potential intersect.

Now, if we change the temperature to $T + dT$ and the pressure to $p + dp$, the chemical potentials of the two phases change according to the Gibbs–Duhem equation:

$$d\mu_{1,2} = -s_{1,2}dT + v_{1,2}dp, \tag{3.68}$$

where the suffices label each phase. In order that the new point $(p + dp, T + dT)$ also lies on the phase boundary, the chemical potentials of the two phases should remain equal. We therefore require that the increments are equal, that is, $d\mu_1 = d\mu_2$, implying that the phase boundary in the $p-T$ diagram is specified by the following relationship between $dT$ and $dp$:

$$(s_1 - s_2)dT = (v_1 - v_2)dp. \tag{3.69}$$

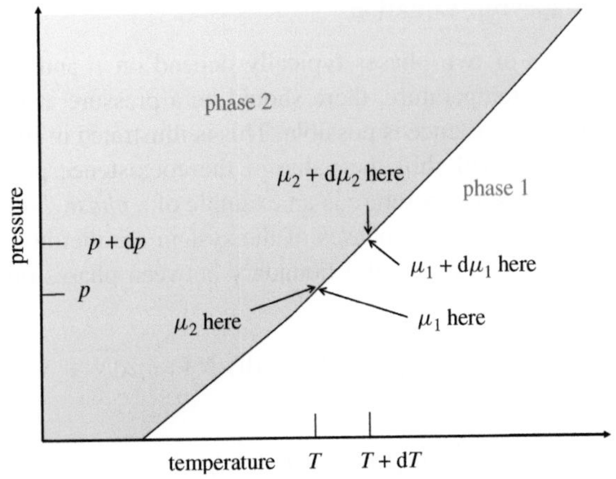


**Figure 3.7** Thermodynamic analysis that leads to the Clausius–Clapeyron equation for a boundary in a phase diagram. The chemical potentials $\mu_1$ and $\mu_2$ are equal under conditions $p$ and $T$, and remain so when they change by $d\mu_1$ and $d\mu_2$, brought about by changes in conditions $dT$ and $dp$.

This leads to the so-called Clausius–Clapeyron equation for the coexistence pressure $p_c(T)$:

$$\frac{dp_c}{dT} = \frac{s_1 - s_2}{v_1 - v_2}. \tag{3.70}$$

The entropy difference per particle in the numerator on the right hand side can be related to the specific latent heat of the transformation $L$. This is the heat that must be provided per particle to drive a quasistatic phase transformation at the temperature $T$. Using the Clausius relationship between quasistatic heat transfer and entropy change, we have $s_1 - s_2 = L/T$. We employ a convention that phase 1 is obtained from phase 2 by the addition of a positive latent heat: phase 1 has the higher entropy per particle under conditions of coexistence. Hence

$$\frac{dp_c}{dT} = \frac{L}{T(v_1 - v_2)}. \tag{3.71}$$

For a gas–liquid phase boundary, with gas labelled as phase 1 and liquid as phase 2, we can make the approximation $v_1 \gg v_2$ to reflect the difference in density (as long as we consider conditions well away from the critical point), and use $v_1 = kT/p$ such that

$$\frac{dp_e}{dT} = \frac{L_e p_e}{kT^2}, \tag{3.72}$$

where the notation now follows that employed in Sections 3.3.3 and 3.3.5 to represent the saturated vapour pressure. Assuming that the specific latent heat of evaporation $L_e$ is independent of temperature and pressure, this may be integrated to give

$$p_e(T) = p_0 \exp\left(-\frac{L_e}{k}\left(\frac{1}{T} - \frac{1}{T_0}\right)\right), \tag{3.73}$$

and so the saturated vapour pressure takes a form reminiscent of (3.32). For a phase boundary between gas (phase 1) and solid (phase 2), the result (3.73) also applies, but

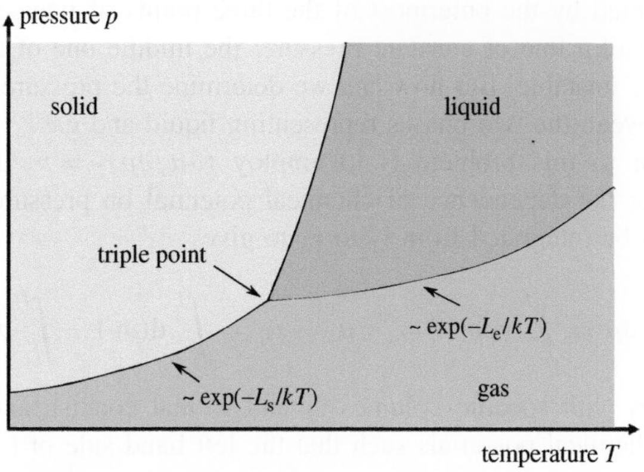


**Figure 3.8** Phase diagram of a typical material, with phase boundaries specified by various applications of the Clausius–Clapeyron equation. $L_e$ and $L_s$ are respectively the latent heats of evaporation and sublimation per particle.

with a specific latent heat of *sublimation* $L_s$ specifying a saturated vapour pressure $p_s(T)$ with respect to the solid phase.

For the phase boundary between solid and liquid, (3.71) takes the form

$$\frac{dp_f}{dT} = \frac{L_f}{T\Delta v}, \tag{3.74}$$

where $L_f$ is the specific latent heat of fusion, and $\Delta v$ is the difference in volume per particle associated with melting solid (phase 2) to liquid (phase 1), which is roughly temperature and pressure independent. The phase boundary is then $p_f(T) = (L_f/\Delta v)\ln T +$ constant, and as $\Delta v$ is usually rather small, it is much steeper than the gas-to-condensed phase boundaries. For ice, which contracts on melting, $\Delta v < 0$ and the slope of the phase boundary is negative, but for most substances it is positive. A phase diagram of a typical material therefore takes the form illustrated in Figure 3.8. The point at the intersection of the three boundaries, at which solid, liquid and gas phases are in coexistence, is the triple point.

### 3.9.3 The Maxwell Equal Areas Construction

Some model equations of state, such as the van der Waals equation encountered in Section 3.4, give rise to an unphysical 'loop' (actually a wiggle) on a $p - V$ plot, as shown in Figure 3.9. We have used dimensionless coordinates $pb/kT$ to represent pressure and $V/Nb$ to represent volume, and chosen parameters such that $a = 4bkT$ to illustrate the behaviour. The ideal gas law is shown using the same coordinates to indicate that the pressure is reduced to reflect the interparticle attraction. Such a wiggle is seen in the equation of state when the temperature lies below a certain threshold and it is clearly nonsense. What kind of gas increases its pressure when the volume is increased, which is suggested for specific volumes $V/N$ between about $2b$ and $5b$? This is an artefact of the model, arising from the assumption that the fluid is homogeneous.

The wiggle actually tells us that for the given temperature, the system separates into phases at two different densities, indicated by two specific volumes at the same pressure.

These are represented by the outermost of the three points of intersection between the equation of state and a line of constant pressure: the middle one of the three turns out to be mechanically unstable. But how can we determine the pressure $p_e$ at which there is equilibrium between the two phases representing liquid and gas?

A neat solution to this problem is to employ $(\partial \mu/\partial p)_T = v$, the Gibbs–Duhem equation (3.68) for the dependence of chemical potential on pressure at constant temperature. This can be integrated from $v_i$ to $v_f$ to give

$$\int_i^f d\mu = \int_i^f v\,dp \quad \Rightarrow \quad \mu_f - \mu_i = \int_i^f d(pv) - \int_i^f p\,dv. \tag{3.75}$$

We consider states with specific volumes $v_f$ and $v_i$ that coexist, and so by definition they have equal chemical potentials such that the left hand side of (3.75) is zero. Also the pressure $p_e$ of the coexisting phases must be the same, so $\int_i^f d(pv) = p_f v_f - p_i v_i = p_e(v_f - v_i) = p_e \int_i^f dv$. Using (3.75) we can therefore conclude that $p_e$, $v_f$ and $v_i$ satisfy

$$\int_{v_i}^{v_f} (p(v) - p_e)dv = 0. \tag{3.76}$$

On a $p - V$ plot this condition is readily interpreted geometrically, as shown in Figure 3.9 for the van der Waals equation of state. The coexistence pressure and associated coexisting specific volumes of the phases are determined by a condition of equality between the two areas defined by the three intersections between the $p - V$ curve provided by the equation of state and the horizontal line at $p = p_e$. This is called the Maxwell equal areas construction. The equilibrium equation of state then properly consists of the horizontal

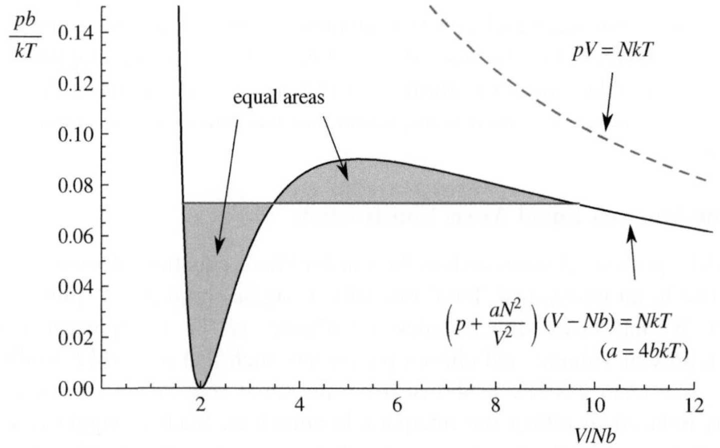


**Figure 3.9** Isotherm of the van der Waals equation of state with parameter choice $a = 4bkT$, together with the Maxwell equal areas construction that determines the specific volumes $v = V/N$ of the coexisting liquid and gas phases at a given temperature. The ideal gas law is shown as a dashed line for comparison.

line at the coexistence pressure, together with the van der Waals equation for $v > v_f$ and $v < v_i$. A system with a volume per particle somewhere between $v_i$ and $v_f$ will consist of an appropriate mixture of the two coexisting phases.

### 3.9.4 Metastability and Nucleation

We have just seen that some model equations of state are inappropriate for the range of specific volumes (or equivalently system densities) between those of the coexisting phases. But in fact the Maxwell construction and phase diagrams such as Figure 3.8 are also misleading. All phase diagrams tell only part of the story because they fail to take into account the physical difficulty a system might have in carrying out a transition between phases as demanded by the second law. A familiar example is the case of diamond: a phase of carbon that is stable at high temperatures and pressures but is also found at room temperature and pressure, where graphite is supposed to be the stable phase.

The diamond phase under such conditions is called *metastable*. It fails to comply with the second law because it cannot easily achieve the necessary atomic rearrangements to move towards a minimum of the free energy. One way to view this is that the necessary initial stages of the rearrangement correspond to structures with a *higher* free energy, and can only be achieved through a thermal fluctuation, apparently a temporary violation of the second law. The system is locked into an inappropriate phase by what is called a *nucleation* barrier. The diamond phase is metastable at room temperature: its transition rate into graphite is negligibly small.

Similarly, it is possible to compress a gas isothermally beyond the coexistence pressure suggested in Figure 3.9. It approximately follows the curve beyond this point and becomes *supersaturated*. Equivalently, the system can pass across the phase boundary from the gas into the liquid region of Figure 3.8, but remain gaseous. Again, the difficulty is that although the liquid phase has a lower chemical potential and is thermodynamically more stable, the tiny droplets that need to be formed in order that the new phase can emerge are *less* stable than the mother phase of supersaturated gas: they have a higher chemical potential. It is therefore improbable that the system should rearrange itself to form small droplets: there is a nucleation barrier. In certain circumstances, gases can be compressed to multiples of the saturated vapour pressure, without condensing.

The freezing of a liquid can be impeded by a nucleation barrier. Water can be supercooled to $-40\,^{\circ}\mathrm{C}$ in spite of the insistence in phase diagrams that it should freeze at zero celsius. In order to achieve this supercooling, the water must be free of impurities or solid surfaces that can nucleate ice. All this is evidence that thermodynamic processes are sometimes controlled by events that take place between the initial and final equilibrium states, and these are not always macroscopic in scale.

## 3.10 Work Processes without Volume Change

Work does not always just involve the mechanical compression of a system: this is the standard example but other cases exist. For solids, work can be performed by distorting the shape without changing the volume. For example, consider the extension $dL$ of a wire under an applied force $\mathcal{F}$ accompanied by a lateral contraction of the wire to conserve volume. If the process is quasistatic, $\mathcal{F}$ describes the equilibrium tension of the system, a state variable analogous to pressure. We write $dW = \mathcal{F}dL$ and add it to the term $-pdV$ that describes a volume change. Similarly, the distortion can produce a change $d\mathcal{A}$ in the surface area of the system without a change in the volume. Work of this kind is performed if we squeeze a spherical droplet into an oblate shape. The corresponding quasistatic work term is written as $dW = \Gamma d\mathcal{A}$ where the coefficient $\Gamma$, known as the surface tension, is again a state variable that describes a system in equilibrium.

The first law $dE = dQ + dW$ and the Clausius expression for an entropy increment then yield an extended fundamental relation:

$$dE = TdS - pdV + \mu dN + \mathcal{F}dL + \Gamma d\mathcal{A}. \tag{3.77}$$

The work terms take the form $Xdx$, where $X$ and $x$ are intensive and extensive variables, respectively. State variables such $E$ and $S$, for example, then acquire a dependence on the new variables $L$ and $\mathcal{A}$. It can become rather complicated: this is why volume work is almost always the standard example considered!

## 3.11 Consequences of the Third Law

Like the other classical laws of thermodynamics, the third law is a statement of empirical observation. We have seen in this chapter how the entropy of a system may be reconstructed through measurements of heat capacities, equations of state, or derived from chemical potentials inferred from phase coexistence. Nearly all this data is compatible with the idea that as $T \to 0$, the entropy of a system goes to a constant, and furthermore, that it is the same constant for all systems. A very few systems appear to have a capacity for nonzero entropy even as the temperature goes to zero, although it is not clear how this behaviour might be exploited. For compatibility with the statistical interpretation of entropy, to be discussed later, the standard reference value of the entropy of systems at $T = 0$ K is taken to be zero.

The principal implication of the third law is that absolute zero temperature is unattainable. This can be illustrated in Figure 3.10, where two possible scenarios involving the entropy function $S(T, V)$ are shown. On the left, the system entropy at $T = 0$ depends on the volume $V$, whereas on the right it does not: the third law is satisfied in the second case. We can imagine sequences of isothermal and adiabatic processes involving changes in volume that drive the system along the zigzag path between functions $S(T, V_1)$ and $S(T, V_2)$ starting from a finite temperature. If the system did not satisfy the third law, then after a few compressions and expansions shown on the left, the system entropy would be reduced to its value at some $V$ between the two extremes, and at a temperature of zero. For the system that satisfied the third law, in contrast, this approach to zero would require an infinite number of steps. Of course there are many technical difficulties in cooling a system, such as the elimination of leakage of heat from the environment, but

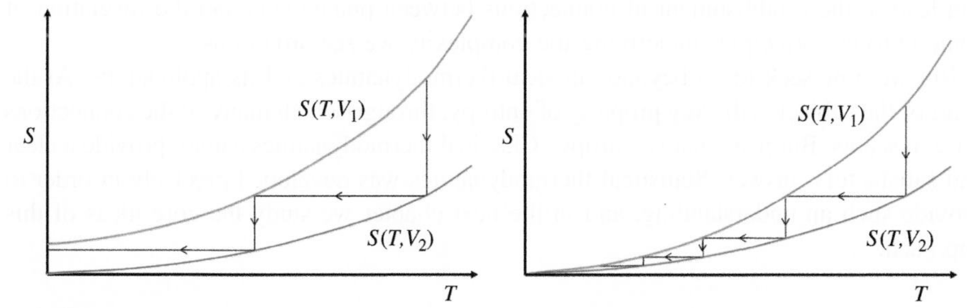


**Figure 3.10** The diagram on the left shows how a temperature of absolute zero can be reached if a system has an entropy function that depends on volume at $T = 0$. The two curves represent the system entropy as a function of temperature at two volumes with $V_1 > V_2$. The zigzag path is followed by a sequence of isothermal compressions (moves downward), followed by quasistatic adiabatic expansions (moves to the left). In contrast, if $S(0, V_1) = S(0, V_2)$ as shown on the right, then an infinite sequence would be needed to reach absolute zero, making it unattainable.

the third law states that we cannot entirely remove all the thermal energy in a system, even in principle.

But if we accept the third law as a necessary boundary condition for models of entropy, then we must conclude that the entropy expression for the ideal gas (2.22), on which we have built much of our intuition, cannot be correct. It quite clearly does not satisfy the third law, tending towards $-\infty$ rather than zero as $T \to 0$! However, the model is *classical*: we need not despair. We might have expected (2.22) to fail in the low temperature limit since we know that classical physics should then be replaced by a quantum mechanical treatment. In later chapters, we shall employ a more appropriate quantum model of the ideal gas, within a framework of statistical mechanics, and find that low temperature behaviour that satisfies the third law emerges, along with some unexpected richness in phenomena.

## 3.12 Limitations of Classical Thermodynamics

Rather than focusing on its limitations, perhaps we should celebrate the successes of classical thermodynamics, only some of which have been addressed in this chapter. The very power of the approach is that it does not particularly rely on specific assumptions about the interactions between the particles in a system. The nature of these interactions is inferred from measurements of macroscopic properties such as an equation of state. The principal value of the discipline is that experimental measurements can be related to one another in ways that are not at first apparent, for example the connection between heat capacities and mechanical properties in (3.48). Indeed this represents the very purpose of theoretical studies, whereby making a few assumptions about the way the world works can lead to the establishment of connections between phenomena and the revelation of some unifying principles underlying the complexity we see around us.

But we now seek to go beyond classical thermodynamics and its applications. At the heart of the subject is the key property of entropy, through which many of the connections arise, it seems. But just what *is* entropy? Classical thermodynamics cannot provide a clear and satisfactory answer. Statistical thermodynamics was developed precisely in order to provide such an understanding, and in the next chapter we study the core ideas of this approach.

## Exercises

**3.1** Taking the enthalpy density of an incompressible liquid to have the form $h = AT + Bp$, where $A$ and $B$ are positive constants, show that the entropy generated per unit volume of fluid flowing through a plug with a drop in pressure from $p_\mathrm{i}$ to $p_\mathrm{o}$ is given by $v(A/B)\ln\,[1 + B(p_\mathrm{i} - p_\mathrm{o})/(AT_\mathrm{i})]$ where $T_\mathrm{i}$ is the initial temperature.

**3.2** The isothermal compressibility $\kappa_T$, the thermal expansion coefficient at constant pressure $\alpha$ and the heat capacity at constant volume $C_V$ of a system are defined as

$$\kappa_T = -\frac{1}{V}\left(\frac{\partial V}{\partial p}\right)_T, \quad \alpha = \frac{1}{V}\left(\frac{\partial V}{\partial T}\right)_p \quad \text{and} \quad C_V = T\left(\frac{\partial S}{\partial T}\right)_V,$$

where constancy of $N$ is understood. Show that the gradient of an adiabat of the system on the $T - V$ plane is given by

$$\left(\frac{\partial T}{\partial V}\right)_S = -\frac{T\alpha}{C_V\kappa_T}.$$

You may use the identities

$$\left(\frac{\partial x}{\partial y}\right)_z\left(\frac{\partial y}{\partial z}\right)_x\left(\frac{\partial z}{\partial x}\right)_y = -1 \quad \text{and} \quad \left(\frac{\partial x}{\partial y}\right)_z = \left[\left(\frac{\partial y}{\partial x}\right)_z\right]^{-1}.$$

Hence determine the equation of an adiabat in the $T - V$ plane for (a) an ideal gas and (b) a substance with constant intensive thermal properties $\alpha_T$ and $\kappa_T$, and a temperature independent but extensive $C_V$.

**3.3** Derive a Maxwell relation involving the quantity $(\partial S/\partial p)_T$ and check that it is satisfied by an ideal gas.

**3.4** Express the Gibbs free energy $G$ and enthalpy $H$ of an ideal gas in terms of $p$, $T$ and $N$. Show that in general $(\partial(G/T)/\partial T)_{p,N} = -H/T^2$ and demonstrate that this result is satisfied by an ideal gas.

**3.5** A system can take two phases, liquid or solid. Above the melting temperature, which phase has the lower chemical potential, and why? At which temperature do both phases have the same chemical potential?

**3.6** Starting from the fundamental relation of thermodynamics, show that

$$\mathrm{d}S = \left[\frac{1}{T}\left(\frac{\partial E}{\partial V}\right)_T + \frac{p}{T}\right]\mathrm{d}V + \frac{1}{T}\left(\frac{\partial E}{\partial T}\right)_V \mathrm{d}T,$$

and derive expressions for $(\partial S/\partial V)_T$ and $(\partial S/\partial T)_V$. Hence show that

$$\left(\frac{\partial E}{\partial V}\right)_T = T^2\left(\frac{\partial(p/T)}{\partial T}\right)_V.$$

Use this relationship to demonstrate that the energy of an ideal classical gas does not change when it is expanded or compressed at constant temperature. Show that the energy of a van der Waals gas does change on isothermal expansion. Does it increase or decrease? Argue physically in support of the direction of change that you deduce. Calculate $(\partial S/\partial V)_T$ for the van der Waals gas.

**3.7** Show that the difference in *specific* heat capacities (the heat capacity per unit mass) can be written as

$$c_p = c_V + \frac{T\alpha^2}{\rho\kappa_T},$$

where $\rho$ is the mass density. Search online for data on the physical properties of liquid water and hence estimate the ratio $(c_p - c_V)/c_V$.

**3.8** Write down the Helmholtz free energy of a neutral plasma of $N_{\rm H}$ atoms of hydrogen, $N_e$ electrons and $N_p = N_e$ free protons at a fixed system volume and temperature, modelling all three components as monatomic ideal classical gases, but taking into account the fact that the energy of the hydrogen atom lies a dissociation energy $\epsilon > 0$ below that of its separated components. Note that the thermal de Broglie wavelength of an electron $\lambda_{{\rm th},e}$ differs from that of a proton, but that the thermal de Broglie wavelengths of a proton and a hydrogen atom are approximately the same. Regarding $N_{\rm H}$ as an unconstrained internal variable, and imposing the constraint $N_{\rm H} + N_p =$ constant, show that the densities of the three components at a given temperature are related by the Saha relation:

$$\frac{n_e n_p}{n_{\rm H}} = \frac{1}{\lambda^3_{{\rm th},e}(T)}\exp\left(-\frac{\epsilon}{kT}\right),$$

corresponding to a balance between the sum of chemical potentials on either side of the reaction $\mathrm{H} \rightleftharpoons e + p$. The fraction of ionised hydrogen in the photosphere (the visible outer layer) of the sun, where the temperature is about 6000 K and the particle density approximately $10^{23}\,\mathrm{m}^{-3}$, is about $10^{-4}$. *Estimate* the ionised fraction if the plasma were heated to $2 \times 10^4$ K at the same density, using $\epsilon = 13.6\,\mathrm{eV}$.

# 4

# Core Ideas of Statistical Thermodynamics

Statistical thermodynamics is an attempt to relate the phenomenological, macroscopic laws of classical thermodynamics to an underlying quantitative picture of molecular behaviour. Statistical mechanics is a broader term that includes dynamical systems not usually treated in thermodynamics, such as star clusters or even crowds of people. But since thermodynamics is applied to macroscopic quantities of gases, liquids and solids, containing of the order of $10^{23}$ particles, this task appears rather daunting. How can we establish the behaviour of this number of particles?

However, we clearly do not need to go so far because thermodynamic systems seem to be well enough characterised by just a handful of macroscopic quantities (energy, pressure, etc.) and the relationships between them. If we were to take the trouble of determining the behaviour of every molecule, then our efforts would be magnificent but pointless[1]: the motion of every molecule surely cannot affect the equation of state. The basic assumption of statistical thermodynamics is that the detail of the molecular motion is irrelevant. This leads to the argument that a study of the *likely* or the *average* behaviour of the particles is good enough. The statistical properties at the microscale are therefore our focus of attention, and this requires us to review our understanding of probability.

## 4.1 The Nature of Probability

It is often stated that if we roll a die, there is a probability of 1/6 that a six will be thrown. Around this apparently simple statement has raged a couple of centuries of philosophical debate.

When we know there to be a variety of possible outcomes of some event, but we cannot determine which will happen, probability is something we use to weight the outcomes in order to define our *expectation*. The probabilities we assign to the outcomes

[1] Rather like Samuel Johnson's view on cucumbers: 'It has been a common saying of physicians in England, that a cucumber should be well sliced, and dressed with pepper and vinegar, and then thrown out, as good for nothing.'

*Statistical Physics: An Entropic Approach*, First Edition. Ian Ford.

could be nothing more than a guess, in which case we would be creating an expectation that might not correlate very well with the actual event. Ideally, we should use some data, or information, to construct a set of probabilities that allow us to make good judgements about the future behaviour. This line of thought is called information theory, wherein probabilities are used as a basis for logical reasoning (see Jaynes in Further Reading). On the other hand, it doesn't sound like a very unique way of proceeding: how can we arrive at the best judgement?

Another point of view is that we should determine the actual frequencies with which each event should turn up in a number of trials, and use these to weight the outcomes. Then it would seem that we could derive an expectation that is tied to real documented behaviour, albeit from the past. However, the problem is that this is a unique approach only if we run an infinite number of trials. Otherwise, the frequencies would only be estimates. If we do not have time for that many trials, there are sophisticated ways of estimating the errors but we essentially revert to making a judgement about the probabilities, on the basis of the limited set of data.

The viewpoint that probabilities represent a distillation of our best judgement has some advantages. If we examine a die and reckon that it is symmetrical or true, we can make a judgement that each face is as likely as any other to come up in a throw. We have a basis for saying that the probability of each outcome is 1/6. This might be wrong: the die might be unfair. If so, then a few trials will provide us with information that will allow us to revise our probabilities in some way. If we were to cast the die a large number of times, then a true die would generate frequencies of the various outcomes that converge towards 1/6, in which case our initial guess was a good one. Or perhaps it is rigged and always throws a 1, indicating that our model of the system is flawed: our judgement was in error, our expectations are wrong and need to be revised.

Whether the numerical values are generated by some sort of judgement, or from trial data, the probabilities that we actually use to weight the events have to satisfy the same rules of arithmetic, so that the distinction need not bother us too much for the present. The basic point, though, is that an average over a set of probabilities might just be a best guess or a hypothesis based on a model, and that the accuracy of the model should be tested.

We can express an intuitive understanding of probability in the form of the following statements:

1. For each possible outcome $i$, there is a positive probability $P(i)$ denoting the statistical weighting (or limiting frequency if you prefer) for it to arise in a trial, such as the rolling of a die.
2. The sum of the $P(i)$ for all possible outcomes of a trial is unity, that is, $\sum_i P(i) = 1$.
3. Intuitively, outcomes related through some symmetry in the system should have the same probability; hence to begin with, we guess that the probabilities for each outcome of the die roll are the same and equal to 1/6.
4. The probability that either outcome $i$ *or* outcome $j$ should occur is given by the sum $P(i) + P(j)$, as long as the two outcomes are mutually exclusive. So the probability of throwing a five or a six is 1/3.
5. There is a probability $P_2(i,j)$ of joint outcomes, for example outcome $i$ as well as outcome $j$. If the events are *uncorrelated*, meaning that the outcome of one trial is

unaffected by the outcome of another, then the joint probability is the product of individual probabilities: $P_2(i,j) = P(i)P(j)$. The trials are said to be *independent*. If the outcomes are correlated, this factorisation does not apply. Thus the probability of throwing two sixes with two normal dice is $(1/6) \times (1/6) = 1/36$. If, on the other hand, the dice are connected in some spooky way, such that they never produce the same number, then the probability of two sixes is zero. The two outcomes in this case are correlated. This idea extends to longer sequences, $P_3(i,j,k)$, and so on. The joint probability can be expressed as a product of the probability of the first outcome times a *conditional* probability of the second outcome given the first, that is, $P_2(i,j) = P(i|j)P(j)$. The vertical line should be read as 'given'. The idea of independence is that $P(i|j) = P(i)$.

The $P(i)$ form a histogram that expresses the relative likelihoods of the outcomes of a trial. The histogram can be generalised to a continuous function if there is a continuum of possible outcomes. A cubic die has six faces and therefore six possible outcomes for a throw. A rolled pound coin, on the other hand, has a continuum of points on the circumference that might be uppermost when it stops (assuming it does not fall over!). We define $p(\theta)\mathrm{d}\theta$ as the probability that the uppermost point on the circumference lies between angles $\theta$ and $\theta + \mathrm{d}\theta$ measured from a vertical with respect to the Queen's head for example, where $p(\theta)$ is a *probability density function* or pdf. If the coin were perfectly circular, then we would guess $p(\theta) = 1/2\pi$, as normalisation in this case would require that $\int_0^{2\pi} p(\theta)\mathrm{d}\theta = 1$. This is analogous to our guess, based on the symmetry of the cube, that a die should roll a six with probability 1/6.

Arguably, the most important statistical properties of a set of probabilities, or probability distribution, are the *mean* and the *standard deviation*. If a variable $n$ is characterised by a discrete probability distribution $P(n)$ over its possible numerical values, then the mean is written using angled brackets as

$$\langle n \rangle = \sum_n nP(n), \tag{4.1}$$

where the sum is over the set of possible outcomes. The standard deviation $\sigma$ is defined as the square root of the variance given by $\sigma^2 = \langle (n - \langle n \rangle)^2 \rangle$; the mean square deviation from the mean. Thus, the mean throw of a true die is 3.5, with a standard deviation of about 1.7. The latter gives an indication of how much deviation from the mean we might expect to see in a typical measurement. This should not, however, mask the fact that the probability of each outcome in this case is the same!

For a continuous variable $x$ characterised by a pdf $p(x)$, the mean and variance are defined by $\langle x \rangle = \int xp(x)\mathrm{d}x$ and $\sigma^2 = \int (x - \langle x \rangle)^2 p(x)\mathrm{d}x$ with the integrals performed over the entire range of possible values of $x$. Note that $\sigma^2 = \int (x^2 - 2x\langle x \rangle + \langle x \rangle^2)\, p(x)\mathrm{d}x = \langle x^2 \rangle - \langle x \rangle^2$. We can define the mean of a function $f(x)$ of the variable $x$ as

$$\langle f(x) \rangle = \int f(x)p(x)\mathrm{d}x, \tag{4.2}$$

where the value of the function for an outcome $x$ is weighted by the probability of producing that outcome. We have already considered cases $f(x) = x$ and $f(x) = (x - \langle x \rangle)^2$.

The die and the coin are characterised by uniform probability distributions over orientation, but in many cases we see distributions with a peak. The most common distribution of this kind is called a Gaussian or normal distribution. It is entirely characterised by the mean and standard deviation. It is the 'bell shaped curve' specified by

$$p_G(x) = \frac{1}{(2\pi)^{\frac{1}{2}}\sigma} \exp\left(-\frac{(x - \langle x \rangle)^2}{2\sigma^2}\right), \tag{4.3}$$

which is normalised such that $\int_{-\infty}^{\infty} p_G(x)\mathrm{d}x = 1$. For example, the distribution of marks in an exam is expected to be approximately Gaussian (for a large class!) so a script selected at random from a stack of scripts should receive a mark in the range $x$ to $x + \mathrm{d}x$ with probability $p_G(x)\mathrm{d}x$, and the statistics are summarised by the two parameters $\langle x \rangle$ and $\sigma$ alone.

In statistical thermodynamics, we aim to establish a probability distribution or a pdf characterising aspects of a system's behaviour. These aspects could be microscopic, such as the values of particle velocities and positions, or macroscopic, such as spatial profiles of particle densities. Such distributions tell us what we need to know about the system's statistical behaviour, such as the mean energy per particle, the mean particle density at a given location or the standard deviation of these quantities. The crux of the matter is to identify such distributions. It seems that this may be done successfully for a system in *equilibrium*, where the properties are time independent and there are no mean flows of heat or particles into or out of the system. The treatment of systems where there are such flows, such as a system in the process of cooling down towards equilibrium, on the other hand, is much more complicated, and not yet fully understood. We shall return to this matter in Chapter 15.

## 4.2 Dynamics of Complex Systems

### 4.2.1 The Principle of Equal a Priori Probabilities

The probability that a system variable should take a value in certain range is clearly something that depends on the dynamics of the system. For example, if we wish to establish the pdf of the momentum of a particular molecule in a sample of gas, we ought to start with a consideration of the molecular dynamics. It turns out, though, that only rudimentary dynamical understanding is needed, at least for a system in thermal equilibrium.

It helps to steer clear of complicated dynamical systems at first and to focus on small discrete systems with simple dynamical rules. Let us start with a dynamical system consisting of several compartments, between which some conserved material is shared. The material comes in the form of indivisible units such that it is distributed in integer quantities between the compartments. The dynamical rule is that every timestep, a specified number of units of material, depending on the prevailing situation, moves from one compartment to another. This is not meant to be a realistic situation, though a version of it approximates to the behaviour of a set of weakly interacting quantised harmonic oscillators.

We consider first a system with two compartments. In fact, let us regard the compartments as two people and that they share a certain number of pound coins. Every couple of seconds one of them gives the other £*x*, according to a definite cashflow rulebook. For example, if A has £5 and B has £10, then at the next exchange, £2 is passed from A to B. From the resulting situation of £3 : £12, the cashflow rulebook says that the next transfer is £3 in the opposite direction. The rules could be of any kind, but let us suppose that they are such that the sequence of situations that is generated includes all possible arrangements of the £15 between the two people. The dynamics would start from an arbitrary point in the sequence, visit all the possibilities, and then it would repeat.

In the case just described, it would seem fair to say that if we observed the situation at an arbitrary later time (and neglected to spot regularities such as the repeat time), we should expect to find the cash distributed in any one of the possible arrangements with equal probability. The two participants generate all arrangements in a regular sequence and at an arbitrary time any one of them might be observed, none with any greater likelihood than another.

We make this more precise now by conceiving of a set of *microstates* of a dynamical system. A microstate in this example might be characterised by the number $q_A$ of units residing in compartment A and $q_B$ in compartment B, with $q_A + q_B = Q$, the fixed total number of units. The microstates are the states of a system we might observe as a snapshot during the evolution and in this case are labelled $(q_A, q_B)$: explicitly they form the set $(Q, 0), (Q-1, 1), \ldots, (1, Q-1), (0, Q)$ illustrated in Figure 4.1. There are $(Q+1)$ such elementary arrangements of the material between the two compartments, and a time series of exchanges would correspond to a path linking them in a sequence.

If the microstates were represented by keys on a piano, a path would consist of a sequence of individual notes, using every one on the keyboard, and then repeating. Each note would represent a snapshot of the system. It might not be music, but it would be dynamic! The probability that a note should be heard at a randomly chosen time is the same for all 88 of them. Explicitly, the probability that a system might be found in a particular microstate would be equal to $1/\Omega$, where $\Omega$ is the number of microstates, and equal to $Q + 1$ in this example.

The set of all microstates of a system, set out as in Figure 4.1, is called the system's *phase space*. This is an unfortunate use of the same word, phase, that is used to specify different states of matter such as solids, liquids and gases. Be aware that a *phase space* is not the same as a *phase diagram*!

Now we distort the logic to apply similar ideas to a dynamical system in the real world. Imagine that A and B exchange cash according to rules that we are not sure about. We want to predict the future, but on the basis of an incomplete knowledge about the rules of the game. How do we proceed?

All we can do is hypothesise about the effect of those rules and see how it works out, and our first guess is that there is no reason to believe the dynamics should favour any one configuration over another, as we do not have enough information to make such a judgement. This is the least biassed thing we can do! It means we should take the probability of finding the system in any individual configuration to be the *same*. Note that this is a probability that is a reflection of our judgement, and is not a frequency over

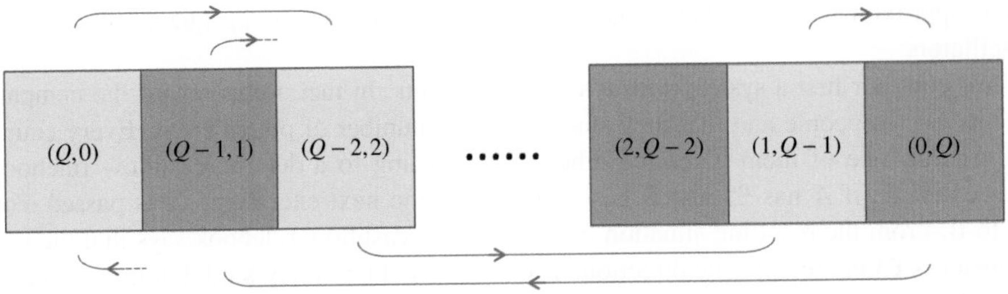


**Figure 4.1** Set of all possible divisions of $Q$ items between two compartments A and B, labelled by $(q_A, q_B)$ with $q_A + q_B = Q$. This forms a *phase space* of microstates of the system. The dynamics takes the system between microstates as shown by the arrows. In this example, the hopping rules are such that all microstates are visited before a return to the starting point.

real trials: we have not done any. Also, note that this is a time-independent assignment of probabilities, and so what we are considering here is a hypothesis about the values of the *equilibrium* microstate probabilities. How they behave for a system that is out of equilibrium, when the probabilities might depend on time, is another matter entirely.

Of course, the hypothesis could easily be flawed. A and B might operate rules whereby they never give their entire wealth to one another: we would never see them in the $(Q, 0)$ and $(0, Q)$ microstates. They might have favourite configurations, to which they return again and again. The nice cashflow rulebook situation that truly gave rise to equal frequencies of visits is a rather special case. Nevertheless, the equal likelihood hypothesis is a way to start thinking about the system, at least until it can be shown to be misleading in some significant way.

A system with unspecified dynamical rules evolves with some uncertainty, but in a similar way the future behaviour of a system with completely specified microscopic rules is also hard to pin down, if it is *large* or *complex* in some sense. The reason is that there will be so many equations of motion to solve and so many initial conditions to specify. Even if we had the rules of cashflow between a group of people labelled A–Z, we would find it tedious, at the very least, to work out the future wealth distribution. Our task would be to judge the probabilities of occurrence of each microstate, given some uncertainty in the initial conditions and practical limits in our ability to compute outcomes. In physical examples, we would have to follow the position and velocity of every particle: we would be at it forever! So rather than give up, perhaps we should make things very simple, and imagine as we did for the uncertain cash exchangers A and B that all of the microstates are equally likely. What a crazy idea.

But this is precisely the assumption upon which we base equilibrium statistical mechanics. The key word here is *equilibrium*, meaning that an appreciable amount of time has elapsed since the last external disturbance to the system, during which the dynamics have a chance to take the system through a reasonable selection of all the available microstates. The system is then considered to have settled down and acquired statistical properties that are time-independent.

Think of a gas of weakly interacting particles, into some of which an amount of kinetic energy (heat) is injected. Before long, the energy will be shared out amongst all the particles as a consequence of collisions, such that there is no mean gradient in temperature across the system and no gross time dependence in mean properties such as density: in short, an equilibrium situation has been reached. We don't quite know the rules of the dynamics and we don't know the initial configuration of all the particles when the energy was injected, but intuition suggests that this does not matter. It is in these situations that we boldly apply our assumption of equal microstate probabilities. This is known as the *principle of equal a priori probabilities*. If we observe an isolated system in equilibrium at an arbitrary time, we claim that it will be found in any one of its microstates with equal likelihood. Notice that part of this hypothesis is that the system is *isolated* and left to sort itself out. The phrase *a priori* refers to the fact that we are basing this hypothesis on the information available to us 'at the outset', which is actually very little. Now we are in a position to make statistical statements about the system properties and then to test them to see how the hypothesis fares.

There are significant problems when it comes to justifying this principle in realistic physical systems. On the positive side, it is known that if a system evolves according to classical or quantum mechanical Hamiltonian dynamics, and if the probability of every microstate is initially the same, then they will remain equal in the future, a result known as Liouville's theorem. Effort has gone into investigating the *ergodic hypothesis*; the idea that such dynamics really do take a system into each and every microstate with equal frequency after an infinite amount of time. This has produced some supportive conclusions, but actually it rather misses the point, as the principle is only a working hypothesis: a simple way of arriving at predictions of behaviour, and we should be prepared to modify it if need be.

In addition, even if the dynamics were ergodic, they would generate equally frequent visits to each microstate only after an *exceedingly* long time; at least as long as the time needed to see the original configuration restored (this is called the Poincaré recurrence time and it can be estimated to be greater than the age of the universe for even quite small physical systems). Only a very small sample of the microstates will actually be visited during a particular observation period. It is seriously crazy to claim that each microstate of a realistic dynamical system in equilibrium will actually turn up with equal *frequency* on making measurements over a finite period. But that does not stop us from using the principle of equal a priori probabilities as a *model* of the world: we just have to remember that the probabilities assigned to the microstates are not frequencies but representations of our best judgement, based on insufficient data, but guided by ideas of even-handedness in the absence of full information.

### 4.2.2 Microstate Enumeration

At this point, we put dynamics aside and focus more on the consequences of the assumption of equal microstate probabilities. Let us extend the model we introduced in the last section. For two compartments, the number of microstates is $Q + 1$, a number that clearly increases with $Q$. If we add more compartments, for a constant $Q$, we increase the number of microstates further. If we have three compartments, for example, the microstates may be labelled by the respective occupancies $(q_1, q_2, q_3)$ of the three. These can be visualised as points on a triangular surface inclined with respect to a 3-d set of Cartesian axes corresponding to the numbers $q_j$, as illustrated in Figure 4.2.

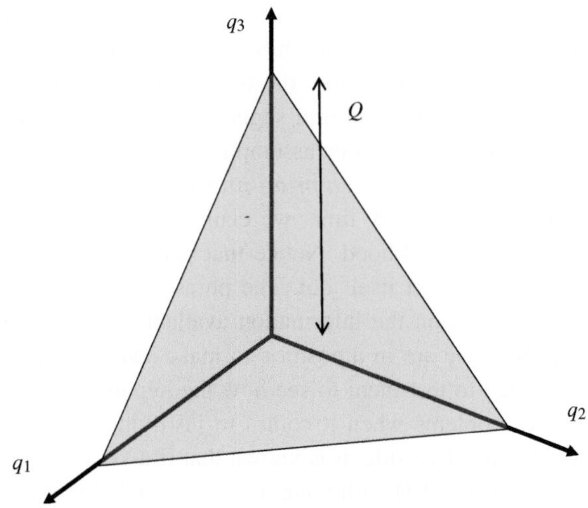


**Figure 4.2** A system of three compartments that together hold $Q$ items has microstates that lie in a triangular pattern on the indicated shaded plane. The axes indicate the number of units $q_j$ in each compartment. The vertices lie at points $(Q, 0, 0)$, $(0, Q, 0)$ and $(0, 0, Q)$.

The sum of the $q_j$ is equal to $Q$, and this locates the vertices of the triangle at points $(Q, 0, 0)$, $(0, Q, 0)$ and $(0, 0, Q)$. If $Q = 1$, these are the only points on the surface. More generally, the number of points goes like the area of the triangle, and in fact is given by the triangular number $1 + 2 + \cdots + (Q + 1) = (Q + 1)(Q + 2)/2$. Notice that this is proportional to $Q^2$ for large $Q$. If we now consider four compartments, the number of microstates with a given $Q$ corresponds to the number of points on or *beneath* this surface. Each plane corresponds to a constant value of $Q_{123} = q_1 + q_2 + q_3$, given that the fourth compartment possesses $Q - Q_{123}$ items. Considering the 3-d geometry, the number of microstates labelled by $(q_1, q_2, q_3, q_4)$ is therefore proportional to $Q^3$ for large $Q$. We begin to see a pattern here and might guess that the number of microstates is proportional to $Q^{N-1}$ for large $Q$. As $N$ and $Q$ increase, this rapidly becomes an absolutely huge number.

## 4.3 Microstates and Macrostates

We have enumerated the microscopic states of our system and judged that when it is isolated, the dynamics will lead to time-independent and equal probabilities of microstate occupation. The next step is to recognise that the microscopic detail is not of interest to us in thermodynamics, but that instead we are chiefly concerned with the behaviour of *macroscopic* properties, gross features of the system that are discernible on a macroscopic scale. They are collective properties of the components of a system.

The principle of equal a priori probabilities can tell us how likely it is that a particular value of a specified macroscopic property is observed. We arrange the microstates into groups according to the available values $\mathcal{M}_\alpha$ of the macroscopic property. Each group would then be called a *macrostate*: a collection of microstates with a common specified property. A macrostate corresponds to our perception of the system on the macroscopic scale. The number of microstates in a particular group, labelled $\alpha$, is called the microstate multiplicity $\Omega_\alpha$ of that macrostate. The microstate multiplicity of the macrostate, divided by the total number of microstates $\Omega = \sum \Omega_\alpha$, is then the probability that the macroscopic property should take value $\mathcal{M}_\alpha$ once the isolated system has come to equilibrium.

Let us illustrate this with our system of three compartments. The labelling of each microstate corresponds to the three numbers $(q_1, q_2, q_3)$ that specify the number of items in each compartment. Let us define the 'spikiness' of the microstate as the difference between the highest and the lowest $q_j$:

$$\mathrm{Sp} = \max(\{q_j\}) - \min(\{q_j\}), \tag{4.4}$$

and as the name suggests, Sp tells us whether the pattern of distribution of the items has highs and lows, or is roughly flat. Thus, the microstate $(1, 3, 5)$ has a spikiness of four and the microstate $(3, 3, 3)$ has a spikiness of zero. Note that these two microstates correspond to a system with total $Q = 9$.

Figure 4.3 illustrates all 55 microstates of this system, colour coded according to spikiness. The microstates form the *phase space* of the system; in this case, it is a triangular plot corresponding to the diagonal plane intersecting the 3-d axes illustrated in Figure 4.2. This is the pattern of points that lie on the shaded triangle. Each corner

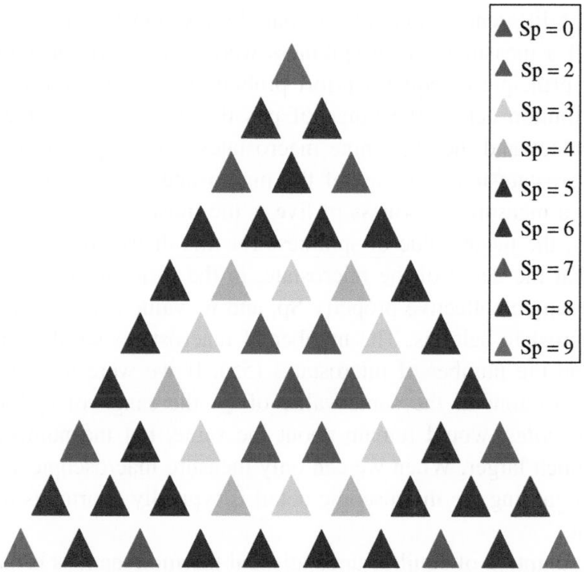


**Figure 4.3** Illustration of the phase space of a system of three compartments possessing nine items. We can imagine laying this pattern over the shaded triangle in Figure 4.2. Each small triangle is a microstate, and the colour coding divides the space into nine macrostates of different spikiness Sp defined in (4.4).

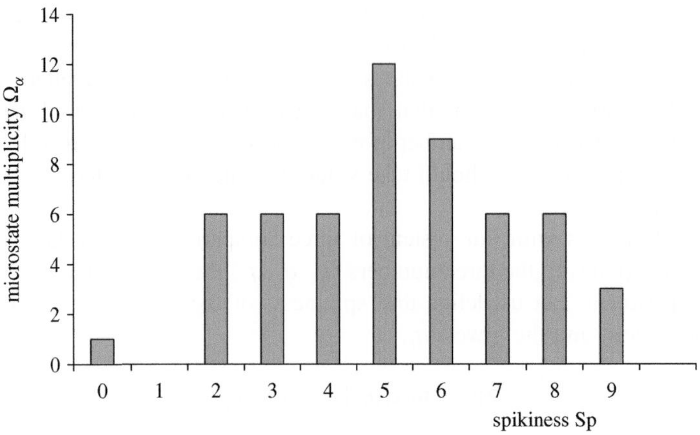


**Figure 4.4** Histogram of microstate multiplicity of spikiness macrostates in the $N = 3$, $Q = 9$ system.

corresponds to microstates labelled $(9, 0, 0)$, $(0, 9, 0)$ and $(0, 0, 9)$ and the point in the centre is microstate $(3, 3, 3)$. At any instant in time, the system is in a configuration represented by a point in this phase space, and as it changes configuration it moves in the phase space according to dynamical rules that we do not need to specify. The groups of microstates that all have the same spikiness, and hence have the same colour in the diagram, are the macrostates $\alpha$ of interest to us.

The contours of different spikiness form roughly hexagonal shells starting at the centre of the triangle. If a measurement of spikiness were to be made of the system when in equilibrium, the principle of equal a priori probabilities would allow us to predict the probability distribution across the range of possible values of spikiness. In this case, there are nine values and therefore nine macrostates with respect to this property. The histogram of microstate multiplicities of the macrostates $\Omega_\alpha$ is given in Figure 4.4. It would seem that a measured spikiness of five is the most likely or modal outcome; and more specifically, the mean value of spikiness for this distribution is 5.18.

We observe that the detail of the microstate, in the form of three numbers, has been subsumed into a single collective property Sp, and its value characterises the macrostate to which the microstate belongs. The number of macrostates so identified (9) is significantly fewer than the number of microstates (55). If we were to increase the number of compartments to four at the same value of $Q$, the range of spikiness, and hence number of *macrostates*, would remain about the same, but the number of *microstates* would become much larger. When we can only measure macroscopic system properties, the uncertainty regarding the microscopic detail is typically enormous for a system with many components.

The central assumption of equilibrium statistical thermodynamics is that if we are concerned with the macroscopic scale, the detail of the dynamics at the microscopic scale is not too important. The laws of motion need only make it likely that the system is found amongst its macrostates in proportion to the relevant microstate multiplicity, in other words, to make the occupation probability of the available microstates roughly equal.

Then statistical information about macroscopic properties can be obtained. The remarkable thing is that real systems seem to behave in just this manner and the implications can deepen our understanding of the principles of classical thermodynamics.

## 4.4 Boltzmann's Principle and the Second Law

We now come to Boltzmann's key insight, and the formula on his gravestone. He suggested that the number of microstates available to a system is related to its entropy $S$. He proposed the connection

$$S = k \ln \Omega, \tag{4.5}$$

where $\Omega$ is the number of microstates and $k$ is the Boltzmann constant. $\Omega$ is sometimes known as the *statistical weight* but the term microstate multiplicity better conveys its meaning as the number of microscale manifestations of the system. An important interpretation that follows is that entropy has something to do with the uncertainty at the microscopic scale when we only specify the state of a system on the macroscopic scale. According to the principle of equal a priori probabilities, $\Omega$ determines the probability that an isolated system might be found in one of the microstates, namely $1/\Omega$.

What is the motivation for this expression? Crucially, Boltzmann recognised that $\Omega$ would increase if a constraint on the dynamics were lifted. Recall that for each available macrostate of a system there is a corresponding microstate multiplicity. Each macrostate would comprise the entire phase space if the system were confined to it through a constraint, brought about perhaps by some aspect of the dynamics. Suppose we were to set up our example system in a particular macrostate, say the $\mathrm{Sp} = 9$ macrostate of the $N = 3$, $Q = 9$ case. In other words, the initial microstate of the system would correspond to one of the corners of the triangle in Figure 4.3. The system phase space would initially comprise just these three microstates. The constraint on the dynamics that confines the system to such a phase space would be that only transfers between compartments of nine units at a time are allowed.

But now consider the lifting of this constraint such that the dynamics can transfer smaller amounts: the system would then be allowed to assume other microstates within the triangle. If the dynamics allowed single unit transfers, for example, the system would move to an adjacent microstate in the triangle at each timestep in a deterministic but complicated manner. After a period of time, according to the principle of equal a priori probabilities, the system would be equally likely to be found in any one of the 55 microstates. It is clear that the release of the constraint, and the associated change in the dynamics, allows the system to explore a phase space with a larger number of microstates. The removal of a constraint cannot reduce the number of accessible states; it can only increase it. This is so reminiscent of the second law that Boltzmann proposed that thermodynamic entropy was related to the microstate multiplicity.

If so, then the requirement that entropy is extensive pins down the functional form of this relationship. If we have a system with microstate multiplicity $\Omega$, then the multiplicity of a combination of two identical systems would be $\Omega^2$: for each microstate of the first replica, there are $\Omega$ microstates of the second. Boltzmann came up with his logarithmic expression because the entropy of the pair of systems would need to be twice that of a single system, and $S_{\rm pair} = k \ln \Omega^2 = 2k \ln \Omega = 2S_{\rm single}$ works as required. Boltzmann's entropy function is extensive, as it needs to be.

Of course, we have to demonstrate that Boltzmann's expression reproduces known results for the thermodynamic entropy of a system such as the ideal gas expression (2.22), otherwise we would have to reject it in spite of these appealing properties. It will take us a while to reach that point in this book, but (spoiler alert!) in Section 9.6, we shall prove it to be so and indeed identify the unknown constant $\hat{c}$ in (2.22). Boltzmann's principle will emerge as a very fruitful hypothesis.

An isolated ideal gas can be used to illustrate the ideas. We describe the gas using the dynamically conserved total energy and number of particles as macrostate variables, and the volume of the container provides the constraint on the dynamics of the particles. The entropy according to Boltzmann is a measure of the multiplicity of microstates associated with choices of these conditions. We can elaborate this by grouping the microstates into a set of macrostates corresponding to various distinct macroscopic configurations of the gas. For example, we might consider macrostates specified by the volume within the container that is actually occupied by the gas. Now, consider a gas confined to one half of the container by a partition. Initially it is in equilibrium, with a uniform mean density throughout its side of the partition. But if the constraining partition were removed, the gas would suddenly find itself in a low multiplicity macrostate with respect to the new constraint: the volume available to the gas has become bigger and macrostates with density profiles with higher multiplicity are newly available. It would begin to evolve.

The gas will expand, of course, after the partition is removed until it is once again homogeneously distributed in the enlarged volume. More precisely, the principle of equal a priori probabilities states that once equilibrium is re-established, the gas will be found in the available macrostates with probability proportional to the multiplicity of the underlying microstates. For the gas released from behind a partition, it turns out that the macrostate with uniform density in the larger container will have the highest microstate multiplicity of all possible density profiles. The microstate multiplicity of this macrostate will make the largest contribution to the total multiplicity. Note carefully that the Boltzmann entropy is related to the *total* number of accessible microstates, not just the number associated with the uniform density macrostate. Other arrangements with lower multiplicity might be observed as an occasional fluctuation away from the uniform density situation. Indeed it is conceivable that this might include the return of the gas into its original half of the container, but the likelihood of this happening is negligible if the number of particles in the gas is large.

If there is a macrostate that dominates the available phase space, then to a good approximation the system will simply adopt that macrostate under the new constraint. Its state variables will be those that characterise that macrostate. Notice that a system in equilibrium is still very dynamic at the microscale and to a lesser extent at the macroscale, but that the probabilities of occupation of micro- or macrostates are independent of time, which is the deeper meaning of the idea of an equilibrium state. This is Boltzmann's theory. A similar picture of new and old phase spaces in another context is given in Figure 4.5.

An alternative viewpoint is to consider that the probabilities of microstate occupation change on the release of a constraint. The probability of observing a density profile of gas that extends across the larger volume was zero before the partition was removed,

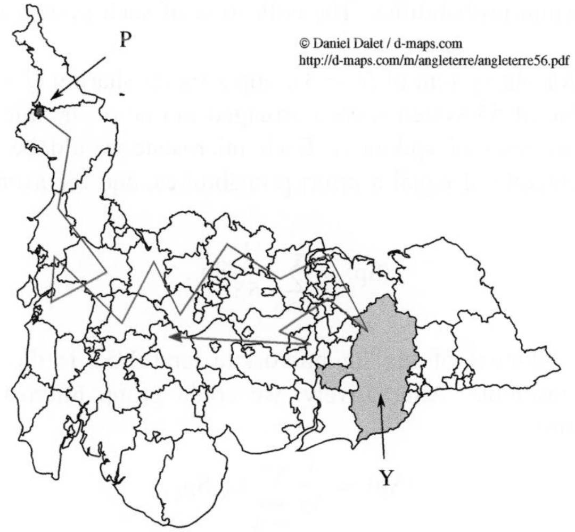


**Figure 4.5** The oddly familiar shapes represent macrostates of a physical system. Initially, the system is constrained to lie within the green macrostate labelled P, but after the removal of the constraint it explores the broader phase space along the blue trajectory. The most likely macrostate visited when in equilibrium is macrostate Y, assuming that statistical weight in that situation is proportional to area. Note that the phase space of a gas of $N$ particles in a given volume would extend across $6N$ dimensions, corresponding to their positions and velocities, not just two as in this representation. Source: Adapted from Daniel Dalet, d-maps.com, http://d-maps.com/m/angleterre/angleterre56.pdf.

and later on it became nonzero. Similarly, the probability of occupation of half the container was unity to begin with, but evolved towards a very small but nonzero value after the partition was removed. This suggests that entropy increase is associated with a change in equilibrium probabilities, equivalent to a change in knowledge or information represented by the macrostate variables and constraints.

The equilibrium probability distribution of any macroscopic quantity follows by identifying the possible macrostates and weighting their contribution to averages according to their respective multiplicity, or statistical weight. In order to develop these ideas further, we now discuss ensemble theory.

## 4.5 Statistical Ensembles

We have seen that a gas released from behind a partition is able to explore various new macrostates, and we proposed that it would be found in each macrostate with a probability proportional to the microstate multiplicity of that macrostate, once it has settled into a new equilibrium. Developing earlier ideas of Maxwell and Boltzmann, Gibbs suggested that the statistics arising from a single system evolving through the macrostates were equivalent to those of a collection of systems, each a realisation of a micro- or macrostate. The time-averaged properties of a single system when in equilibrium are equivalent to a weighted average of the properties of these copies of the system, the weighting being the appropriate equilibrium probabilities. The collection of such system snapshots is called an *ensemble.*

As an example, for our system of $N = 3$ compartments sharing $Q = 9$ units, we could imagine an ensemble of 55 systems, each arranged in one of the microstates, and work out the ensemble average of spikiness. Each microstate would be equally weighted, according to the principle of equal a priori probabilities, and the average would be

$$\langle \mathrm{Sp} \rangle = \sum_{i=1}^{55} \frac{1}{55} \mathrm{Sp}_i, \tag{4.6}$$

where $\mathrm{Sp}_i$ is the spikiness of the $i$th microstate, and 1/55 is the weighting of that microstate in the ensemble. Alternatively, we could group microstates into the nine macrostates and write

$$\langle \mathrm{Sp} \rangle = \frac{1}{55} \sum_{\alpha=1}^{9} \Omega_\alpha \mathrm{Sp}_\alpha, \tag{4.7}$$

where $\Omega_\alpha$ denotes the multiplicity of the $\alpha$th macrostate. Clearly, the system is assumed to be in microstate $i$ with probability $1/\Omega$ or in macrostate $\alpha$ with probability $\Omega_\alpha/\Omega$. Here, $\Omega$ is the total multiplicity of 55 given by $\Omega = \sum_\alpha \Omega_\alpha$ where the $\Omega_\alpha$ are specified in the histogram in Figure 4.4.

The assumed equivalence of the ensemble average and the time average of a single system is actually quite hard to justify. As we have already stated, a time average corresponding to a physical measurement of a system at equilibrium corresponds to an average over a small sample of the totality of snapshots available in the ensemble. The equivalence would be valid as long as these microstates are sampled from the macrostate groups in proportion to the size of those groups; attempts to prove this are called ergodic theory.

But if we accept the ensemble approach and the underlying principle of equal a priori probabilities, we obtain a very simple strategy for calculating the statistics of system properties in equilibrium. We establish the statistical weighting across the micro- or macrostates, carry out a weighted average of the system properties, and regard this as a prediction of an empirical measurement. We need not concern ourselves with the dynamics. We place our faith in the principle of equal a priori probabilities, and proceed from there, hoping that comparison with experiment will justify the approach.

## 4.6 Statistical Thermodynamics: the Salient Points

In summary, the core ideas upon which statistical thermodynamics is based are as follows:

- Probability is a means of weighting outcomes in order to work out an expectation of future behaviour.
- The principle of equal a priori probabilities for the occupation of microstates is an attempt to capture the effect of the complicated dynamics of isolated systems in equilibrium, when the statistical properties are independent of time.

- Macrostates of a system are characterised by values of macroscopic state variables, but the actual microstate taken at any moment could be one of many possibilities, the number of which is called the microstate multiplicity of the macrostate.
- It is proposed that values of macrostate variables are observed with a probability proportional to the corresponding microstate multiplicity, for an isolated system.
- We can evaluate equilibrium averages of micro- or macroscopic variables by performing ensemble averages over all possible microstates or macrostates.
- Boltzmann's principle is that the entropy of an isolated system in equilibrium is given by $k \ln \Omega$, where $\Omega$ is the total microstate multiplicity consistent with the macroscopic state variables and dynamical constraints.
- Following the relaxation of a constraint, the number of available microstates increases, and this is the underlying rationale for the second law.

## Exercises

**4.1** Two people A and B together possess £9 and exchange cash between them in units of £1. Determine the change in Boltzmann entropy that arises when a third person C joins in, according to the assumptions of complex system behaviour we have discussed.

**4.2** A, B and C share £3. List the 10 microstates in terms of labels $(q_1, q_2, q_3)$, where $q_j$ is the number of pound coins possessed by the $j$th participant. Plot a histogram of the microstate multiplicity of each spikiness macrostate. Calculate the mean and standard deviation of the spikiness, assuming the principle of equal a priori microstate probabilities.

**4.3** A, B and C share £9 but exchange it in batches of £3. Determine the Boltzmann entropy of the system and the probability of observing the most likely, or *modal*, spikiness. They change the rules to allow exchanges of £1. Calculate the new entropy and the probability of observing the new modal spikiness.

**4.4** A has £3 but then starts sharing it with B, C and D. Calculate the probability under the new situation that A should hold the £3 again.

# 5

# Statistical Thermodynamics of a System of Harmonic Oscillators

We now apply the principles developed in the last chapter to a reasonably realistic physical system: a set of $N$ weakly coupled quantum harmonic oscillators that collectively hold $Q$ quanta of energy. We will construct ensembles and obtain statistical information about the system when it is in equilibrium.

A quantum harmonic oscillator with natural frequency $\omega$ has energy levels $E_q = (q + 1/2)\hbar\omega$ where $q$ is a non-negative integer. We shall ignore the zero point energy $\hbar\omega/2$, and denote the number of quanta held by the $j$th oscillator by the integer $q_j$. We then have total energy $E = \hbar\omega\sum_j q_j$, such that $Q = \sum_j q_j$ is proportional to $E$. The oscillators are weakly coupled, in the sense that quanta can be exchanged between them, but without complicating the specification of the system energy. The system is very similar to the example of compartments sharing units of material introduced in the last chapter and we illustrate four oscillators, represented as springs, in Figure 5.1.

## 5.1 Microstate Enumeration

The number of microstates with total number of quanta $Q$ may be evaluated for this system using a neat trick. Unfortunately, it is not so easy for other systems! We have to distribute $Q$ quanta amongst $N$ oscillators. Each microstate of the system might be pictured as a set of $N$ groups of objects, the $j$th group consisting of $q_j$ objects, with the total number of objects equal to $Q$. If we line up the $Q$ objects next to each other and insert $N - 1$ dividers to separate them into $N$ groups, the situation might resemble the arrangement shown in Figure 5.1. The key point is that each possible microstate of the system (a specific set of labels $\{q_j\}$) corresponds to an arrangement of these $Q + N - 1$ elements. A different choice of positions of the dividers would correspond to a different set of $q_j$. For example, in Figure 5.1, there are seven objects and three dividers, making ten elements in all. The pattern corresponds to $q_1 = 3$, $q_2 = 4$ and $q_3 = q_4 = 0$, roughly represented by the amplitudes of oscillation shown for the springs.

*Statistical Physics: An Entropic Approach*, First Edition. Ian Ford.

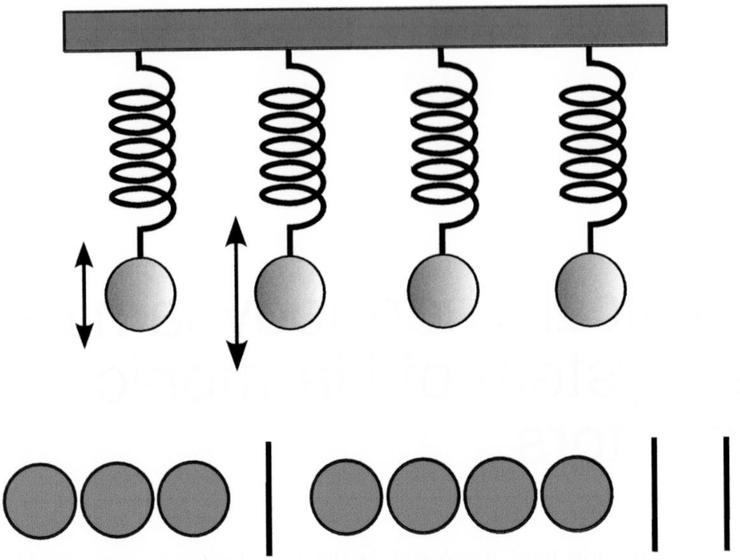

**Figure 5.1** Illustration of four quantum oscillators holding 3, 4, 0 and 0 quanta, respectively. The microstate is represented by the sequence of seven objects (circles) and three dividers (lines) shown. The possible microstates of seven quanta held by four oscillators correspond to different linear arrangements of these ten objects.

So how many arrangements might there be? We can solve this problem by working out the number of ways we can place the $N-1$ dividers on a set of $Q+N-1$ positions. The objects then take the positions not occupied by a divider, and can be ignored. For example, one such way is to have the $N-1$ dividers in the first $N-1$ positions, and the $Q$ objects in the rest. This corresponds to the microstate $(0,0,\cdots,Q)$ in the notation $(q_1,q_2,\cdots,q_N)$.

There are $Q+N-1$ positions available for the first divider, $Q+N-2$ positions for the second and $Q+N-(N-1)$ positions for the last one, giving a multiplicity of arrangements of $(Q+N-1)(Q+N-2)\cdots(Q+N-(N-1))$. But this is the number of ways a set of *different* dividers can be arranged and therefore overcounts the true multiplicity since the dividers are in fact indistinguishable. We can correct the result by dividing by the number of possible arrangements of the $N-1$ dividers on the $N-1$ occupied positions, which is $(N-1)(N-2)\cdots 1$, and so the correct microstate multiplicity of the system is

$$\Omega(N,Q)=\frac{(Q+N-1)(Q+N-2)\cdots(Q+N-(N-1))}{(N-1)(N-2)\cdots 1}=\frac{(Q+N-1)!}{Q!(N-1)!}. \tag{5.1}$$

We can check this with $N=3$, $Q=9$ to obtain the multiplicity of 55 considered in the previous chapter.

Factorials crop up naturally in calculations of multiplicity and they are often inconvenient in analysis. However, using *Stirling's approximation*, factorials of large numbers are readily converted into powers. Stirling's formula states that for large $m$:

$$\ln m! \approx m \ln m - m. \tag{5.2}$$

This may be derived by noting that $\ln m! = \ln m + \ln(m-1) + \cdots + \ln 1 = \sum_{n=1}^{m} \ln n \approx \int_1^m \ln x \, dx = [x \ln x - x]_1^m \approx m \ln m - m$ for large $m$. So, if both $Q$ and $N$ are large, then

$$\begin{aligned} \ln \Omega(N,Q) &= \ln(Q+N-1)! - \ln Q! - \ln(N-1)! \\ &\approx (Q+N-1)\ln(Q+N-1) - (Q+N-1) \\ &\quad -Q\ln Q + Q - (N-1)\ln(N-1) + N - 1 \\ &= (Q+N-1)\ln(Q+N-1) - Q\ln Q - (N-1)\ln(N-1). \end{aligned} \tag{5.3}$$

This implies that

$$\ln \Omega(N,Q) \approx Q \ln\left(1 + \frac{N-1}{Q}\right) + (N-1)\ln\left(1 + \frac{Q}{N-1}\right), \tag{5.4}$$

and by a further approximation, we get

$$\Omega(N,Q) \approx \left(1 + \frac{N}{Q}\right)^Q \left(1 + \frac{Q}{N}\right)^{N-1}. \tag{5.5}$$

This indicates that $\Omega$ can be a very large number, for any ratio of $Q$ to $N$. Notice that the multiplicity of the system for $Q \gg N$ is

$$\Omega(N,Q) \approx \left(\frac{Q}{N}\right)^{N-1}, \tag{5.6}$$

and this is consistent with the result we guessed in Section 4.2.2.

## 5.2 Microcanonical Ensemble

The set of microstates available to a system is known as its *phase space*. This may often be visualised as a coordinate space of high dimension, such as the $N$-dimensional space labelled by the harmonic oscillator quantum numbers $\{q_j\}$. A snapshot of the system is represented by a single point in this space, and the evolution of the system in time corresponds to the motion of that point along a trajectory through the phase space.

The dynamics of an *isolated* system are constrained by the conservation of energy, and the phase space is then characterised by a specific value of energy. According to the principle of equal a priori probabilities, the statistics of an isolated system are to be found by giving equal weight to every possible microstate in this phase space. Equivalently, we can imagine an ensemble of systems each of which takes one of the microstates, and then obtain averages of properties across this ensemble, with each member of the ensemble given equal weight in the averaging. The collection of configurations under consideration here is called the *microcanonical* ensemble and is designed for studying an isolated system.

For example, a microcanonical ensemble of the $N = 3$, $Q = 9$ oscillator system consists of 55 copies of the system, in each of the possible microstates. In order to calculate the mean spikiness, we would calculate Sp for each microstate and average over the ensemble with equal weighting for each copy. This is exactly what was done in Section 4.5. The calculation might be represented as

$$\langle \mathrm{A} \rangle = \sum_i \mathrm{A}_i P_i, \tag{5.7}$$

where $i$ labels the microstates, $P_i$ is the equilibrium probability that the system is found in microstate $i$ and $\mathrm{A}_i$ is the value of a system property of interest when in microstate $i$. For the microcanonical ensemble, the microstate probabilities are all equal to $1/\Omega$.

## 5.3 Canonical Ensemble

The canonical ensemble is a method for calculating the statistical properties of a system that is *not* isolated. It is able to exchange energy with its environment. Consequently, it is able to explore a phase space that includes microstates with different energies, in contrast to the microcanonical case just considered. However, the weightings of the system microstates when performing ensemble averages are now *not* equal. The word *canonical* is used in the sense of *standard*. The word microcanonical used to describe an ensemble of an isolated system is derived from this and although the terminology is not particularly revealing, it has become established.

Let us see how the canonical ensemble arises by considering a system of $N$ oscillators weakly coupled to a larger system of $N_\mathrm{r} = N_\mathrm{tot} - N$ oscillators that we shall call the reservoir or heat bath. The total energy of the combined system of $N_\mathrm{tot}$ oscillators is $Q_\mathrm{tot}$. The reservoir has a microstate multiplicity $\Omega_\mathrm{r}(N_\mathrm{r}, Q_\mathrm{r})$ when it has $Q_\mathrm{r}$ quanta. The label r reminds us that it refers to the reservoir. The combination of system and reservoir is allowed to explore the $(N_\mathrm{tot} - 1)$ dimensional phase space of the combined system, characterised by a constant energy, and will reach equilibrium corresponding to equal probabilities of occupation of any individual combined microstate. The system plus reservoir can therefore be studied using a microcanonical ensemble.

We are interested in the average properties of the *system*, not the combination of system and reservoir. We want to construct an ensemble of just microstates of the system. It is therefore sensible to divide the phase space of the combined system and reservoir into macrostates labelled by the microstate of the system. We then deduce that each of these macrostates has a multiplicity equal to that of the reservoir, given that the system is in a particular microstate.

For example, consider again our $N_\mathrm{tot} = 3$, $Q_\mathrm{tot} = 9$ system, and regard this as a combination of a reservoir (two oscillators) and a system (one oscillator, with label $j = 3$). The combined microcanonical ensemble consists of the 55 microstates of the three oscillators. An ensemble of the single oscillator, in contrast, would comprise all possible configurations of the single oscillator, allowing it to take different energies, specifically the ten microstates corresponding to the values of $q_3$ from 0–9.

Now, the macrostate of combined system and reservoir with the system microstate specified by $q_3 = 0$ has a microstate multiplicity of ten. This is because the nine quanta

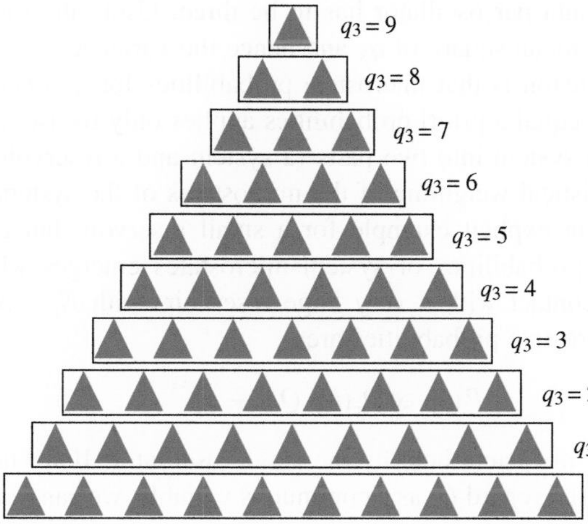


**Figure 5.2** Illustration of the construction of the canonical ensemble. The phase space of three oscillators with nine quanta is carved up into a ziggurat of macrostates labelled by the microstate of oscillator 3, namely by $q_3$. The statistics of oscillator 3 (the system) are to be found by weighting its microstates by the size of the phase space of oscillators 1 and 2 (the reservoir) when $q_3$ takes a specific value. This weighting is $10 - q_3$.

can be shared amongst the two reservoir oscillators in ten ways (corresponding to $\Omega_r(N_r = 2, Q_r = 9) = 10!/(1!9!) = 10$). For $q_3 = 5$, the four remaining quanta are to be shared between the two reservoir oscillators and the multiplicity of this macrostate of the combined system and reservoir is now $\Omega_r(N_r = 2, Q_r = 4) = 5!/(1!4!) = 5$. The relative weightings of the $q_3 = 0$ and $q_3 = 5$ system microstates of the single oscillator in this ensemble are therefore 10 and 5. The weightings for other system microstates are found to be $(10 - q_3)$ by similar reasoning. The argument is summarised in Figure 5.2.

We may then write mean system properties as:

$$\langle \mathrm{A} \rangle = \sum_{q_3} \mathrm{A}(q_3) P(q_3), \tag{5.8}$$

which is a weighted sum over system microstates labelled by $q_3$. The probability $P(q_3)$ is proportional to the multiplicity of the *reservoir* when it possesses $Q_{\mathrm{tot}} - q_3$ quanta, namely $\Omega_r(N_r, Q_{\mathrm{tot}} - q_3)$. The proportionality constant is chosen to ensure the normalisation of the probabilities, so in the case considered, $P(q_3) = (10 - q_3)/55$. The average number of quanta in the system according to this ensemble is then clearly

$$\begin{aligned} \langle q_3 \rangle &= \sum q_3 P(q_3) = \sum_{q_3=0}^{9} q_3 \frac{(10 - q_3)}{55} \\ &= \frac{1}{55}(0 + 9 + 16 + 21 + 24 + 25 + 24 + 21 + 16 + 9) = 3. \end{aligned} \tag{5.9}$$

This answer should be no surprise, since the combination of system and reservoir has nine quanta distributed between three oscillators with equal microstate probabilities: the average number of quanta per oscillator has to be three. Using the same probabilities, we could work out the mean square of $q_3$ and hence the variance.

The important conclusion is that microstate probabilities for an *open* system are not equal. The principle of equal a priori probabilities applies only for isolated systems. But by splitting an isolated system into two parts (a system and a reservoir), we have been able to deduce the statistical weighting of the microstates of the system.

We have just seen an explicit example for a small reservoir, but a very important universal form of the probabilities of system microstates emerges when we consider a single oscillator in contact with a very large reservoir, with $N_r = N_{tot} - 1 \gg 1$. As before, the system microstate probabilities are

$$P(q) \propto \Omega_r(N_r, Q_{tot} - q), \tag{5.10}$$

where $q$ is the number of quanta held by the single oscillator. If we take $q \ll Q_{tot}$ for almost all microstates, and regard $Q_r$ as a continuous variable, we can make the following Taylor expansion:

$$\ln \Omega_r(N_r, Q_{tot} - q) \approx \ln \Omega_r(N_r, Q_{tot}) - q \left.\frac{\partial \ln \Omega_r(N_r, Q_r)}{\partial Q_r}\right|_{Q_r=Q_{tot}} + \cdots, \tag{5.11}$$

and ignore terms in $q^2$ and beyond. We expand the logarithm of $\Omega_r$ and not $\Omega_r$ itself for two reasons. First, we wish to obtain an approximate representation where the reservoir multiplicity and hence the system microstate probability $P(q)$ is always positive, and this is guaranteed if we expand the logarithm. Secondly, $\Omega_r(N_r, Q_{tot} - q)$ is a very rapidly varying function of $q$ and so an expansion of the more slowly changing logarithm is likely to be accurate over a wider range of $q$.

From (5.3), we can determine the derivative in (5.11) to be

$$\begin{aligned}\hat{\beta} &= \left.\frac{\partial \ln \Omega_r(N_r, Q_r)}{\partial Q_r}\right|_{Q_r=Q_{tot}} = \frac{\partial \ln \Omega_r(N_r, Q_{tot})}{\partial Q_{tot}} \\ &= \ln (Q_{tot} + N_r - 1) + 1 - \ln Q_{tot} - 1 \approx \ln \left(1 + \frac{N_{tot}}{Q_{tot}}\right),\end{aligned} \tag{5.12}$$

where we have introduced a symbol $\hat{\beta}$ to denote the derivative. Since $\hat{\beta}$ depends on $N_{tot}/Q_{tot}$, it is clearly a property of the combined system and reservoir, and effectively of the reservoir alone, to a good approximation, since the system is a tiny component of the combination. From (5.10)–(5.12) we are then able to write the system microstate probabilities as

$$P(q) \propto \exp(-\hat{\beta}q). \tag{5.13}$$

When we considered the mini-reservoir with $N_r = 2$, we deduced a linear relationship between the probability of each single oscillator microstate and the number of quanta held, as employed in (5.9). For very large $N_r$, this goes over to an exponential decrease. This is the so-called *canonical probability distribution* and it is illustrated in Figure 5.3.

Canonical ensemble averages of system properties can now be obtained by performing summations such as

$$\langle \mathrm{A} \rangle = \frac{1}{Z} \sum_q \mathrm{A}(q) \exp(-\hat{\beta}q), \tag{5.14}$$

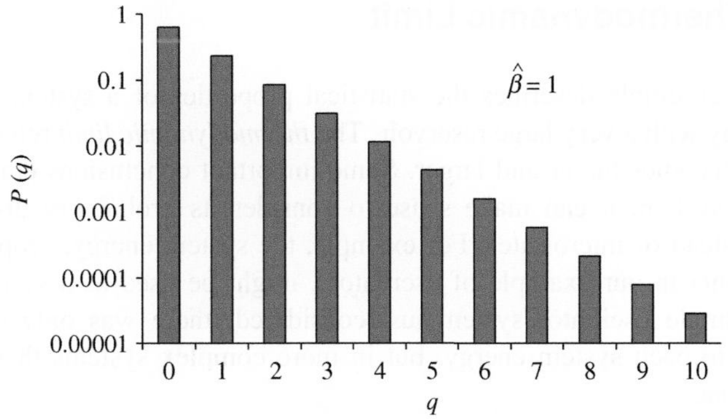


**Figure 5.3** The canonical probability distribution $P(q) \propto \exp(-\hat{\beta}q)$ of quanta $q$ held by a single oscillator in thermal contact with a large reservoir characterised by $\hat{\beta} = 1$. Note the logarithmic scale.

where

$$Z = \sum_q \exp(-\hat{\beta}q) \tag{5.15}$$

is a normalising factor; A($q$) is the value of a system property when the system is in the microstate $q$, and the sum is taken over all microstates of the system labelled by $q$.

For example, consider a single oscillator that is part of a large system of $N_{\text{tot}}$ oscillators characterised by parameters $N_{\text{tot}} \gg 1$, $Q_{\text{tot}} \gg 1$ and $Q_{\text{tot}}/N_{\text{tot}} = 3$. Then from (5.12), we find that $\hat{\beta} = \ln(4/3)$ and the average number of quanta held by the single oscillator is

$$\langle q \rangle = \frac{1}{Z} \sum_{q=0}^{Q_{\text{tot}}} q \exp(-\hat{\beta}q). \tag{5.16}$$

The system has $Q_{\text{tot}} + 1$ microstates, and if $Q_{\text{tot}} \gg 1$, the upper limit in the sum can effectively be replaced with $\infty$. We define $x = \exp(-\hat{\beta})$ such that $Z = \sum_0^\infty x^q = (1-x)^{-1}$. Similarly

$$\sum_0^\infty q \exp(-\hat{\beta}q) = -\frac{\mathrm{d}}{\mathrm{d}\hat{\beta}} \sum_0^\infty \exp(-\hat{\beta}q) = -\frac{\mathrm{d}Z}{\mathrm{d}\hat{\beta}} = -\frac{\mathrm{d}Z}{\mathrm{d}x}\frac{\mathrm{d}x}{\mathrm{d}\hat{\beta}} = -(1-x)^{-2}(-x), \tag{5.17}$$

and so

$$\langle q \rangle = Z^{-1} \sum_0^\infty q \exp(-\hat{\beta}q) = \frac{(1-x)x}{(1-x)^2} = \frac{\exp(-\hat{\beta})}{(1-\exp(-\hat{\beta}))}. \tag{5.18}$$

For $\hat{\beta} = \ln(4/3)$, $\langle q \rangle = 3$. Once again this is as we expect, since the system plus reservoir is simply $N_{\text{tot}}$ oscillators carrying $Q_{\text{tot}} = 3N_{\text{tot}}$ quanta. We have recovered the mean number of quanta per oscillator, but this time by considering the *canonical distribution* of a single oscillator over its microstates.

## 5.4 The Thermodynamic Limit

The canonical ensemble describes the statistical properties of a system that is able to exchange energy with a very large reservoir. The *thermodynamic limit* refers to behaviour as the *system* becomes larger and larger. Some important conclusions emerge.

For a large system, it can make sense to consider its probability distribution over macrostates instead of microstates. For example, the system energy, proportional to the number of quanta in our example of oscillators, might be used as a system macrostate label. In the single oscillator system just considered, there was only one microstate corresponding to each system energy, but in more complex systems there will usually be more than one.

Let us consider a system consisting of $N$ oscillators in contact with a reservoir characterised by parameter $\hat{\beta}$. Each system macrostate is labelled by the number of quanta $Q$ it holds, and has a microstate multiplicity of $\Omega(N,Q)$. The system moves between macrostates as time progresses as a consequence of exchanges of quanta with the reservoir that alter $Q$. The canonical ensemble average over *macrostates* may then be written in the form

$$\langle \mathrm{A} \rangle = \frac{1}{Z} \sum_{Q=0}^{\infty} \mathrm{A}(Q)\Omega(N,Q) \exp\,(-\hat{\beta}Q), \tag{5.19}$$

where

$$Z = \sum_{Q=0}^{\infty} \Omega(N,Q) \exp\,(-\hat{\beta}Q). \tag{5.20}$$

Note that this is simply a reorganisation of the sum over microstates. Each *microstate* with $Q$ quanta has canonical weighting $\exp(-\hat{\beta}Q)$ and we have grouped all such microstates ($\Omega(N,Q)$ of them) to form one of the macrostate terms in the sum in (5.19).

We saw in (5.5) that $\Omega$ is a power-like, rapidly increasing function of $Q$ for an oscillator system. On the other hand, the factor $\exp(-\hat{\beta}Q)$, with $\hat{\beta} > 0$, is a rapidly decreasing function of $Q$. As a consequence, the macrostate weighting factor $\Omega(N,Q)\exp(-\hat{\beta}Q)$ increases with $Q$ until it reaches a peak and then falls, as exponential suppression at large $Q$ is stronger than power-law amplification.

It is very instructive to expand the weighting factor around the peak in the distribution at $Q = Q^*$. This is the largest or *modal* probability over the macrostates. We write the probability distribution as

$$P(Q) = Z^{-1} \exp\,(\ln\,\Omega(N,Q) - \hat{\beta}Q), \tag{5.21}$$

and determine $Q^*$ from the condition $\partial P/\partial Q = 0$, equivalent to $\partial \ln P/\partial Q = 0$, or

$$\frac{\partial}{\partial Q}(\ln\,\Omega(N,Q) - \hat{\beta}Q)\bigg|_{Q=Q^*} = 0, \tag{5.22}$$

from which we deduce that

$$\frac{\partial \ln\,\Omega(N,Q)}{\partial Q}\bigg|_{Q=Q^*} = \hat{\beta}. \tag{5.23}$$

Let us recall that $\hat{\beta}$ is effectively a property of the reservoir and that it is defined in (5.12) as a derivative of the logarithm of *reservoir* multiplicity with respect to the number of quanta it holds. The system property that appears on the left hand side in (5.23) has a similar form, and we might call it the system $\hat{\beta}$-parameter, or $\hat{\beta}_s$, for the system macrostate at the modal number of quanta $Q^*$. Each macrostate of the system has a property $\hat{\beta}_s(Q) = \partial \ln \Omega/\partial Q$ that is a function of $Q$. We conclude from (5.23) that when it is in equilibrium, the system is most likely to be found in the macrostate that has the same $\hat{\beta}$ parameter as the reservoir. This is very reminiscent of the equality of temperature between a system and reservoir in thermal equilibrium, which we will consider later.

For this particular case, and assuming the system parameters $N$ and $Q$ are large, (5.12) and (5.23) may be used to identify $Q^*$ in terms of $\hat{\beta}$:

$$\frac{\partial}{\partial Q} \ln \Omega(N, Q) \approx \frac{\partial}{\partial Q}((Q + N - 1) \ln (Q + N - 1) - Q \ln Q) = \ln \left(\frac{Q + N - 1}{Q}\right), \tag{5.24}$$

so from (5.23)

$$\hat{\beta}_s(Q^*) = \ln \left(\frac{Q^* + N - 1}{Q^*}\right) = \hat{\beta} \quad \Rightarrow \quad Q^* = \frac{N - 1}{\exp (\hat{\beta}) - 1}. \tag{5.25}$$

Next we look at the extent of fluctuations away from this most likely or modal macrostate. We expand the first term in the exponent in (5.21) about $Q^*$:

$$\begin{aligned} \ln \Omega(N, Q) \;\approx\; & \ln \Omega(N, Q^*) + (Q - Q^*) \left.\frac{\partial \ln \Omega(N, Q)}{\partial Q}\right|_{Q=Q^*} \\ & + \frac{(Q - Q^*)^2}{2} \left.\frac{\partial^2 \ln \Omega(N, Q)}{\partial Q^2}\right|_{Q=Q^*}, \end{aligned} \tag{5.26}$$

and recognising through (5.23) that the second term on the right hand side is $(Q - Q^*)\hat{\beta}$ we get

$$\ln \Omega(N, Q) - \hat{\beta} Q \approx \ln \Omega(N, Q^*) - \hat{\beta} Q^* + \frac{(Q - Q^*)^2}{2} \left.\frac{\partial^2 \ln \Omega(N, Q)}{\partial Q^2}\right|_{Q=Q^*}, \tag{5.27}$$

and so we can approximate the system macrostate probability as

$$P(Q) \propto \Omega(N, Q) \exp (-\hat{\beta} Q) \approx \Omega(N, Q^*) \exp (-\hat{\beta} Q^*) \exp \left(-\frac{(Q - Q^*)^2}{2\sigma^2}\right), \tag{5.28}$$

where we define

$$\frac{1}{\sigma^2} = -\frac{\partial^2 \ln \Omega(N, Q^*)}{\partial Q^{*2}}. \tag{5.29}$$

According to this approximation, the canonical pdf over macrostates is *Gaussian* and the modal macrostate labelled by $Q^*$ is also the mean macrostate characterised by $\langle Q \rangle$, since $P(Q)$ is symmetric about $Q = Q^*$. The normalisation sum $Z$ may be written

$$Z \approx \int_0^\infty dQ\, \Omega(N,Q) \exp\left(-\hat{\beta} Q\right) \approx \Omega(N,\langle Q\rangle) \exp\left(-\hat{\beta}\langle Q\rangle\right)$$
$$\times \int_{-\infty}^{\infty} dQ \exp\left(-\frac{(Q-\langle Q\rangle)^2}{2\sigma^2}\right) \approx \Omega(N,\langle Q\rangle) \exp\left(-\hat{\beta}\langle Q\rangle\right)(2\pi)^{\frac{1}{2}}\sigma. \quad (5.30)$$

$Z$ is therefore well approximated by the largest term in the sum in (5.20), multiplied by a factor of $(2\pi)^{1/2}\sigma$.

Let us examine further the standard deviation $\sigma$ of the distribution. We find from differentiating (5.24) that

$$\sigma^2 = -\left[\frac{\partial^2 \ln \Omega(N,Q^*)}{\partial Q^{*2}}\right]^{-1} = \frac{Q^*(Q^*+N-1)}{N-1}, \quad (5.31)$$

and for $N \gg 1$, the ratio of the standard deviation to the mean is proportional to $N^{-1/2}$:

$$\frac{\sigma}{Q^*} = \left(\frac{Q^*+N}{NQ^*}\right)^{\frac{1}{2}} \approx \left(\frac{\exp(\hat{\beta})}{N}\right)^{\frac{1}{2}}, \quad (5.32)$$

having used (5.25). This means that the distribution becomes very sharply peaked as $N \to \infty$. This is illustrated in Figure 5.4, where for clarity $Q$ is scaled by its mean to ensure that the peak does not move as $N$ changes. In this so-called *thermodynamic limit* where $N$ is large, the fluctuations in macrostate are unlikely to be observable, and the system can be regarded as having a constant number of quanta $Q^*$ related to the $\hat{\beta}$ parameter of the reservoir through (5.25). Thus, on the macroscopic scale, a system described by the canonical ensemble seems to be static, although underlying this apparent quiescence is a continual exploration of microstates. It is simply that the exploration rarely strays away from the $Q = Q^*$ macrostate.

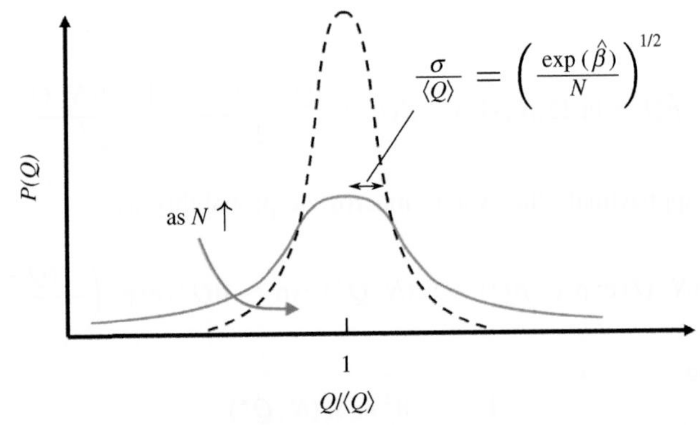


**Figure 5.4** Probability distribution function of system parameter $Q$ for two system sizes $N$, with $Q$ scaled by the mean of the distribution, indicating how the relative width of the distribution shrinks as the system size increases. In the thermodynamic limit $N \to \infty$, the distribution becomes sharply peaked and fluctuations are extremely unlikely.

## 5.5 Temperature and the Zeroth Law of Thermodynamics

We have seen that the microstate probabilities of a system of harmonic oscillators in the canonical ensemble depend on a property $\hat{\beta}$ that characterises the large reservoir of oscillators with which it is in thermal contact. We now define an analogous property of systems or reservoirs that are more general than collections of harmonic oscillators. We write

$$\beta = \frac{\partial \ln \Omega(N, E)}{\partial E}, \tag{5.33}$$

where $\Omega$ is the microstate multiplicity of a macrostate of a system labelled by energy $E$ and number of particles $N$. For oscillators, the number of quanta was proportional to energy and so the $\hat{\beta}$ parameter of (5.12) matched this form. The $\beta$ parameter has dimensions of inverse energy, while $\hat{\beta}$ was dimensionless. But following a similar development, we can show that the most likely energy macrostate of a system has the same value of $\beta$ as the reservoir with which it is in equilibrium. In the thermodynamic limit, this modal macrostate is essentially the *only* macrostate ever explored by the system. Therefore, a large system and reservoir, when in equilibrium under heat exchange, are characterised by the same parameter $\beta$. Other systems in equilibrium with the reservoir will all be characterised by this parameter. This suggests that $\beta$ is an indicator of thermal equilibrium between macroscopic objects, and must therefore correspond to some function of temperature.

Let us now insert Boltzmann's formula $S = k \ln \Omega$, implying that

$$\beta = \frac{1}{k}\frac{\partial S(N, E)}{\partial E} = \frac{1}{kT}, \tag{5.34}$$

as long as $S(N, E)$ is indeed equivalent to the thermodynamic entropy, as Boltzmann claimed, and where we have employed (2.45). Now we see that $\beta k$ is the inverse temperature, and so it is no surprise that it plays the role of indicator of thermal equilibrium. On the basis of statistical thermodynamics, we seem to have found a justification of the zeroth law of thermodynamics.

Thus, when a system is in equilibrium with a reservoir or heat bath, the fixed temperature $T$ of the heat bath imposes two related conditions. Firstly, all microstate probabilities of the system take a canonical form proportional to $\exp(-E/kT)$, where $E$ is the microstate energy. This is called the *Boltzmann factor*. Secondly, the most likely energy macrostate of a large system has the same temperature as the heat bath, where macrostate temperature is defined in terms of a derivative, with respect to energy, of the microstate multiplicity of the macrostate. Furthermore, energy and temperature fluctuations of a system usually become unobservable as the system gets very large: this is the thermodynamic limit, so called because such behaviour ties in with the concept of time-independent macrostate variables in classical thermodynamics.

## 5.6 Generalisation

The example of a system and reservoir of harmonic oscillators studied in this chapter has given us an illustration of the following general approach, regarded as the central doctrine of statistical physics. The canonical ensemble is a collection of all possible microstates of a system, of which there might be an infinite number, each weighted by a canonical probability $P(E_i) = Z^{-1} \exp(-E_i/kT)$, where $Z = \sum_i \exp(-E_i/kT)$ and $E_i$ is the energy of the microstate. Such an ensemble provides the statistical properties of the system when in thermal equilibrium with an environment at temperature $T$. We can also construct a canonical ensemble over macrostates, such that $P(E_\alpha) = Z^{-1}\Omega(E_\alpha)\exp(-E_\alpha/kT)$, where $Z = \sum_\alpha \Omega(E_\alpha)\exp(-E_\alpha/kT)$ and $E_\alpha$ is the energy of the macrostate. Such an approach is consistent with various features of the thermodynamics of macroscopic systems, but also extends to systems that are much smaller. This entire picture follows from the principle of equal a priori probabilities, and is a remarkably simple proposition for describing the equilibrium behaviour of a complex system. We investigate some of its implications in the next few chapters.

## Exercises

**5.1** A system of $N$ weakly coupled quantum oscillators contains $Q = 2N$ quanta. Show that the total microstate multiplicity of the system macrostate in which oscillator 1 possesses $q_1$ quanta is given by $\omega(N, q_1) = (3N - 2 - q_1)!/[(2N - q_1)!(N - 2)!]$.

**5.2** For cases $N = 2$, 3 and 4: (a) specify the allowed range of $q_1$, (b) evaluate the microstate multiplicity $\omega(N, q_1)$ for each value of $q_1$ in this range and (c) calculate the mean and standard deviation of $q_1$, assuming that the principle of equal a priori probabilities holds.

**5.3** This question examines the approach towards the canonical ensemble description of the properties of a single oscillator. (a) Use Stirling's approximation to simplify $\ln \omega(N, q_1)$ for $N \gg q_1$. (b) Expand $\ln \omega(N, q_1)$ to first order in $q_1$ and take the limit $N \to \infty$ to show that $\omega(\infty, q_1) \propto \exp(-\hat{\beta} q_1)$. Demonstrate that the $\hat{\beta}$ parameter is equal to $\ln(3/2)$ for this system. (c) Sketch the $q_1$ dependence of the weighting factors $\omega(2, q_1)$, $\omega(3, q_1)$, $\omega(4, q_1)$ and $\omega(\infty, q_1)$.

**5.4** A system of $N = 4$ oscillators holding $Q = 8$ quanta is initially maintained in equilibrium in a macrostate with $q_1 = 4$ quanta held by oscillator 1. Calculate the Boltzmann entropy of the system in this initial state. The constraint on the value of $q_1$ is removed, and the system assumes a new equilibrium state. What is the change in Boltzmann entropy of the system?

**5.5** Two coupled quantum harmonic oscillators share nine quanta of energy according to unspecified dynamical rules. The system is then allowed to exchange quanta with a third oscillator. Initially, the third oscillator possesses no quanta. Determine the change in Boltzmann entropy once equilibrium has been restored.

**5.6** A system of three oscillators sharing four quanta is coupled to a system of seven oscillators sharing eight quanta such that they can exchange energy. Calculate the Boltzmann entropy of each separate system, and of the coupled system once it has reached equilibrium. Determine the change in entropy brought about by the coupling.

**5.7** Consider a system of $N$ oscillators sharing $Q$ quanta, with $N \gg 1$, $Q \gg 1$ and $Q = \alpha N$, where $\alpha$ is a constant. Use Stirling's formula to obtain an approximate expression for the Boltzmann entropy of the system and show that it is an extensive quantity.

**5.8** Calculate the mean and standard deviation of the number of quanta $q$ held by a single oscillator in a canonical ensemble when the reservoir parameter is $\hat{\beta} = \ln(3/2)$.

**5.9** Calculate the mean and standard deviation of the number of quanta $Q$ held by three oscillators in a canonical ensemble when the reservoir parameter is $\hat{\beta} = \ln(3/2)$.

# 6

# The Boltzmann Factor and the Canonical Partition Function

The canonical ensemble discussed in the previous chapter is designed to represent the statistics of a system that is open to energy exchange with a large environment. It is mathematically easier to employ in practice than the microcanonical ensemble designed for isolated systems. Let us now see some examples of its use, where we will also encounter the important concept of the canonical partition function.

## 6.1 Simple Applications of the Boltzmann Factor

The Boltzmann factor is $\exp(-E/kT)$, where $E$ is the energy of the micro- or macrostate in question. It plays a role in the statistical weighting of such a state in a canonical ensemble. It can be applied in a variety of ways.

### 6.1.1 Maxwell–Boltzmann Distribution

We consider a gas of atoms of mass $m$, and regard one atom as the system and the remainder of the gas as the heat bath or reservoir. The system exchanges energy with the reservoir through occasional atomic collisions. To a very good approximation, the system energy is just the kinetic energy of the atom, if interatomic interactions are weak. Let us now consider macrostates of the system labelled by the magnitude $v$ of the velocity of the atom. We want to determine the canonical probability density function $p(v)$ and so we write

$$p(v)\mathrm{d}v \propto \mathrm{d}\Omega(v)\exp\left(-\frac{E(v)}{kT}\right) \propto \rho(v)\exp\left(-\frac{mv^2}{2kT}\right)\mathrm{d}v. \tag{6.1}$$

For $E$ we have used the kinetic energy $mv^2/2$. The factor $\mathrm{d}\Omega(v)$ is the number of microstates of the atom, each specified by a 3-d velocity $\mathbf{v}$, that have a speed in the range $v$ to $v + \mathrm{d}v$. It is written as an increment to reflect the fact that it is a multiplicity of microstates within an incremental range of a microscopic variable, and has

*Statistical Physics: An Entropic Approach*, First Edition. Ian Ford.

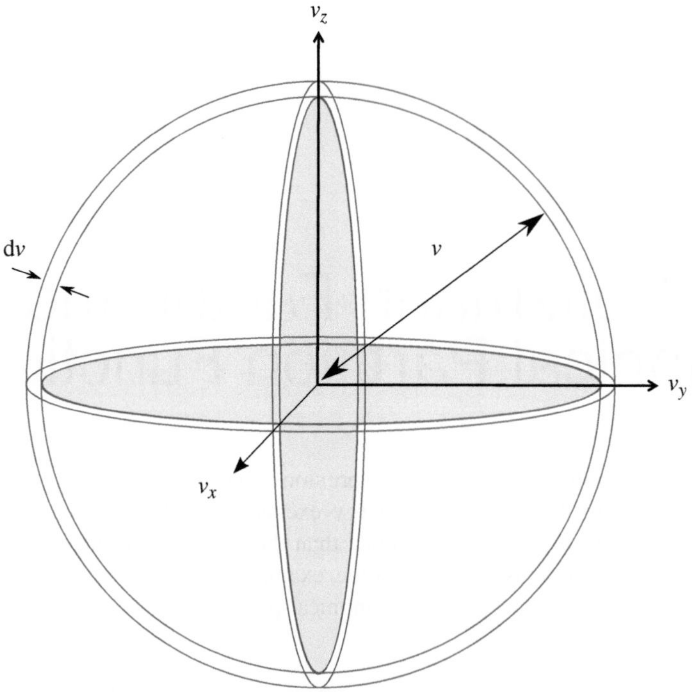


**Figure 6.1** The number of microstates with speed in a range d$v$ about a value $v$ is proportional to the volume of a shell of radius $v$ and thickness d$v$, assuming a uniform density of microstates in 3-d velocity space.

been expressed as $\rho(v)\mathrm{d}v$ using a *density of microstates* $\rho(v)$ per unit range of speed. Assuming that velocity microstates are to be found with equal density across the 3-d velocity space, an assumption that will be justified in Chapter 9 using quantum mechanics, the multiplicity $\mathrm{d}\Omega$ is proportional to the volume $4\pi v^2\mathrm{d}v$ of a shell of radius $v$ and thickness d$v$ in 3-d velocity space, as illustrated in Figure 6.1. The density of microstates is therefore proportional to $4\pi v^2$. This produces the celebrated Maxwell–Boltzmann speed distribution function:

$$p(v) \propto v^2 \exp\left(-\frac{mv^2}{2kT}\right), \tag{6.2}$$

and this may be employed to calculate a number of system properties.

The mean energy of the atom, for example, is

$$\langle E\rangle = \left\langle \frac{1}{2}mv^2\right\rangle = \frac{1}{Z}\int_0^\infty \frac{1}{2}mv^4 \exp\left(-\frac{mv^2}{2kT}\right)\mathrm{d}v \tag{6.3}$$

where $Z = \int_0^\infty v^2 \exp(-mv^2/2kT)\mathrm{d}v$, such that the pdf is normalised, that is, $\int_0^\infty p(v)\mathrm{d}v = 1$. After an integration by parts, we find that

$$\left\langle \frac{1}{2}mv^2 \right\rangle = \frac{1}{Z}\frac{kT}{2}\left( -\left[ v^3 \exp\left(-\frac{mv^2}{2kT}\right)\right]_0^\infty + 3\int_0^\infty v^2 \exp\left(-\frac{mv^2}{2kT}\right)\mathrm{d}v \right) = \frac{3}{2}kT, \tag{6.4}$$

having noticed that the remaining integral is proportional to $Z$. Thus the temperature of the reservoir controls the mean kinetic energy of the atom.

Each atom in the gas must have the same statistical properties, so the mean energy of a gas of $N$ atoms at temperature $T$ is $(3/2)NkT$. The constant volume heat capacity of such a gas is the temperature derivative of the mean energy when the volume is held constant and we find that

$$C_V = \frac{\mathrm{d}\langle E\rangle}{\mathrm{d}T} = \frac{3}{2}Nk. \tag{6.5}$$

These results may be compared with (2.6) and (2.10) and suggest that averages of quantities in statistical thermodynamics correspond to the values of corresponding state variables in classical thermodynamics.

### 6.1.2 Single Classical Oscillator and the Equipartition Theorem

Similarly, we make the assumption that the microstates of a classical 1-d harmonic oscillator, labelled by position $x$ and velocity $v$, are distributed with uniform density $\rho(x, v) = \rho_0$ over the phase space spanned by these two coordinates. The pdf over these variables, according to the canonical ensemble, would then be

$$p(x, v) \propto \rho(x, v) \exp\left(-\frac{E(x, v)}{kT}\right) \propto \exp\left(-\frac{(\kappa x^2 + mv^2)}{2kT}\right), \tag{6.6}$$

where $\kappa$ is the spring constant of the oscillator, and this distribution provides us with statistical averages of system properties. The canonical average of system quantity $\mathrm{A}(x, v)$ is

$$\langle \mathrm{A}\rangle = \frac{1}{Z}\int_{-\infty}^{\infty}\int_{-\infty}^{\infty} \mathrm{A}(x, v)\rho_0 \exp\left(-\frac{(\kappa x^2 + mv^2)}{2kT}\right)\mathrm{d}x\,\mathrm{d}v, \tag{6.7}$$

where $Z = \int\int \rho_0 \exp(-(\kappa x^2 + mv^2)/2kT)\mathrm{d}x\mathrm{d}v$. Using the standard Gaussian integral $\int_{-\infty}^{\infty} \exp(-\alpha x^2)\mathrm{d}x = (\pi/\alpha)^{1/2}$, the normalising factor is

$$Z = \rho_0 \left(\frac{2\pi kT}{\kappa}\right)^{\frac{1}{2}} \left(\frac{2\pi kT}{m}\right)^{\frac{1}{2}}. \tag{6.8}$$

For example, the average energy, using integration by parts, is

$$\begin{aligned}
\langle E\rangle &= \frac{1}{Z}\int_{-\infty}^{\infty}\int_{-\infty}^{\infty}\frac{1}{2}(\kappa x^2+mv^2)\rho_0\exp\left(-\frac{(\kappa x^2+mv^2)}{2kT}\right)\mathrm{d}x\,\mathrm{d}v\\
&= \frac{1}{2}\left[\left(\frac{\kappa}{2\pi kT}\right)^{\frac{1}{2}}\int_{-\infty}^{\infty}\kappa x^2\exp\left(-\frac{\kappa x^2}{2kT}\right)\mathrm{d}x\right.\\
&\quad\left.+\left(\frac{m}{2\pi kT}\right)^{\frac{1}{2}}\int_{-\infty}^{\infty}mv^2\exp\left(-\frac{mv^2}{2kT}\right)\mathrm{d}v\right]\\
&= \frac{1}{2}\left[kT\left(\frac{\kappa}{2\pi kT}\right)^{\frac{1}{2}}\int_{-\infty}^{\infty}\exp\left(-\frac{\kappa x^2}{2kT}\right)\mathrm{d}x+kT\left(\frac{m}{2\pi kT}\right)^{\frac{1}{2}}\int_{-\infty}^{\infty}\exp\left(-\frac{mv^2}{2kT}\right)\mathrm{d}v\right]\\
&= kT.
\end{aligned} \tag{6.9}$$

It is a general result that for each term in the energy that is quadratic in a microstate variable, such as $x$ or $v$, the mean energy in the canonical ensemble is $kT/2$. We saw this in the case of the ideal gas in the last section, when there were three such terms (proportional to $v_x^2$, $v_y^2$ and $v_z^2$). This result is called the *equipartition theorem*, and it is very powerful since quadratic terms in the energy are frequently seen. For a 3-d oscillator, the energy is $(1/2)\kappa(x^2+y^2+z^2)+(1/2)m(v_x^2+v_y^2+v_z^2)$, using Cartesian position and velocity components, and the mean energy in a canonical ensemble would be $3kT$, a contribution of $(1/2)kT$ for each quadratic term, or equivalently for each so-called *degree of freedom*.

### 6.1.3 Isothermal Atmosphere Model

Now we consider an ideal gas in a gravitational field. Once again, we take the system to be a single molecule, and the reservoir to be the rest of the gas. Imagine the gas is isothermal, that is, the temperature is uniform with height $z$. Let us consider *macrostates* of the single molecule labelled by height and we seek a pdf $p(z)$ such that the probability of finding the single molecule in the height range $z \to z+\mathrm{d}z$ is $p(z)\mathrm{d}z$ with a normalisation $\int_0^\infty p(z)\mathrm{d}z = 1$. The potential energy of the macrostate is $mgz$, where $m$ is the mass of the molecule and $g$ the acceleration due to gravity.

The microstate multiplicity of such a macrostate is the number of microstates corresponding to different velocities and various states of rotation and internal vibration available to the molecule at height $z$, but this does not depend on height. Thus, the $z$-dependence of the pdf for the macrostate variable $z$ is simply proportional to $\exp(-mgz/kT)$. By extension, the *density* of the gas is proportional to this factor, as it will mirror the probability distribution function for the position of a single molecule. The gas pressure $p$ (not to be confused here with the pdf $p(z)$) is proportional to density, for an isothermal ideal gas, and hence

$$p \propto \exp\left(-\frac{mgz}{kT}\right). \tag{6.10}$$

Applying this to the atmosphere, the pressure should fall exponentially with height, reducing by a factor of e with each ascent through a distance $z_s = kT/mg$. For the Earth, this height is about 16 km and is the right order of magnitude, but the real terrestrial atmosphere is not isothermal. The temperature varies with height, and so the pressure profile differs from the isothermal profile in practice.

### 6.1.4 Escape Problems and Reaction Rates

As well as providing a basis for detailed calculations of statistical averages, as we have seen in the above examples, the Boltzmann factor can be used to *estimate* the likelihood of rare events. Consider a particle in a potential well, but weakly interacting with a reservoir of other particles at temperature $T$. Imagine that the particle can escape from the well if it can acquire a threshold energy $E_t$ equal to its binding energy. An example might be an electron bound to a nucleus but interacting with other particles too. What is the thermal ionisation rate?

The probability that the particle might acquire sufficient energy to reach the escape threshold is estimated to be proportional to $\exp(-E_t/kT)$. This is only an approximation, of course, because a system such as this is not in equilibrium. The statistical properties are time dependent as the particle is able to escape: as time goes on the probability that it remains in the well might go to zero. Nevertheless, the Boltzmann factor provides a rule of thumb, and we conclude that a particle with a binding energy of many $kT$ will find it difficult to escape by thermal excitation.

An escape problem on a larger scale concerns the rate of loss of a planetary atmosphere. The potential energy of a molecule in the gravitational field of a body of mass $M$ is $-GMm/r$, where $G$ is the gravitational constant, $m$ is the molecular mass and $r$ is the radial distance from the centre of the body. If the atmosphere is at a temperature $T$, molecules at a radius $r$ will escape at a significant rate if $kT$ is greater than some specified fraction of the molecular gravitational energy. Hot or low mass planets therefore lose their atmospheres more quickly than massive or cool planets.

By a similar argument, liquids should evaporate at a rate that resembles a Boltzmann factor. The exponential form of the Clausius–Clapeyron equation for equilibrium vapour pressure in (3.73) confirms this expectation. A molecule will escape from the condensed phase if it acquires an amount of energy equal to the latent heat of evaporation per particle $L_e$. It will condense from the gas at a rate proportional to the pressure of the vapour. In equilibrium, there is a balance between the rates of escape and condensation and therefore the saturated vapour pressure will be proportional to $\exp(-L_e/kT)$.

As a final example, consider that the temperature dependence of chemical reactions is often found to be proportional to an Arrhenius factor $\exp(-E_c/kT)$ with some characteristic energy $E_c$. The interpretation, in a similar spirit, is that the reaction requires an energy barrier to be surmounted, and that a Boltzmann factor expresses the likelihood that the reactants acquire this energy thermally from their surroundings.

## 6.2 Mathematical Properties of the Canonical Partition Function

The normalising factor $Z$ that appears in the canonical probability distribution $P(E) = Z^{-1}\exp(-E/kT)$ is far more central to the application of the canonical ensemble than it might appear. It is so prominent that it has been given a name. It is called the *canonical partition function*, terminology that expresses its role in describing the partitioning of probability amongst the available possibilities.

Consider a system with microstates labelled $i$, each with energy $E_i$. The canonical partition function is

$$Z = \sum_i \exp(-\beta E_i), \tag{6.11}$$

where the sum is over all microstates, and where it is convenient to use the reservoir parameter $\beta$ instead of $1/kT$. The derivative of $Z$ with respect to $\beta$ is

$$\frac{\mathrm{d}Z}{\mathrm{d}\beta} = -\sum_i E_i \exp(-\beta E_i), \tag{6.12}$$

and hence the mean energy in the canonical ensemble is

$$\langle E\rangle = \sum_i E_i P(E_i) = \frac{1}{Z}\sum_i E_i \exp(-\beta E_i) = -\frac{1}{Z}\frac{\mathrm{d}Z}{\mathrm{d}\beta} = -\frac{\mathrm{d}\ln Z}{\mathrm{d}\beta} = kT^2\frac{\mathrm{d}\ln Z}{\mathrm{d}T}, \tag{6.13}$$

using $\mathrm{d}\beta/\mathrm{d}T = -1/kT^2$. This demonstrates the mathematical convenience of the canonical ensemble and the central role of $Z$: statistical properties can often be obtained just by taking suitable derivatives of the partition function.

Moreover, the partition function plays a role in making the connection between classical and statistical thermodynamics. Let us consider the temperature derivative of the Helmholtz free energy divided by temperature. Using methods familiar from Chapter 3, we have

$$\begin{aligned}\mathrm{d}\left(\frac{F}{T}\right) &= \frac{1}{T}\mathrm{d}F - \frac{F}{T^2}\mathrm{d}T = \frac{1}{T}(-S\,\mathrm{d}T - p\mathrm{d}V + \mu\mathrm{d}N) - \frac{E-TS}{T^2}\mathrm{d}T \\ &= -\frac{E}{T^2}\mathrm{d}T - \frac{p}{T}\mathrm{d}V + \frac{\mu}{T}\mathrm{d}N,\end{aligned} \tag{6.14}$$

having employed (3.20), so that

$$E = -T^2\left(\frac{\partial(F/T)}{\partial T}\right)_{V,N}. \tag{6.15}$$

If we compare this with (6.13), and once again regard the thermodynamic state variable $E$ as equivalent to the *average* system energy in the canonical ensemble, we conclude that

$$F = -kT\ln Z = \langle E\rangle - TS, \tag{6.16}$$

which is an extremely important result that we shall meet again in Section 8.2.

Another powerful result is

$$\begin{aligned}\sigma_E^2 &= \langle(E-\langle E\rangle)^2\rangle = \langle E^2\rangle - \langle E\rangle^2 = \frac{1}{Z}\sum_i E_i^2\exp(-\beta E_i) - \langle E\rangle^2 \\ &= \frac{1}{Z}\frac{\mathrm{d}^2Z}{\mathrm{d}\beta^2} - \frac{1}{Z^2}\left(\frac{\mathrm{d}Z}{\mathrm{d}\beta}\right)^2 = \frac{\mathrm{d}^2\ln Z}{\mathrm{d}\beta^2} = -\frac{\mathrm{d}\langle E\rangle}{\mathrm{d}\beta} = kT^2\frac{\mathrm{d}\langle E\rangle}{\mathrm{d}T},\end{aligned} \tag{6.17}$$

that establishes a connection between the variance of the energy and $\mathrm{d}\langle E\rangle/\mathrm{d}T$, the heat capacity at constant volume.

We can show in general that the relative magnitude of thermal fluctuations is inversely proportional to the square root of the number of particles. We saw a specific illustration of this for the oscillator system in Section 5.4. From the expression for the variance in (6.17), we evaluate the ratio of the standard deviation in energy to the mean:

$$\frac{\sigma_E}{\langle E\rangle} = \frac{(\langle E^2\rangle - \langle E\rangle^2)^{\frac{1}{2}}}{\langle E\rangle} = \left(\frac{kT^2}{\langle E\rangle^2}\frac{\mathrm{d}\langle E\rangle}{\mathrm{d}T}\right)^{\frac{1}{2}}. \tag{6.18}$$

Both the heat capacity $\mathrm{d}\langle E\rangle/\mathrm{d}T$ and the mean energy $\langle E\rangle$ are proportional to the number of particles in the system and hence are extensive: if the system size is doubled, they both double. Inserting such dependence, it is clear that the ratio in (6.18) is proportional to $N^{-1/2}$. The mean and the standard deviation both increase as the thermodynamic limit is approached, but in a relative sense, the distribution over energy becomes sharper. The thermodynamic limit is characterised by the complete neglect of fluctuations.

Note, however, that there are circumstances where statistical fluctuations can become apparent even in macroscopic systems. An example is the phenomenon of *critical opalescence* of a fluid near its so-called critical conditions of pressure and temperature. These conditions correspond to the right hand end of the gas–liquid coexistence line in Figure 3.8, where the distinction between a gas and a liquid is lost. Near this point, the fluid experiences strong local fluctuations in fluid density, since the surface tension characterising the interface between such patches becomes very small, giving rise to variations in optical properties and a consequent cloudiness in the fluid. Equation (6.17) would suggest that the heat capacity near the critical point becomes anomalously large as well, and this is borne out experimentally.

In the next few sections, we study several examples of the role of the partition function in establishing the statistical properties of simple systems.

## 6.3 Two-Level Paramagnet

One of the simplest systems that can be studied in a canonical ensemble is a two-level paramagnet. A paramagnet is a material that can acquire a magnetisation when exposed to an external magnetic field, but loses it when the field is turned off. The basic element of the material is an atomic magnetic dipole that can orient itself with respect to the external field. Quantum mechanics tells us that the number of possible dipole orientations is finite, not continuous, and in the very simplest case, there might be just two microstates labelled by $m_s$: the dipole can be aligned with ($m_s = +1$) or against ($m_s = -1$) the field. The microstates are characterised by magnetic dipole moments $m_s\mu_B$ and energies $-m_s\mu_B B$, where $B$ is the magnitude of the external magnetic field and $\mu_B = e\hbar/2m_e$ is the Bohr magneton, with $e$ the elementary charge and $m_e$ the mass of the electron.

We are interested in the mean magnetisation $M$ of the paramagnet. If there are $N$ dipoles in the material, then $M = N\mu_B\langle m_s\rangle$, where the brackets denote a canonical average under the influence of a heat bath at temperature $T$. The partition function is

$$Z = \sum_i \exp\left(-\frac{E_i}{kT}\right) = \sum_{m_s=-1}^{+1} \exp\left(\frac{m_s \mu_B B}{kT}\right)$$
$$= \exp\left(\frac{\mu_B B}{kT}\right) + \exp\left(-\frac{\mu_B B}{kT}\right) = 2\cosh\left(\frac{\mu_B B}{kT}\right). \quad (6.19)$$

The average may be constructed as

$$\langle m_s \rangle = \frac{1}{Z}\sum_{m_s} m_s \exp\left(\frac{m_s \mu_B B}{kT}\right) = \frac{1}{Z}\left[\exp\left(\frac{\mu_B B}{kT}\right) - \exp\left(-\frac{\mu_B B}{kT}\right)\right]$$
$$= \tanh\left(\frac{\mu_B B}{kT}\right), \quad (6.20)$$

but the same result can be obtained by differentiating the partition function:

$$\langle m_s \rangle = \frac{1}{Z}\sum_{m_s} m_s \exp\left(\frac{m_s \mu_B B}{kT}\right) = \frac{1}{Z}\frac{\mathrm{d}Z}{\mathrm{d}\left(\mu_B B/kT\right)} = \frac{2\sinh\left(\mu_B B/kT\right)}{2\cosh\left(\mu_B B/kT\right)}$$
$$= \tanh\left(\frac{\mu_B B}{kT}\right). \quad (6.21)$$

Thus, $\langle m_s \rangle = 0$ if $B = 0$ (the absence of an external field), but for a strong field such that $|B| \gg kT/\mu_B$, the dipole is fully aligned in the direction of the field, corresponding to $|\langle m_s \rangle| \to 1$. For $|B| \ll kT/\mu_B$, $\tanh(\mu_B B/kT) \approx \mu_B B/kT$ and the paramagnetisation of the material is then approximately

$$M \approx \frac{N\mu_B^2 B}{kT}. \quad (6.22)$$

This linear proportionality between magnetisation and external field, and inverse proportionality to temperature, is a well-established experimental result known as Curie's law. The mean magnetisation of the two-level paramagnet over a range of fields is illustrated in Figure 6.2.

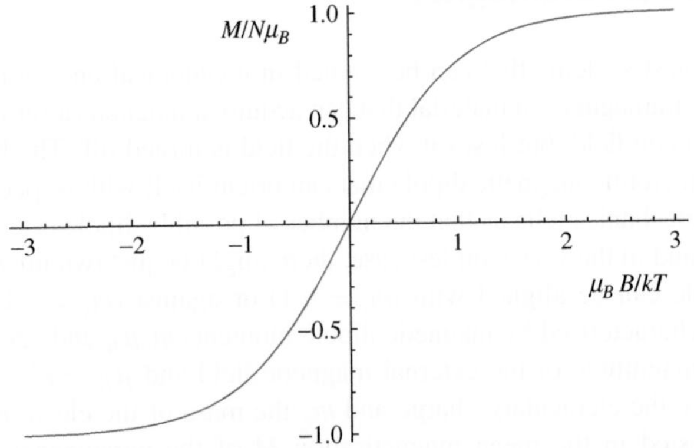


**Figure 6.2** Magnetisation $M$ of a two-level paramagnet as a function of external magnetic field $B$. Linear dependence at low external field goes over to saturation at higher fields.

## 6.4 Single Quantum Oscillator

We based our discussion of statistical thermodynamics on quantum oscillators in Chapter 5, and it is instructive to consider now the canonical statistical properties of a 1-d quantum harmonic oscillator using the partition function. The energies of the microstates are $E_n = [n + (1/2)]\hbar\omega$, such that the partition function is

$$Z = \exp\left(-\frac{x}{2}\right)\sum_{n=0}^{\infty}\exp(-nx), \tag{6.23}$$

where $x = \hbar\omega/kT$. This is a geometric series, and sums to

$$Z = \frac{\exp\left(-\frac{x}{2}\right)}{1-\exp(-x)} = \frac{1}{2\sinh\left(\frac{x}{2}\right)}, \tag{6.24}$$

and the mean vibrational energy $E_{\text{vib}}$ is given by

$$\langle E_{\text{vib}}\rangle = -\frac{1}{Z}\frac{\text{d}Z}{\text{d}\beta} = -\frac{\hbar\omega}{Z}\frac{\text{d}Z}{\text{d}x} = 2\hbar\omega\sinh\left(\frac{x}{2}\right)\frac{2\cosh\left(\frac{x}{2}\right)}{\left(2\sinh\left(\frac{x}{2}\right)\right)^2}\frac{1}{2} = \frac{\hbar\omega}{2\tanh\left(\frac{\hbar\omega}{2kT}\right)}. \tag{6.25}$$

This has the correct behaviour at high and low temperature: it is equal to the zero point energy $\hbar\omega/2$ for $T \ll \hbar\omega/k$, where thermal excitation is very weak, and goes to the classical equipartition result $\langle E_{\text{vib}}\rangle = kT$ as $T \to \infty$, as shown in Figure 6.3. The agreement with the equipartition result obtained for a single classical oscillator in Section 6.1.2 suggests that the constant density of states across the $(x, v)$ phase space used in that derivation was appropriate.

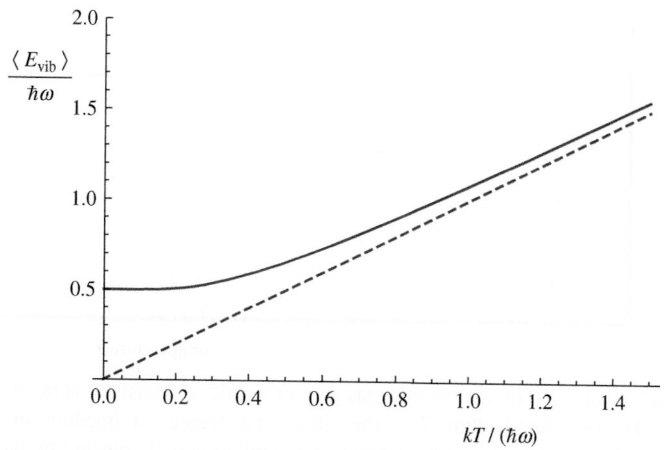


**Figure 6.3** Mean energy of a 1-d quantum oscillator as a function of temperature. The classical result $kT$ is obtained for $T \gtrsim \hbar\omega/k$. Below this temperature, the mean energy tends towards the zero point energy of the oscillator $\hbar\omega/2$.

## 6.5 Heat Capacity of a Diatomic Molecular Gas

The classical and quantum limits of the canonical statistical behaviour of an oscillator can be observed in the temperature dependence of the heat capacity of a diatomic molecular gas such as $O_2$. The vibrational motion of the two atoms with respect to one another may be modelled as a 1-d quantum oscillator. Using (6.25), the heat capacity of molecular vibration is

$$C_{\text{vib}} = \frac{\mathrm{d}\langle E_{\text{vib}}\rangle}{\mathrm{d}T} = \frac{\hbar\omega}{2}\frac{\mathrm{d}}{\mathrm{d}T}\left[\frac{1}{\tanh\left(\frac{\hbar\omega}{2kT}\right)}\right] = \frac{\hbar^2\omega^2}{4kT^2\sinh^2\left(\frac{\hbar\omega}{2kT}\right)}. \tag{6.26}$$

The classical expression emerges for $T \gg \hbar\omega/k$, in which case, we can write $\hbar\omega/2kT \ll 1$ and $\sinh(\hbar\omega/2kT) \approx \hbar\omega/2kT$, such that $C_{\text{vib}} \to k$. In the other extreme, $T \ll \hbar\omega/k$ and $\sinh(\hbar\omega/2kT) \approx (1/2)\exp(\hbar\omega/2kT)$ and so $C_{\text{vib}} \to 0$.

This behaviour is apparent in experimental data for diatomic molecules such as $O_2$, illustrated in Figure 6.4. On the basis of the equipartition theorem considered in Section 6.4, a contribution of $k/2$ per degree of freedom would be expected. There are three components of centre of mass linear momentum, two degrees of freedom for rotation of the molecule about the centre of mass (a third does not arise for a diatomic molecule) and then two oscillator degrees of freedom from the vibrational motion, one each for the potential energy and kinetic energy contributions. Therefore, we would expect to see a constant volume heat capacity per molecule of $7k/2$, and indeed at temperatures above 1500 K for $O_2$, this is what we see experimentally. However, as the temperature is reduced, the heat capacity falls to $5k/2$, as sketched in Figure 6.4. This is taken as evidence that the mean energy due to vibration is suppressed at lower temperatures corresponding to the approach to the quantum limit. This is referred to as the 'freezing out' of the vibrational degrees of freedom, and the temperature $T_{\text{vib}}$

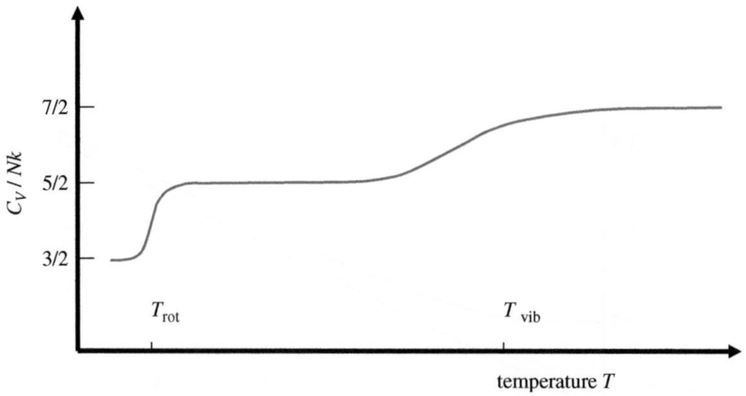


**Figure 6.4** Heat capacity of a diatomic gas per molecule at constant volume as a function of temperature. Below a temperature $T_{\text{vib}}$, the vibrational degrees of freedom are progressively 'frozen out' until the vibrational energy of molecular oscillation is dominated by the temperature-independent zero point energy. Below a temperature $T_{\text{rot}}$, rotational degrees of freedom are also frozen out.

below which this sets in should correspond to the condition $\hbar\omega/kT_{\text{vib}} \approx 1$, where $\omega$ is the vibrational frequency of the molecular bond.

Furthermore, below a temperature $T_{\text{rot}}$, there is a suppression of the heat capacity from $5k/2$ to $3k/2$, which is interpreted as the freezing out of the rotational degrees of freedom, which is explored in an exercise at the end of this chapter. The early pioneers of statistical thermodynamics were very concerned at the apparent failure of the equipartition theorem for diatomic gases; the anomalies actually provided evidence for the quantisation of energy.

The separation of the total energy of a systems into translational, vibrational and rotational energy has the important implication that the partition function can be factorised. Consider a complex molecule with an energy that separates as

$$E = E_{\text{trans}} + E_{\text{vib}} + E_{\text{rot}} + E_{\text{elec}}, \tag{6.27}$$

where a term for the energy associated with the electrons is included as well. The partition function for such a molecule is

$$\begin{aligned} Z &= \sum_{\text{microstates}} \exp\left(-\frac{E}{kT}\right) \\ &= \sum_{\text{trans}} \exp\left(-\frac{E_{\text{trans}}}{kT}\right) \sum_{\text{vib}} \exp\left(-\frac{E_{\text{vib}}}{kT}\right) \sum_{\text{rot}} \exp\left(-\frac{E_{\text{rot}}}{kT}\right) \sum_{\text{elec}} \exp\left(-\frac{E_{\text{elec}}}{kT}\right) \\ &= Z_{\text{trans}}\, Z_{\text{vib}}\, Z_{\text{rot}}\, Z_{\text{elec}} \end{aligned} \tag{6.28}$$

as long as the energies of the microstates of vibrational motion do not depend on the value of the translational energy and so on. The factorisation of the partition function when describing independent modes of excitation appears in other contexts later in this book.

## 6.6 Einstein Model of the Heat Capacity of Solids

The analysis of a quantum oscillator can be used to understand the heat capacity of a solid. Albert Einstein (1879–1955) proposed a model of the vibrational energy of a solid based on the idea that each particle vibrates about its rest position at the same angular frequency $\omega_{\text{E}}$, known as the Einstein frequency. Each particle therefore may be represented by a 3-d quantum oscillator. The vibrational frequency spectrum of a solid is much broader and richer than this, but the single frequency approximation makes the analysis quite easy as each particle makes the same contribution to the heat capacity.

Using (6.26), the Einstein vibrational heat capacity of a solid consisting of $N$ particles is

$$C_{\text{E}} = \frac{3N\hbar^2\omega_{\text{E}}^2}{4kT^2\sinh^2\left(\frac{\hbar\omega_{\text{E}}}{2kT}\right)}, \tag{6.29}$$

which goes to $3Nk$ for $T \gg T_{\text{E}} = \hbar\omega_{\text{E}}/k$, where $T_{\text{E}}$ is called the Einstein temperature. $C_{\text{E}}$ approximates to the form $3Nk(T_{\text{E}}^2/T^2)\exp(-T_{\text{E}}/T)$ for $T \ll T_{\text{E}}$, as illustrated in Figure 6.5. Although experimental data deviates somewhat from this particular expression

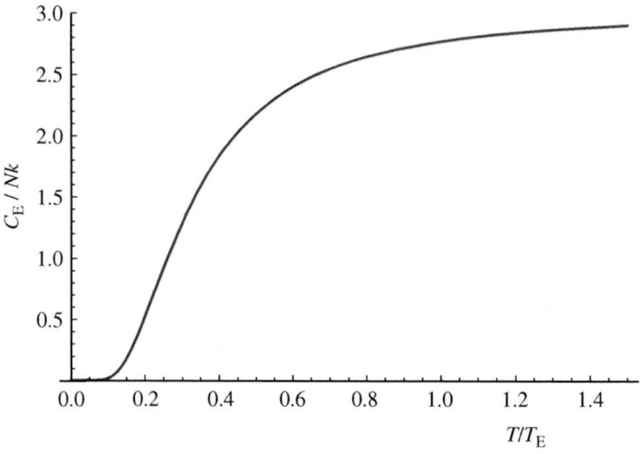


**Figure 6.5** Temperature dependence of the Einstein vibrational heat capacity of a solid of $N$ atoms, each oscillating at a common frequency $\omega_E$ related to the Einstein temperature $T_E = \hbar\omega_E/k$.

at low temperature, the rough agreement was used by Einstein as evidence that the vibrational energy of a solid is indeed quantised, and that it progressively becomes frozen out as the temperature is decreased. Such an application offers strong support for the validity of statistical thermodynamics.

## 6.7 Vacancies in Crystals

Our next example concerns the presence of defects in crystals known as vacancies. In a perfect crystal, the $N$ atoms form a precisely ordered spatial array called a lattice. But real crystals will contain the occasional unoccupied lattice site, and these are called vacancies. The displaced atom can either squeeze in between its neighbours somewhere, when it is called an interstitial atom, or migrate to the crystal surface or some other interface.

Let us make things simple and consider each atom to have two available positions: in its proper place with energy zero, or displaced elsewhere with positive energy $E_v$. The configuration of the atoms, and hence of the crystal, is a microstate labelled by the state of occupancy of each site in the lattice. The partition function is

$$Z = \sum_{\{n_j\}} \exp\left( -\frac{\sum_j n_j E_v}{kT} \right), \tag{6.30}$$

where the set $\{n_1, n_2, n_3, \cdots, n_N\}$ denotes the state of occupancy of each of the $N$ sites: $n_j = 0$ means site $j$ is occupied and $n_j = 1$ means that it is vacant. The energy of the configuration is therefore $\sum_{j=1}^{N} n_j E_v$. The partition sum is over microstates $\{n_j\}$, that is, over all possible values of all the $n_j$.

This model is a useful illustration of how the statistical properties of a system of independent components can be regarded as the aggregate of the statistical properties of each component. As the energy is a sum of separate contributions, we can factorise the partition function as follows

$$Z = \sum_{n_1=0}^{1} \exp\left(-\frac{n_1 E_v}{kT}\right) \sum_{n_2=0}^{1} \exp\left(-\frac{n_2 E_v}{kT}\right) \cdots \sum_{n_N=0}^{1} \exp\left(-\frac{n_N E_v}{kT}\right), \tag{6.31}$$

and evaluate each part. The first is just $\left[1+\exp\left(-E_v/kT\right)\right]$. In fact, every factor is the same; so the partition function is simply

$$Z = \left[1+\exp\left(-\frac{E_v}{kT}\right)\right]^N. \tag{6.32}$$

We can interpret $N_v = \sum_j n_j$ as the total number of vacancies in the microstate $\{n_j\}$. The mean number of vacancies $\langle N_v \rangle$ is therefore given by $\langle \sum_j n_j \rangle = Z^{-1}\sum_{\{n_j\}}\left(\sum_j n_j\right)\exp(-\sum_j n_j E_v/kT)$, but considering (6.30), this is equivalent to $-\mathrm{d}\ln Z/\mathrm{d}y$ where $y = E_v/kT$. Thus

$$\langle N_v \rangle = -\frac{\mathrm{d}}{\mathrm{d}y}\ln\left(1+\exp(-y)\right)^N = -N\frac{\mathrm{d}}{\mathrm{d}y}\ln(1+\exp(-y)) = \frac{N}{1+\exp\left(\frac{E_v}{kT}\right)}. \tag{6.33}$$

Let us consider the implications of this expression. As the temperature is raised, $\langle N_v \rangle$ increases towards an upper limit of $N/2$. In contrast, as $T \to 0$, we find that $\langle N_v \rangle \approx N\exp(-E_v/kT)$ such that vacancies become rare as the temperature is reduced. Notice that this result is in accord with the rule of thumb that the Boltzmann factor $\exp(-E_v/kT)$ determines the likelihood of a rare event such as vacancy formation at low temperature.

The mean energy of the crystal associated with its defects is the mean number of vacancies times the energy of vacancy formation, $\langle N_v \rangle E_v$. The contribution to the heat capacity of the crystal is therefore $\mathrm{d}(\langle N_v \rangle E_v)/\mathrm{d}T$. This is

$$\begin{aligned} C_v &= \frac{NE_v^2}{kT^2}\frac{\exp\left(\frac{E_v}{kT}\right)}{\left[1+\exp\left(\frac{E_v}{kT}\right)\right]^2} \\ &= \frac{NE_v^2}{kT^2}\frac{1}{\left[\exp\left(-\frac{E_v}{2kT}\right)+\exp\left(\frac{E_v}{2kT}\right)\right]^2} = \frac{NE_v^2}{4kT^2}\mathrm{sech}^2\left(\frac{E_v}{2kT}\right). \end{aligned} \tag{6.34}$$

This supplements the heat capacity due to vibrations of the crystal discussed in Section 6.6.

As the temperature increases, expression (6.34) goes through a peak at around $T = E_v/k$, as sketched in Figure 6.6. As $T$ rises, the increasing value of $\mathrm{sech}^2(E_v/2kT)$, which behaves like $\exp(-E_v/kT)$ for $T \ll E_v/k$, is countered by the decreasing value of $T^{-2}$. However, as a typical vacancy formation energy is around 1 eV, the peak will lie at $T \sim 1.6\times10^{-19}/1.38\times10^{-23} \sim 10^4$ K, and the crystal will have melted before then. Nevertheless, any system containing components that each have two possible microstates

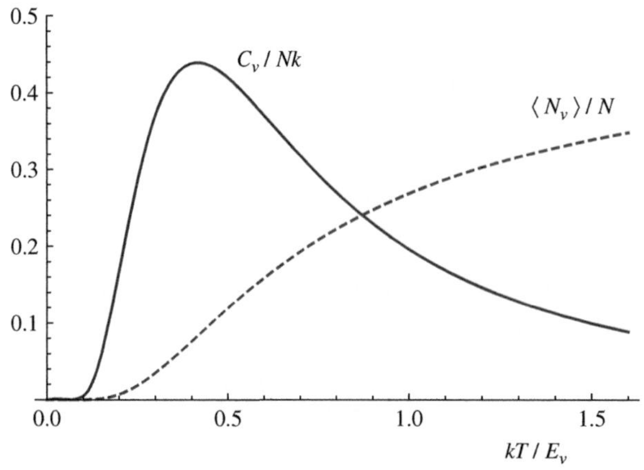


**Figure 6.6** Heat capacity $C_v$ of a crystal due to a temperature-dependent mean population $\langle N_v \rangle$ of vacancies. The peak is known as a Schottky anomaly, which might be visible above the monotonic temperature behaviour of the heat capacity arising from solid vibrations sketched in Figure 6.5, assuming that the solid has not melted.

will behave in a similar manner, and as long as the energy scale is suitable then such a peak, known as a Schottky anomaly, will be observable above the vibrational heat capacity background. For example, the two-level paramagnet discussed in Section 6.3 has a mean energy of $\langle E \rangle = -N\mu_B B \tanh(\mu_B B/kT)$ for $N$ dipoles, and a heat capacity of

$$C_M = -N\mu_B B \frac{\mathrm{d}}{\mathrm{d}T} \tanh\left(\frac{\mu_B B}{kT}\right) = \frac{N\mu_B^2 B^2}{kT^2} \mathrm{sech}^2\left(\frac{\mu_B B}{kT}\right), \qquad (6.35)$$

which takes the same form as (6.34). The temperature $T_M = \mu_B B/k$ for a magnetic field of order 1 tesla is a few kelvin, and Schottky anomalies can be readily detected in paramagnetic materials at such temperatures, providing further confidence in the methods of the canonical ensemble.

## Exercises

**6.1** The canonical partition function of a classical 1-d harmonic oscillator of mass $m$ and spring constant $\kappa$ may be written as

$$Z = h_0^{-1} \int_{-\infty}^{\infty} \int_{-\infty}^{\infty} \mathrm{d}x\, \mathrm{d}p\, \exp\left(-\frac{p^2}{2mkT}\right) \exp\left(-\frac{\kappa x^2}{2kT}\right),$$

where $x$ and $p$ are the oscillator position and momentum, respectively, and $h_0$ is a constant. (a) Evaluate $Z$ and hence the mean energy $\langle E \rangle$ of the oscillator in equilibrium with a heat bath at temperature $T$. You may assume that

$\int_{-\infty}^{\infty} \exp(-\alpha x^2) dx = (\pi/\alpha)^{1/2}$. (b) Calculate the Helmholtz free energy of the oscillator, and show that the entropy is given by $S = k(1 + \ln[2\pi kT/(h_0\omega)])$ where $\omega = (\kappa/m)^{1/2}$ is the natural angular frequency of the oscillator. (c) Demonstrate that the oscillator entropy does not satisfy the third law of thermodynamics and explain physically why this is so.

**6.2** The canonical partition function of a 1-d quantum harmonic oscillator is

$$Z = [2\sinh(\hbar\omega/(2kT))]^{-1}.$$

(a) Evaluate the mean energy of the quantum oscillator when in equilibrium with a heat bath at temperature $T$. (b) Evaluate the Helmholtz free energy and entropy of the quantum oscillator. (c) Does the oscillator entropy satisfy the third law of thermodynamics? (d) By considering the high temperature limit of the partition function of the quantum oscillator, identify the constant $h_0$ employed in the classical treatment in the previous question.

**6.3** The spring constant of a *classical* harmonic oscillator is very slowly changed from $\kappa$ to $2\kappa$ while the oscillator remains in thermal equilibrium with a heat bath at temperature $T$. Making use of results obtained in question 6.1: (a) Calculate the change in mean energy of the oscillator. (b) Calculate the change in entropy of the oscillator. (c) Calculate the heat delivered to the heat bath during the process. (d) Calculate the change in Helmholtz free energy of the oscillator. (e) Calculate the work done on the oscillator during the process. (f) If the process were repeated more rapidly, would your answer to part (e) be greater than, less than, or the same as in the very slow process?

**6.4** The magnetic moment of an atom may take two orientations with respect to an external magnetic field $B$: aligned with the field, with energy $-\mu_B B$, or against the field with energy $+\mu_B B$. (a) Calculate the canonical partition function of an array of $N$ atoms in equilibrium with a heat bath at temperature $T$. (b) Show that the mean energy of the array is given by $\langle E\rangle = -N\mu_B B \tanh(\mu_B B/kT)$. (c) Show that the entropy of the array is given by

$$S = -N\mu_B B/T \;\tanh(\mu_B B/kT) + Nk\ln(2\cosh(\mu_B B/kT)).$$

(d) Evaluate the entropy of the array for $B = 0$ and $B \to \infty$. (e) Show that the standard deviation of the energy is given by $\sigma_E = \mu_B B N^{1/2}/\cosh(\mu_B B/kT)$.

**6.5** Demonstrate that the following expression

$$kTV^2\left(\frac{\partial(V^{-1}\ln Z)}{\partial V}\right)_{T,N}$$

is equal to the Gibbs free energy of a system.

**6.6** The rotational energy of a molecule is quantised according to $E_{\mathrm{rot}}(\ell) = \ell(\ell+1)\Theta$ where $\Theta$ is a constant, and with $\ell = 0, 1, 2$ and so on. The number of rotational quantum states at a given value of $\ell$ is $2\ell + 1$. Write down the rotational canonical partition function of the molecule as a sum over $\ell$. State the probability that the molecule might be found with a particular value of $\ell$. At high temperatures, it is possible to regard the general term in the sum as a function of a continuous variable $\ell$. Show that the most probable value of $\ell$ in these circumstances is

$$\ell_{\text{mode}} = \frac{1}{2}\left[\left(\frac{2kT}{\Theta}\right)^{\frac{1}{2}} - 1\right],$$

and interpret this in terms of the suppression of rotational energy as the temperature is reduced.

# 7

# The Grand Canonical Ensemble and Grand Partition Function

The behaviour of a system in contact with a heat bath has been investigated using the ideas of statistical thermodynamics in previous chapters and we have developed a scheme, the canonical ensemble, that provides a framework for a treatment of such behaviour. The next step is to develop the statistical thermodynamics of a system in contact with a reservoir that provides *particles* as well as energy. The ensemble of system microstates that emerges is designed to capture the statistical properties of a system coupled to a heat and particle bath. It is called the *grand canonical ensemble*. This is very suitable for a description of coexistence between phases such as a gas and a liquid. We shall also find that vacancy formation in crystals, discussed in Section 6.7, can be naturally treated with these methods.

## 7.1 System of Harmonic Oscillators

As in Chapter 5, it is instructive to use an example involving quantised harmonic oscillators as the basis for discussion. Consider a set of $N_{tot}$ harmonic oscillators holding $Q_{tot}$ energy quanta. Our system of interest is now a variable size *group* of these oscillators. How it might be that oscillators are counted as in or out of the group is not important, but the idea is akin to a situation where people associate together, and make or break their social bonds to each other in a rather complicated way. A physical example that might help intuition is to imagine the oscillators floating around inside a container, and as they pass in and out of a defined subvolume, we regard them as entering and exiting our system of interest. The ensemble we plan to develop would tell us the likelihood that the subvolume should contain a certain number of oscillators.

Alternatively, we could be more abstract and simply consider the creation and removal of network connections between the oscillators, without their having to move, and

*Statistical Physics: An Entropic Approach*, First Edition. Ian Ford.

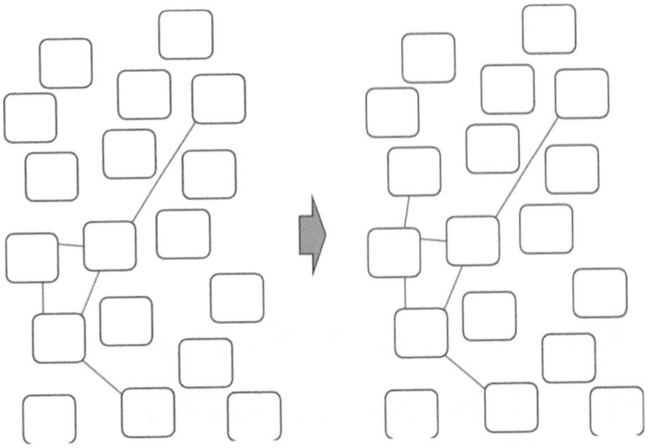

**Figure 7.1** A system of oscillators is defined by a set of connections between them that appear and disappear as time evolves according to given rules. In the illustration, the system grows in size from $N = 5$ to $N = 6$. The disconnected oscillators form the reservoir. If the oscillators also exchange quanta of energy, then we have a framework for describing a system in contact with a heat and particle bath. The grand canonical ensemble is the collection of all available conformations of the system.

imagine that the pattern of connections evolves in time in some complex manner. A system would then correspond to the set of interconnected oscillators, and the oscillators that are not part of the network would be regarded as components of the reservoir. The dynamics that modify connections would run alongside the dynamics of the exchange of quanta between oscillators considered in Section 4.2. Such a network is illustrated in Figure 7.1, undergoing a change in size as a new connection is made.

The system moves from microstate to microstate in a complicated phase space that is now labelled by variable $N$ as well as the distribution of quanta $\{q_1, \cdots, q_N\}$. We need to establish the statistical weighting of these system microstates. We expect to find that the average number of oscillators in the system, as well as the average number of quanta, will depend on the properties of the reservoir, and we would also like to construct an analogue of the chemical potential of the reservoir. Fluctuations around the mean population will emerge from the statistical treatment.

We start as in the canonical case by considering the microcanonical ensemble of the system plus the reservoir. We define macrostates of this larger phase space labelled by the number of oscillators $N$ in the system and their total number of quanta $Q$. The microstate multiplicity of a macrostate would be

$$\Omega(N, Q)\Omega_r(N_{tot} - N, Q_{tot} - Q), \tag{7.1}$$

formed from the product of the separate microstate multiplicities of system and reservoir when characterised by parameters $(N, Q)$ and $(N_{tot} - N, Q_{tot} - Q)$, respectively. As in Section 5.3, we have denoted the multiplicity of the reservoir using suffix r for added clarity.

If $N \ll N_{\text{tot}}$ and $Q \ll Q_{\text{tot}}$, we can expand $\ln \Omega_r$ to first order in $Q$ and $N$:

$$\ln \Omega_r(N_{\text{tot}} - N, Q_{\text{tot}} - Q) \approx \ln \Omega_r(N_{\text{tot}}, Q_{\text{tot}}) - Q\frac{\partial \ln \Omega_r(N_{\text{tot}}, Q_{\text{tot}})}{\partial Q_{\text{tot}}} - N\frac{\partial \ln \Omega_r(N_{\text{tot}}, Q_{\text{tot}})}{\partial N_{\text{tot}}}. \quad (7.2)$$

We again define

$$\hat{\beta} = \frac{\partial \ln \Omega_r(N_{\text{tot}}, Q_{\text{tot}})}{\partial Q_{\text{tot}}}, \quad (7.3)$$

and now introduce a parameter $\hat{\mu}$ such that

$$\hat{\mu}\hat{\beta} = -\frac{\partial \ln \Omega_r(N_{\text{tot}}, Q_{\text{tot}})}{\partial N_{\text{tot}}}. \quad (7.4)$$

For $N_{\text{tot}} \gg 1$ and $Q_{\text{tot}} \gg 1$, we can employ (5.3) and write $\ln \Omega_r(N_{\text{tot}}, Q_{\text{tot}}) \approx (Q_{\text{tot}} + N_{\text{tot}}) \ln(Q_{\text{tot}} + N_{\text{tot}}) - Q_{\text{tot}} \ln Q_{\text{tot}} - N_{\text{tot}} \ln N_{\text{tot}}$. Then, we find that

$$\hat{\beta} = \ln(Q_{\text{tot}} + N_{\text{tot}}) + 1 - \ln Q_{\text{tot}} - 1 = \ln\left(1 + \frac{N_{\text{tot}}}{Q_{\text{tot}}}\right), \quad (7.5)$$

as in (5.12), and

$$\hat{\mu} = -\hat{\beta}^{-1} \ln\left(1 + \frac{Q_{\text{tot}}}{N_{\text{tot}}}\right) = -\frac{\ln\left(1 + \frac{Q_{\text{tot}}}{N_{\text{tot}}}\right)}{\ln\left(1 + \frac{N_{\text{tot}}}{Q_{\text{tot}}}\right)}. \quad (7.6)$$

The parameters $\hat{\beta}$ and $\hat{\mu}$ are clearly properties of the combined system and reservoir, as they depend on $N_{\text{tot}}$ and $Q_{\text{tot}}$. If the number of oscillators and quanta in the reservoir are typically much larger than those in the system, we may also regard them as properties of the reservoir on its own. Clearly, we intend the $\hat{\mu}$ parameter to represent the chemical potential of classical thermodynamics, though since energy in our example is measured in units of $\hbar\omega$, both $\hat{\beta}$ and $\hat{\mu}$ are dimensionless. It is interesting to notice that $\hat{\mu}$ is negative, in line with the claims made in Section 2.11.

In this case, the parameters $\hat{\beta}$ (inverse temperature) and $\hat{\mu}$ (chemical potential) are not independent of each other, in contrast to the independence of $T$ and $\mu$ for a physical heat and particle bath in classical thermodynamics. The reason for this is that the reservoir is specified by only two parameters, $N_{\text{tot}}$ and $Q_{\text{tot}}$, and we lack a third parameter, analogous to the volume occupied by a gas, to add further flexibility. Nevertheless, the example is still instructive.

We now see from (7.2) that the microstate multiplicity of a macrostate characterised by $N$ and $Q$ can be expressed as

$$\Omega(N, Q)\Omega_r(N_{\text{tot}} - N, Q_{\text{tot}} - Q) \approx \Omega(N, Q)\Omega(N_{\text{tot}}, Q_{\text{tot}}) \exp\left(-\hat{\beta}(Q - \hat{\mu}N)\right), \quad (7.7)$$

having used (7.2–7.4). This provides the statistical weighting for an ensemble of all system macrostates. This is illustrated for a range of $N$ and $Q$ with $\hat{\beta} = \ln(3/2)$

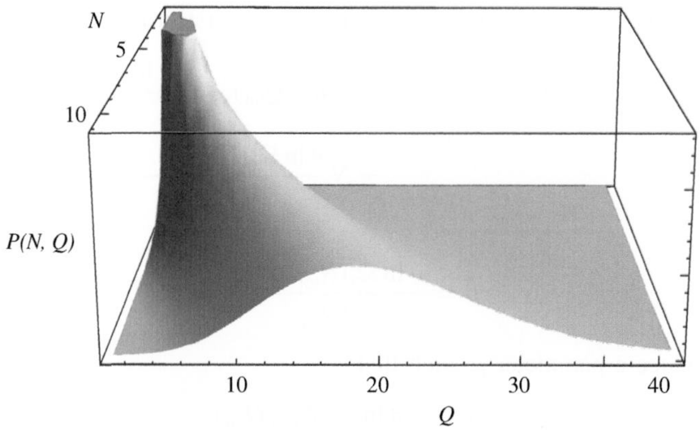


**Figure 7.2** Surface indicating the grand canonical weightings of $(N,Q)$ macrostates of a system of oscillators in contact with a heat and particle bath at $\hat{\beta} = \ln(3/2)$ and $\hat{\mu} = -\ln 3/\ln(3/2)$. The ranges chosen are $0 \leq Q \leq 40$ and $0 \leq N \leq 10$. The peak rises further around $N = 0$, $Q = 0$, but is cut off for clarity.

and $\hat{\mu}\hat{\beta} = -\ln 3$, arising from an underlying choice $Q_{\rm tot} = 2N_{\rm tot}$, in Figure 7.2. The distribution over quanta $Q$ is peaked, as we might expect from the considerations in Section 5.4, and the distribution over $N$ falls steadily. The probability distribution over the ensemble of system macrostates is

$$P(N,Q) = Z_G^{-1}\Omega(N,Q)\exp\left(-\hat{\beta}(Q - \hat{\mu}N)\right), \tag{7.8}$$

where we have introduced a new kind of partition function playing the role of the normalising factor in the probability distribution. It is related to a sum over canonical partition functions $Z(\hat{\beta},N)$ of systems with fixed $N$:

$$\begin{aligned} Z_G(\hat{\mu},\hat{\beta}) &= \sum_N \sum_Q \Omega(N,Q)\exp\left[-\hat{\beta}(Q - \hat{\mu}N)\right] \qquad (7.9)\\ &= \sum_N \exp\left(\hat{\mu}\hat{\beta}N\right)\left[\sum_Q \Omega(N,Q)\exp\left(-\hat{\beta}Q\right)\right] = \sum_N Z(\hat{\beta},N)\exp\left(\hat{\mu}\hat{\beta}N\right). \end{aligned}$$

The ensemble of system macrostates labelled by $N$ and $Q$ could also be written as an ensemble of system *microstates* labelled by the specific populations of quanta $\{q_1, q_2, \cdots, q_N\}$ residing in the system oscillators, in which case we would write

$$P(N,\{q_j\}) = Z_G^{-1}\exp\left(-\hat{\beta}(Q - \hat{\mu}N)\right), \tag{7.10}$$

and

$$Z_G = \sum_N \sum_{\{q_i\}} \exp\left(-\hat{\beta}(Q - \hat{\mu}N)\right), \tag{7.11}$$

where $Q = \sum_1^N q_j$. The ensemble is a collection of copies of the system, prepared in every possible microstate, where both the number of quanta and the particle number can vary. It is called the *grand canonical ensemble* and $Z_G$ is the *grand partition function*: a summation of the so-called Gibbs factor $\exp(-\hat{\beta}(Q - \hat{\mu}N))$ over all microstates. It is a function of the dimensionless reservoir parameters $\hat{\beta}$ and $\hat{\mu}$.

The principal role for the grand canonical ensemble is to enable us to understand how the reservoir chemical potential controls the mean number of particles in a system, and how that number might fluctuate. Just as we discussed in Section 5.4, we expect there to be a thermodynamic limit where the distribution over particle number in the system has a sharp peak. The expression in (7.8) is a product of $\Omega(N, Q)$, which is a rapidly increasing function of $N$, and, recalling that $\hat{\mu} < 0$, a decreasing function $\exp(\hat{\beta}\hat{\mu}N)$. The location of the peak with respect to $N$ increases as $\hat{\mu}$ increases, or more precisely as $\hat{\mu}$ is made less negative, when the rate of exponential decrease in the second factor is weakened.

For example, considering $Q$ to be constant, we have $\Omega(N, Q) \sim N^Q$ for $N \gg Q$ from (5.5) and the peak in $\Omega(N, Q)\exp(\hat{\beta}\hat{\mu}N)$ lies at $N = N^*$, found from the condition

$$\frac{\mathrm{d}}{\mathrm{d}N}\ln\left(\Omega(N, Q)\exp\left(\hat{\beta}\hat{\mu}N\right)\right) \approx \frac{\mathrm{d}}{\mathrm{d}N}(Q\ln N + \hat{\beta}\hat{\mu}N) = \frac{Q}{N} + \hat{\beta}\hat{\mu} = 0, \tag{7.12}$$

such that $N^* \propto -\hat{\mu}^{-1}$. In the thermodynamic limit brought about by $\hat{\mu} \to 0$ from below, in which $N^*$ becomes macroscopic, the standard deviation in the particle number grows less quickly than the mean, such that fluctuations about the mean essentially disappear.

## 7.2 Grand Canonical Ensemble for a General System

Having introduced the concept of the grand canonical ensemble using the example of the system of oscillators, we now generalise. We consider systems specified by macrostate variables energy $E$ and number of particles $N$, and propose the following equilibrium probability distribution over macrostates:

$$P(N, E) = \frac{1}{Z_G}\Omega(N, E)\exp\left(-\frac{(E - \mu N)}{kT}\right), \tag{7.13}$$

where the grand partition function is

$$Z_G(\mu, T) = \sum_{\text{microstates}} \exp\left(-\frac{(E - \mu N)}{kT}\right) = \sum_N \sum_E \Omega(N, E)\exp\left(-\frac{(E - \mu N)}{kT}\right). \tag{7.14}$$

The system is considered to be coupled to a reservoir with a temperature $T$ and a parameter $\mu$, with dimensions of energy, defined as

$$\mu = -kT\frac{\partial \ln \Omega_\mathrm{r}(N_\mathrm{r}, E_\mathrm{r})}{\partial N_\mathrm{r}}. \tag{7.15}$$

Using Boltzmann's expression for the entropy $S = k\ln\Omega$, this definition is consistent with that of the chemical potential $\mu = -T\partial S/\partial N$ in classical thermodynamics, given by (2.48).

The main purpose of the grand partition function, as we found with the canonical partition function, is that it allows ensemble averages to be obtained by differentiation. The mean number of oscillators in the system, for example, is a weighted sum of macrostate variable $N$ over the grand canonical ensemble:

$$\langle N\rangle = \sum_N \sum_E NP(N,E) = \frac{1}{Z_G}\sum_N \sum_E N\ \Omega(N,E)\exp\left(-\frac{(E-\mu N)}{kT}\right)$$
$$= \frac{1}{Z_G}\left(\frac{\partial Z_G}{\partial(\frac{\mu}{kT})}\right)_T = kT\left(\frac{\partial \ln Z_G}{\partial \mu}\right)_T, \qquad (7.16)$$

where we employ partial differentiation since $Z_G$ is a function of both $\mu$ and $T$. We shall use the grand canonical ensemble, and this result in particular, in Chapters 11 and 12 when we examine quantum gases. We shall find that the grand partition functions for such cases are actually quite straightforward to evaluate. But as a preliminary example, let us examine once again the problem of vacancies in crystals.

## 7.3 Vacancies in Crystals Revisited

In Section 6.7, we determined the mean population of vacancies at a lattice site using a canonical ensemble. A vacancy is not a particle, but rather the absence of one, but nevertheless with some bending of the concepts, we can approach the same problem using the grand canonical ensemble.

We consider a system consisting of a single lattice site and imagine it to be exposed to a reservoir of vacancies at a chemical potential $\mu$, in addition to a heat bath at temperature $T$. Conceptually, the rest of the crystal acts as a gas of vacancies, in a state of motion as atoms hop between the various lattice sites, and at a certain concentration and chemical potential determined by the temperature and the vacancy formation energy $E_v$.

Let us construct the grand partition function for the single lattice site. There are only two values of the population $N$ of vacancies in the system, namely zero and one, and for each population, the energy takes just one value, zero and $E_v$, respectively. There are just two microstates and we write

$$Z_G = \sum_{\text{microstates}} \exp\left(-\frac{(E-\mu N)}{kT}\right) = 1 + \exp\left(-\frac{(E_v-\mu)}{kT}\right). \qquad (7.17)$$

The mean number of vacancies per site then follows:

$$\langle N\rangle = kT\left(\frac{\partial \ln Z_G}{\partial \mu}\right)_T = kT\frac{(kT)^{-1}\exp\left(-\frac{(E_v-\mu)}{kT}\right)}{1+\exp\left(-\frac{(E_v-\mu)}{kT}\right)} = \frac{1}{\exp\left(\frac{(E_v-\mu)}{kT}\right)+1}. \qquad (7.18)$$

This is similar to (6.33), but the appearance of $\mu$ in (7.18) needs further comment. It is inserted since we are employing the concept of a vacancy as a real particle that can be injected from or lost to a reservoir. In Section 6.7, we were calculating the total energy of a system with a fixed number of real particles, but we interpreted the outcome in terms of the occupation of lattice sites and hence the population of vacancies. The grand canonical approach is a more powerful approach than this since there are occasions when a crystal can effectively be exposed to sources and sinks of vacancies, such that the population of vacancies might be changed without altering the temperature. Essentially, atoms can migrate out of the system, controlled by the properties of the environment, and the idea of a chemical potential for the supply of vacancies is a useful one. But in order that the previous result (6.33) might be recovered, we must conclude that the chemical potential of vacancies in an isolated crystal is zero. A similar outcome in different circumstances will appear in Chapter 13 when we consider the statistical thermodynamics of electromagnetic radiation, and its associated quantised particle, the photon.

## Exercises

**7.1** A system is in contact with a heat and particle bath at temperature $T$ and chemical potential $\mu$. If there are only two microstates available to the system, one with $N = 0$, $E = 0$; and the other with $N = 1$, $E = \epsilon > 0$, derive an expression for the grand partition function and show that $\langle N \rangle = [\exp((\epsilon - \mu)/kT) + 1]^{-1}$. Sketch the dependence of $\langle N \rangle$ on the bath chemical potential in the range $-\infty < \mu < \infty$.

**7.2** Show that the variance in the population of a system $\sigma_N^2 = \langle N^2 \rangle - \langle N \rangle^2$ is given by

$$\sigma_N^2 = (kT)^2 \left( \frac{\partial^2 \ln Z_G}{\partial \mu^2} \right)_T.$$

Determine $\sigma_N$ for the case in question 7.1, plotting it in the range $-\infty < \mu < \infty$.

**7.3** A system can hold zero, one or two particles, and a particle in the system can take an energy of zero or $E_p = kT \ln 2$. Calculate the grand partition function and hence the mean particle number for a given $\mu$ and $T$.

**7.4** Determine the relationship between the mean energy of a system and the grand partition function.

# 8

# Statistical Models of Entropy

Caution: Entropy

Several chapters have passed without much mention of entropy, and as this book is *an entropic approach*, it is time to put this right.

It is quite acceptable to use the tools of statistical ensembles to study system behaviour without delving more deeply into explanations, just as it is feasible in quantum mechanics to master the *mechanics* of performing calculations without necessarily enquiring deeply into the *physics* of quantum phenomena. In statistical thermodynamics, the equally deep matters are the justification for the principle of equal a priori probabilities, and the relationship between statistical ideas and the concept of thermodynamic entropy. This is where most of the trouble in understanding the subject lies, and in this chapter, we discuss the various views that have developed.

In Section 4.4, we briefly discussed Boltzmann's insight that entropy is a measure of the number of underlying microstates that are compatible with a macroscopically measurable state. This is the key principle of statistical thermodynamics, but it does not quite stand up to detailed scrutiny for systems, coupled to an environment, that could potentially access an infinite number of microstates. We also referred to the possibility that Boltzmann's expression was actually a statement of a connection between entropy and microstate probabilities, in the context of an isolated system where the probabilities were to be specified by the principle of equal a priori probabilities. We now investigate this underlying connection further, which of course could raise the question of what we mean by probability, matters that we discussed in Section 4.1. The outcome of such considerations is a multifaceted view of entropy, a collage of mutually supportive ideas, some of which can extend to systems out of equilibrium.

## 8.1 Boltzmann Entropy

Boltzmann's model of entropy was introduced in Section 4.4, and we repeat it here with a suffix B for clarity:

$$S_B = k \ln \Omega, \tag{8.1}$$

*Statistical Physics: An Entropic Approach*, First Edition. Ian Ford.

where $\Omega$ is the total multiplicity of microstates available to an isolated system. $\Omega$ depends on system parameters that are conserved by the dynamics, such as energy and the number of particles. It is often sensible to divide the available microstates into groups, each characterised by a common feature that is not dynamically conserved, and therefore changes with time, and to refer to the groups as macrostates. The idea of a macroscopic state in classical thermodynamics maps onto the concept of macrostates in statistical thermodynamics. The multiplicity $\Omega_\alpha(N, E)$ of such a macrostate depends on the conserved quantities, and on the value of a non-conserved quantity $\alpha$ that serves to define the macrostate, such as the oscillator macrostate spikiness that we encountered in Section 4.3.

### 8.1.1 The Second Law of Thermodynamics

The second law of thermodynamics emerges in statistical thermodynamics if we set up an isolated system under an initial constraint that it should take only a certain range of values of a property $\alpha$, and then remove this restriction on the dynamics. If the system remains isolated, the parameters $N$ and $E$ do not change, but the range of available values of $\alpha$ is typically increased, and after a relaxation period the probabilities that the system might take values of $\alpha$ are assumed to be proportional to the microstate multiplicity $\Omega_\alpha$ of the associated macrostate.

In such discussions, the concept of a macrostate can be employed in two ways: in a narrow and in a broad sense. We can state that the system is in a macrostate where it takes a specific value of the non-conserved quantity $\alpha$, but the collection of all such macrostates can also be viewed as a macrostate, this time labelled by the whole range of available values of $\alpha$ and corresponding to the complete accessible phase space. So in a thermodynamic process, we start off in one macrostate, labelled by the initial range of values of $\alpha$. We release the constraint on $\alpha$, allowing the system to visit new (narrow sense) macrostates that were inaccessible before. Eventually the system settles into a new equilibrium, taking a final (broad sense) macrostate labelled by the entire range of $\alpha$ made possible under the new dynamics.

This is illustrated in Figure 8.1, where we see a phase space divided into white regions characterised by different values of $\alpha$. The initial macrostate might be one or more of the white regions, and after a release of a constraint the final macrostate would be the entire white sector. Beyond that are pink regions that are inaccessible under the remaining dynamical constraints. As the microstate multiplicity of the final (broad sense) macrostate is clearly larger than the microstate multiplicity of any initial (narrow sense) macrostates, Boltzmann's expression clearly explains the second law. This is essentially the same argument we gave in Section 4.4.

As entropy in thermodynamics is traditionally regarded as an equilibrium property, Boltzmann's expression should perhaps only apply before the constraint is removed, or some time in the future when the statistical properties of the system have become time independent. These situations are where the principle of equal a priori probabilities is supposed to apply, and so Boltzmann's expression could also be taken to be a connection between entropy and the equilibrium probabilities of microstate occupation, which we shall explore later.

Boltzmann proposed that the dynamics of particles in a gas, for example, were capable of establishing equal, time independent microstate probabilities. An extreme case would

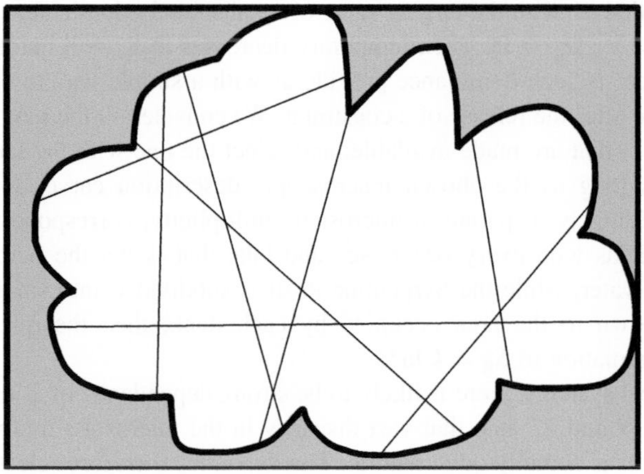

**Figure 8.1** A white phase space that is accessible to a system, divided into macrostates corresponding to values of a non-conserved quantity $\alpha$. The area of each patch is a representation of its microstate multiplicity.

be that the dynamics should allow a system in equilibrium to visit every microstate for an equal length of time. This is called the assumption of *ergodicity*: unfortunately, it cannot be proved in general, and would appear to be a rather demanding requirement for a finite measurement interval. So in Boltzmann's approach, the fundamental axiom is that the dynamics of an isolated system are ergodic, or something approaching it.

It is possible to extend these dynamical ideas to the nonequilibrium behaviour seen just after the release of the dynamical constraint. We can define the microstate multiplicity $\Omega_\alpha$ associated with each of the macrostates $\alpha$, and we could take the view that $S_\alpha = k \ln \Omega_\alpha$ represents something similar to entropy as the system moves through the macrostates during its relaxation towards equilibrium. As a system explores its phase space, it passes from (narrow sense) macrostate to macrostate, and so the value of $S_\alpha$ changes. Although some directions of travel in phase space can conceivably lead to a decrease in the value of $S_\alpha$, most lead to its increase, as long as the dynamics are of a kind that explores phase space broadly and efficiently. Such a picture gives us a quantity that tends to increase after the release of a constraint, and to reach a peak at a value corresponding to the macrostate with the largest microstate multiplicity, thereafter taking only rare excursions into neighbouring macrostates accompanied by decreases in $S_\alpha$. This resembles the presumed developing behaviour of a nonequilibrium entropy.

Using a geographical analogy, it is like starting a journey from London that involves a complicated schedule of flights to destinations all over the world. During the journey, a traveller looks out of the window and sees water. The chances are that she will be over the Pacific Ocean, as it is the largest in area. Sometimes she might see the Atlantic or the Indian Oceans. The Caspian Sea is visited rarely. She might spot the Serpentine in London's Hyde Park, and this might even be quite likely during the period just after take-off, but later on it would become extremely improbable. On the whole, she need not bother to check the sat-nav: the water she sees is most likely the Pacific Ocean.

Such a view does not quite correspond to the second law since occasional decreases in total entropy are not supposed to happen. But it is a very good approximation when the macrostate labelled $\alpha^*$ with the largest value of $S_\alpha$ absolutely dominates the phase space, such that $\Omega_{\alpha*} \approx \sum_\alpha \Omega_\alpha = \Omega$. Then temporary decreases in $S_\alpha$ will hardly ever be seen. Cases where there is such dominance provide us with a simple way to determine a new equilibrium state after the release of a constraint. We consider all the new macrostates (in the narrow sense) that are made available, and select the one with the largest microstate multiplicity. As long as the chosen macroscopic description carves phase space into patches that are hugely disparate in microstate multiplicity, corresponding in our flight example to an atlas with every ocean, sea and lake that is *not* the Serpentine labelled as one body of water, while the Serpentine itself is subdivided into smaller and smaller patches right down to the area occupied by each duck, then the quantity $S_{\alpha*}$ is an excellent approximation to $S_\text{B} = k \ln \Omega$.

In real physical systems, there is likely to be strong dependence of $\Omega$ on the parameter $\alpha$ as well as on $N$ and $E$, such that vast disparity in the microstate multiplicities of different macrostates is actually quite natural. The dynamics therefore select the macrostate with maximum multiplicity, or more colloquially (but with a slight misuse of terminology) they select the *maximum entropy macrostate*. The isolated system evolves to maximise its 'entropy' $S_\alpha$. This provides a rationale for the second law as a variational principle. We now examine some examples.

### 8.1.2 The Maximum Entropy Macrostate of Oscillator Spikiness

The phase space of the three oscillators with nine quanta studied in Section 4.5 has 55 available microstates divided into nine groups, or spikiness macrostates, each with a specific multiplicity. We start with a constraint that holds the system in the $\text{Sp} = 9$ macrostate of the phase space in Figure 4.3 and then release it. The total multiplicity $\Omega(N, Q)$ of available microstates changes instantaneously from 3 to 55, but our interest is in the value of $\Omega_\text{Sp}(N, Q)$ as the system moves in phase space and Sp changes.

What emerges is a time sequence of values sampled from the histogram shown in Figure 4.4. We would recognise the $\text{Sp} = 5$ macrostate as the maximum entropy macrostate, and identify $k \ln \Omega_{\text{Sp}=5}(N = 3, Q = 9) = k \ln 12$ as an estimator of the total Boltzmann entropy $k \ln 55$. We would frequently see $k \ln \Omega_\text{Sp}(N, Q)$ fluctuate below this value as Sp changed with time, but for systems with larger $N$ and $Q$, departure from the maximum entropy macrostate becomes much less frequent, as indicated in Figure 8.2. In such a thermodynamic limit, the histogram for Sp becomes sharp, as suggested by Figure 5.4, and most of the statistical weight is concentrated in a narrow region around the peak. Nevertheless, the estimate $\text{Sp} = 5$ of the mean spikiness $\langle \text{Sp} \rangle = 5.18$ in the example is a rather good approximation; so the procedure can be instructive even for a small system.

### 8.1.3 The Maximum Entropy Macrostate of Oscillator Populations

Now for a more complicated example that will have some value to us later on. We consider a set of $N$ oscillators and divide the system phase space into macrostates according to a rather different kind of label: the set of numbers $\{n_k\}$ of oscillators that possess $k$ quanta. It is a population distribution that satisfies the conditions $\sum_k n_k = N$ and $\sum_k k n_k = Q$. For a system with $Q$ quanta, there will be $Q + 1$ oscillator populations: the number $n_0$ of oscillators possessing no quanta, the number $n_1$ that hold one quantum

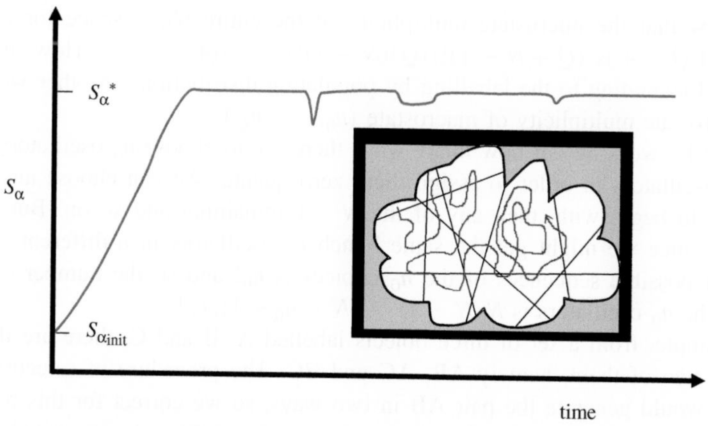


**Figure 8.2** Time dependence of $S_\alpha = k \ln \Omega_\alpha$ as a system evolves from an initial macrostate into a larger set of macrostates labelled $\alpha$, each with multiplicity $\Omega_\alpha$, as illustrated in the inset. In the thermodynamic limit, fluctuations away from the maximum entropy macrostate $\alpha^*$ become rare and $S_{\alpha^*}$ may be taken to be a good estimate of the final entropy $S_B = k \ln \Omega$, where $\Omega = \sum_\alpha \Omega_\alpha$.

and so on, up to a population $n_Q$ that holds all $Q$ quanta. So for the case $Q = 4$, there are five populations $n_{0-4}$, some of which can be zero.

There will be a certain number of manifestations of the population distribution, and these can be taken to define the different macrostates of the system. For $N = 3$ and $Q = 4$, there are four macrostates of population distribution. In the notation $(n_0, n_1, n_2, n_3, n_4)$, these are (2,0,0,0,1), (1,1,0,1,0), (1,0,2,0,0) and (0,2,1,0,0). The last of these, for example, is the macrostate with two oscillators holding one quantum and one holding two quanta, accounting for the four quanta in all. The division of the phase space is illustrated in Figure 8.3.

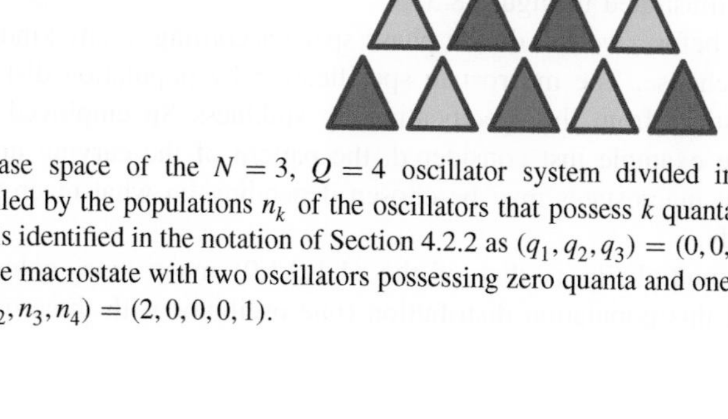

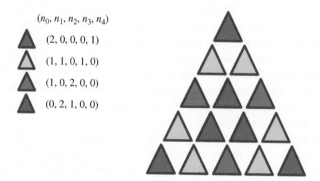


**Figure 8.3** Phase space of the $N = 3$, $Q = 4$ oscillator system divided into colour-coded macrostates labelled by the populations $n_k$ of the oscillators that possess $k$ quanta. The microstate at the top vertex is identified in the notation of Section 4.2.2 as $(q_1, q_2, q_3) = (0, 0, 4)$, which makes it a member of the macrostate with two oscillators possessing zero quanta and one possessing four, namely $(n_0, n_1, n_2, n_3, n_4) = (2, 0, 0, 0, 1)$.

We know that the microstate multiplicity of the entire phase space for parameters $N = 3$ and $Q = 4$ is $(Q + N - 1)!/(Q!(N - 1)!) = 6!/(4!2!) = 15$. How might these be grouped according to the labelling by population distribution? In other words, what is the microstate multiplicity of macrostate $(n_0, \cdots, n_Q)$?

We start by working out how many ways there are to choose $n_0$ oscillators from the set of $N$ oscillators in order to assign them zero quanta. We can choose any of the $N$ oscillators to begin with, then any of the $N - 1$ remaining and so on. But this is an overcount since we might get the same bunch of oscillators in a different order. The number of possible sequences of the $n_0$ choices is $n_0!$ and so the number of ways of selecting the $n_0$ oscillators is $N(N - 1) \cdots (N - n_0 + 1)/n_0!$.

For example, from a set of three objects labelled A, B and C, there are three ways to choose two of them, namely AB, AC and BC. The procedure of selection we just described would generate the pair AB in two ways, so we correct for this by dividing by two. The formula with $N = 3$ and $n_0 = 2$ gives $3 \times 2/2! = 3$ as required.

Similarly there are $(N - n_0)(N - n_0 - 1) \cdots (N - n_0 - n_1 + 1)/n_1!$ ways to choose the $n_1$ oscillators from the remaining $N - n_0$, in order to assign them one quantum, and so on. The final $n_Q$ oscillators are chosen in $(N - n_0 - n_1 - \cdots - n_{Q-1}) \cdots (N - n_0 - n_1 - \cdots - n_{Q-1} - n_Q + 1)/n_Q! = (N - n_0 - n_1 - \cdots - n_{Q-1}) \cdots 1/n_Q!$ ways since $N = n_0 + \cdots + n_Q$.

Multiplying together these factors for all $Q + 1$ populations, we find the number of ways in which the oscillators can be assigned a specified population distribution $\{n_k\}$ of quanta is written

$$\Omega_{\{n_k\}}(N, Q) = \frac{N!}{n_0! \cdots n_Q!}, \tag{8.2}$$

and by definition this is the microstate multiplicity of the macrostate $\{n_k\} \equiv (n_0, \cdots, n_Q)$.

As a check, let us work out the multiplicity of the four macrostates of the $N = 3$, $Q = 4$ system. The (2,0,0,0,1) macrostate corresponds to one oscillator holding all four quanta ($n_4 = 1$), while the other two have none ($n_0 = 2$) and there are no oscillators with one, two or three quanta ($n_1 = n_2 = n_3 = 0$). According to (8.2) this population macrostate has multiplicity $3!/(2!0!0!0!1!) = 3$. Note that we take $0!$ to be equal to unity. The three microstates in the macrostate correspond to the vertices of the triangular phase space, where each oscillator in turn takes possession of all four quanta.

By a similar reasoning, the (1,1,0,1,0), (1,0,2,0,0) and (0,2,1,0,0) population macrostates have multiplicities 6, 3 and 3, respectively. The sum of all the macrostate multiplicities, 15, is equal, as expected, to the total number of microstates in the phase space. We have successfully carved up the phase space into macrostates of population distribution, as illustrated in Figure 8.3.

As discussed before, we can divide phase space according to any kind of macrostates that we might choose. The macrostate specification by population distribution $\{n_k\}$ is in principle distinct from the specification by spikiness Sp employed in Section 4.3 (although in the example just considered, the pattern of the carving up is exactly the same). Different macrostates may be chosen depending on what macroscopic property we wish to study.

The largest multiplicity is that of the (1,1,0,1,0) macrostate, coloured yellow in Figure 8.3, and this population distribution (one oscillator with three quanta, one with one, and one with none), denoted $\{n_k^*\}$, is the most likely to be seen during the dynamics, assuming ergodicity. If the system were set up in macrostate (2,0,0,0,1) (the blue macrostate) and then allowed to evolve according to the dynamics, then much later the system would be presumed most likely to be found in the yellow macrostate (1,1,0,1,0).

The estimate of the final equilibrium entropy $k \ln \Omega_{\{n_k^*\}}(N,Q) = k \ln 6$ is somewhat short of the correct value $k \ln 15$, but for a system approaching the thermodynamic limit, where the difference in multiplicity between different macrostates becomes vast, identifying the maximum entropy macrostate gives a good approximation of the Boltzmann entropy $S_{\mathrm{B}} = k \ln \Omega(N,Q)$, and provides an excellent estimate of system properties averaged over the complete phase space.

Let us investigate this thermodynamic limit and identify the population macrostate $\{n_k^*\}$ that has the largest multiplicity $\Omega_{\{n_k\}}(N,Q)$. We start with (8.2) and write

$$\ln \Omega_{\{n_k\}}(N,Q) = \ln N! - \sum_k \ln n_k!, \tag{8.3}$$

and if we assume that all the $n_k$ are large, such that Stirling's formula can be used, this becomes

$$\begin{aligned} \ln \Omega_{\{n_k\}}(N,Q) &\approx N \ln N - N - \sum_k n_k \ln n_k + \sum_k n_k \\ &= N \ln N - \sum_k n_k \ln n_k = -\sum_k n_k \ln\left(\frac{n_k}{N}\right), \end{aligned} \tag{8.4}$$

where we have used $N = \sum_k n_k$.

Now we find the maximum entropy macrostate by maximising $\ln \Omega_{\{n_k\}}(N,Q)$, while ensuring that the conditions $N = \sum_k n_k$ and $Q = \sum_k k n_k$ are met. We use Lagrange's method of undetermined multipliers and maximise

$$I = -\sum_k n_k \ln\left(\frac{n_k}{N}\right) - \lambda \sum_k n_k - \tilde{\beta} \sum_k k n_k, \tag{8.5}$$

over the populations $\{n_k\}$, where $\lambda$ and $\tilde{\beta}$ are constants. We assume that the populations are large, allowing us to treat them as continuous variables. Taking a partial derivative of $I$ with respect to $n_{k'}$ then gives

$$-\ln\left(\frac{n_{k'}^*}{N}\right) - 1 - \lambda - \tilde{\beta} k' = 0, \tag{8.6}$$

and so the population of oscillators holding $k$ quanta in the maximum entropy macrostate is

$$n_k^* = N \exp\,(-1 - \lambda - \tilde{\beta} k). \tag{8.7}$$

We evaluate $\lambda$ through the condition $\sum_k n_k^* = N$ such that

$$n_k^* = \frac{N}{Z} \exp\,(-\tilde{\beta} k), \tag{8.8}$$

where $Z = \sum_k \exp\,(-\tilde{\beta} k)$.

The maximum entropy population macrostate is therefore $\{n_k^*\} = (n_0^*, n_0^* x, n_0^* x^2, n_0^* x^3 \cdots)$ where $n_0^* = N/Z$, $x = \exp(-\tilde{\beta})$ and $Z = \sum_0^\infty x^k = (1-x)^{-1}$, as long as $\tilde{\beta} > 0$ such that $x \leq 0$. Notice that we extend the upper limit of the sum to infinity since we are approaching the thermodynamic limit.

Finally, the constant $\tilde{\beta}$ can be identified through the condition

$$Q = \sum_k k n_k^* = \frac{N}{Z} \sum_k k \exp\left(-\tilde{\beta} k\right) = -\frac{N}{Z}\frac{dZ}{d\tilde{\beta}} = -\frac{N}{Z}\frac{dZ}{dx}\frac{dx}{d\tilde{\beta}}$$

$$= \frac{N}{Z}\frac{x}{(1-x)^2} = N\frac{x}{(1-x)}, \tag{8.9}$$

which leads to $\tilde{\beta} = \ln(1 + N/Q)$. Notice that the form of (8.8) resembles the canonical distribution, and that the parameter $\tilde{\beta}(N, Q)$ is identical to the parameter $\hat{\beta}(N, Q)$ of a system of $N$ oscillators holding $Q$ quanta, derived in (5.12), and which we came to regard as the dimensionless inverse temperature of such a system.

The reason for this correspondence is that the fraction of the $N$ oscillators that have $k$ quanta in the thermodynamic limit, namely, $n_k^*/N$ since the properties of the system are then those of the maximum entropy population macrostate $\{n_k^*\}$, will correspond to the probability that a single oscillator in the group should possess $k$ quanta. The latter probability was shown to be $Z^{-1}\exp(-\hat{\beta} k)$ in (5.13) using the canonical ensemble. The populations of oscillators with $k$ quanta in the maximum entropy population macrostate appear to be distributed canonically. Indeed an argument similar to this is often employed as an alternative derivation of the canonical distribution, and it can be used in the context of a collection of general systems, not just a collection of oscillators.

### 8.1.4 The Third Law of Thermodynamics

The next use we have for Boltzmann's entropy expression is to justify the third law of thermodynamics, namely that the entropy of a system should approach a universal constant (taken to be zero) as the temperature approaches zero. This limit corresponds to the removal of all energy from a system, if we ignore zero point energy, and this is the context in which we shall discuss the third law.

A system with little energy to distribute amongst its various degrees of freedom has a very reduced multiplicity of microstates. We have already argued that the multiplicity is an increasing function of energy. The ultimate situation is that the smallest number of available microstates of a system applies when the energy is zero, and this number is unity if there is a unique ground state. The Boltzmann entropy of the system would then be zero. If there are several configurations of the system with zero energy, then the Boltzmann entropy at $E = 0$ would be nonzero, but physical systems do not commonly have such degeneracy. The situation is illustrated by the example of $N$ oscillators with $Q$ quanta, for which the microstate multiplicity is $\Omega(N, Q) = (Q + N - 1)!/(Q!(N - 1)!)$, and this goes to unity if $Q = 0$. There is just one microstate with zero energy, the one where every oscillator has zero quanta.

## 8.2 Gibbs Entropy

The Gibbs model of entropy differs from the Boltzmann approach in some important ways. It is necessary to have another model in order to describe the entropy of a system in contact with an environment. If energy transfers are possible, then a system might be able to access an unlimited number of microstates, and the Boltzmann expression would be inappropriate. The key feature of the Gibbs approach is that it connects thermodynamic entropy directly to statistical ideas through the use of the equilibrium microstate occupation probabilities. It turns out that this yields a finite entropy for an open system.

The microstate probabilities for a system in equilibrium are established using the principle of equal a priori probabilities, either for an isolated system or for the combination of system with a reservoir, and in the latter case, we would expect to obtain canonical system microstate probabilities, if the reservoir is large. The justification of the principle could be argued on the grounds of the ergodic hypothesis, the fundamental basis according to Boltzmann, but the principle could very well be regarded as the prime axiom in itself. This would follow from an interpretation of probability from the viewpoint of information theory: the microstate probabilities are not frequencies generated by the dynamics, but are rather a best judgement of the statistical weight accorded to each microstate. We might feel that no microstate of an isolated system should be weighted differently from any other, when we lack any information by which we might distinguish them. This automatically leads to the canonical and grand canonical equilibrium microstate probabilities $P_i$ for a system in contact with different reservoirs, as we saw in Chapters 5 and 7, and we need not worry about ergodic dynamics.

The Gibbs entropy $S_{\mathrm{G}}$ of a system is defined by the important formula:

$$S_{\mathrm{G}} = -k \sum_i P_i \ln P_i, \tag{8.10}$$

where $i$ labels the microstates of the system. Let us insert the canonical probabilities $P_i = Z^{-1} \exp\,(-E_i/kT)$ for a system in contact with a heat bath at temperature $T$ to see what this implies. We find that

$$S_{\mathrm{G}} = -k \sum_i P_i \left( -\frac{E_i}{kT} - \ln Z \right) = \frac{1}{T}\langle E \rangle + k \ln Z, \tag{8.11}$$

where the brackets indicate a canonical average. A clear indication that the Gibbs entropy differs from the Boltzmann entropy is that $S_{\mathrm{G}}$ is a function of the reservoir property $T$ and the number of particles $N$. It therefore naturally applies to a system in contact with a heat bath. In contrast, $S_{\mathrm{B}}$ is a function of conserved quantity $E$, and is designed to apply to an isolated system. Of course, in the thermodynamic limit, when fluctuations about mean properties become negligible, there is an essentially unique equilibrium system energy associated with a given reservoir temperature, and the mathematical distinction between functions $S_{\mathrm{G}}(N,T)$ and $S_{\mathrm{B}}(N,E)$ becomes less important.

The Gibbs expression is compatible with the maximum entropy macrostate estimate of Boltzmann entropy derived in Section 8.1.3. According to a generalisation of (8.4) we write

$$S_{\mathrm{B}} = k \ln \Omega \approx k \ln \Omega_{\alpha *} = -Nk \sum_i \left(\frac{n_i^*}{N}\right) \ln \left(\frac{n_i^*}{N}\right), \tag{8.12}$$

for a system consisting of $N$ similar subsystems (oscillators in that earlier discussion) each able to assume one of a set of subsystem microstates now labelled by $i$. The number of subsystems in microstate $i$ is $n_i^*$, for the maximum entropy population macrostate, and this is found to take the canonical form (8.8). The probability $P_i$ that a particular subsystem should assume microstate $i$ is $n_i^*/N$; so we can equate the Boltzmann entropy of a set of $N$ similar subsystems, in the thermodynamic limit, to be $N$ times the Gibbs entropy (defined in (8.10)) of each one: namely, $S_{\mathrm{B}}(N, E) = N S_{\mathrm{G}}(1, T)$, with $E$ and $T$ related uniquely.

Turning to more straightforward matters, note that (8.11) can be rewritten as

$$-kT \ln Z = \langle E \rangle - T S_{\mathrm{G}}, \tag{8.13}$$

and so with reference to (3.13), and if we accept a correspondence between the thermodynamic state variable $E$ and the statistical mean $\langle E \rangle$, we appear to be able to identify $-kT \ln Z$ with the Helmholtz free energy:

$$F = -kT \ln Z, \tag{8.14}$$

just as we found in (6.16). This correspondence lends support to the form of entropy proposed by Gibbs. Note that $Z$ in (8.14) is a function of $T$ and $N$, and that together with the system volume, these are the natural variables for the Helmholtz free energy in classical thermodynamics.

The Gibbs entropy in (8.10) is a sum over microstates, but we can group the microstates into energy macrostates $\alpha$ with equilibrium probabilities $P_\alpha = \Omega_\alpha \exp\,(-\beta E_\alpha)/Z$ to obtain the expression

$$S_{\mathrm{G}} = -k \sum_\alpha P_\alpha \ln \left(\frac{P_\alpha}{\Omega_\alpha}\right), \tag{8.15}$$

where $\Omega_\alpha$ is the microstate multiplicity of the macrostate $\alpha$. Whether expressed in terms of microstate or macrostate probabilities, the Gibbs entropy is independent of the choice of macrostate patchwork, because like the Boltzmann entropy we wish it to be a measure of the size of the whole available phase space.

If we employ the grand canonical microstate probabilities $P_i^G = Z_G^{-1} \exp\,[-(E_i - \mu N_i)/kT]$ in the Gibbs expression for the entropy, we obtain

$$S_{\mathrm{G}} = -k \sum_i P_i^G \left(-\frac{E_i}{kT} + \frac{\mu N_i}{kT} - \ln Z_G\right) = \frac{1}{T}\langle E \rangle - \frac{\mu \langle N \rangle}{T} + k \ln Z_G, \tag{8.16}$$

where the angled brackets now represent a grand canonical average. This can be rearranged into

$$-kT \ln Z_G = \langle E \rangle - T S_{\mathrm{G}} - \mu \langle N \rangle = \Phi, \tag{8.17}$$

where $\Phi(\mu, V, T)$ is the grand potential defined in (3.31), interpreting the system energy and particle number in that classical thermodynamic expression as grand canonical averages. This is another connection between a quantity from statistical thermodynamics (the left hand side) and a quantity from classical thermodynamics (the right hand side), similar to the connection established in (8.14).

### 8.2.1 Fundamental Relation of Thermodynamics and Thermodynamic Work

If we accept the Gibbs expression for the entropy, then it allows us to derive familiar classical thermodynamic relationships starting from the ideas of statistical thermodynamics. Consider the change in mean system energy in the canonical ensemble brought about by a change in the canonical probabilities, brought about by an alteration to the reservoir temperature, for example, together with a variation in a parameter $x$ that affects the microstate energies $E_i$. From $\langle E\rangle = \sum_i E_i P_i$, where the $P_i$ are canonical probabilities, we write

$$\begin{aligned}\mathrm{d}\langle E\rangle &= \sum_i E_i \mathrm{d}P_i + \sum_i P_i \mathrm{d}E_i = -kT\sum_i (\ln P_i + \ln Z)\mathrm{d}P_i + \sum_i P_i \frac{\partial E_i}{\partial x}\mathrm{d}x \\ &= -kT\sum_i \ln P_i \mathrm{d}P_i - f_x \mathrm{d}x, \end{aligned} \qquad (8.18)$$

where the condition $\sum_i \mathrm{d}P_i = 0$ has been employed in order to maintain normalisation $\sum_i P_i = 1$, and where we define a thermodynamic force $f_x$ associated with $x$:

$$f_x = -\sum_i P_i \frac{\partial E_i}{\partial x} = -\left\langle \frac{\partial E_i}{\partial x}\right\rangle. \qquad (8.19)$$

Next we note that

$$\mathrm{d}S_{\mathrm{G}} = -k\left[\sum_i \ln P_i \mathrm{d}P_i + \sum_i P_i\left(\frac{1}{P_i}\right)\mathrm{d}P_i\right] = -k\sum_i \ln P_i \mathrm{d}P_i, \qquad (8.20)$$

again having used $\sum_i \mathrm{d}P_i = 0$. Combining this with (8.18), we get

$$\mathrm{d}\langle E\rangle = T\mathrm{d}S_{\mathrm{G}} - f_x \mathrm{d}x. \qquad (8.21)$$

This resembles the fundamental relation of thermodynamics $\mathrm{d}E = T\mathrm{d}S - p\mathrm{d}V$, if we associate the classical state variable $E$ with the canonical average $\langle E\rangle$. The second term on the right hand side of (8.21) should therefore correspond to the *quasistatic work* done on the system, which we can define in statistical thermodynamics to be the contribution to the change in mean system energy brought about by a change in a parameter that affects microstate energies, on the assumption that the microstate probabilities remain canonical in form throughout the process.

For example, if the system microstate energies depended on system volume $V$, then $-f_V \mathrm{d}V$ is just the change in mean system energy brought about by increasing the volume by $\mathrm{d}V$. The thermodynamic force $f_V$ would correspond to the system pressure $p$.

Notice that (8.19) implies that

$$p = -\frac{1}{Z}\sum_i \frac{\partial E_i}{\partial V}\exp\left(-\frac{E_i}{kT}\right) = \frac{kT}{Z}\frac{\partial}{\partial V}\sum_i \exp\left(-\frac{E_i(V)}{kT}\right) = kT\left(\frac{\partial \ln Z}{\partial V}\right)_{T,N}, \qquad (8.22)$$

which is compatible with (3.35), namely $p = -(\partial F/\partial V)_T$, if we again employ $F = -kT\ln Z$. This is a powerful result that enables us to calculate the pressure in general systems. We shall check that it reproduces the pressure of an ideal classical gas when we develop the statistical thermodynamics of such a system in Chapter 9, and we shall employ it in Section 13.3.3 to calculate the pressure of electromagnetic radiation.

### 8.2.2 Relationship to Boltzmann Entropy

The machinery of classical thermodynamics seems to emerge naturally if we base statistical thermodynamics on the Gibbs entropy and on canonical probabilities, noting that some of the classical state variables then correspond to canonical averages of microscopic quantities. Nevertheless, the Gibbs and Boltzmann entropies are closely related to each other in the thermodynamic limit. The way to demonstrate this is to start with the Gibbs entropy expressed in terms of macrostate probabilities

$$S_{\mathrm{G}} = -k \sum_{\alpha} P_{\alpha} \ln \left( \frac{P_{\alpha}}{\Omega_{\alpha}} \right), \tag{8.23}$$

and to employ an expression for $P_{\alpha}$ that is appropriate as the thermodynamic limit is approached. We label the macrostates by energy $E_{\alpha}$ and write $S_{\mathrm{G}} = -k \sum_{\alpha} P(E_{\alpha}) \ln(P(E_{\alpha})/\Omega(E_{\alpha}))$. Following (5.28), the $P(E_{\alpha})$ may be taken to be approximately Gaussian:

$$P(E_{\alpha}) = Z^{-1} \Omega(\langle E \rangle) \exp\left(-\frac{\langle E \rangle}{kT}\right) \exp\left(-\frac{(E_{\alpha} - \langle E \rangle)^2}{2\sigma_E^2}\right) = \frac{1}{\mathcal{N}} \exp\left(-\frac{(E_{\alpha} - \langle E \rangle)^2}{2\sigma_E^2}\right), \tag{8.24}$$

where $\langle E \rangle$ is the canonical mean energy of the system, $\sigma_E^2$ is the variance in system energy, and $\mathcal{N} = \sum_{\alpha} \exp\left(-(E_{\alpha} - \langle E \rangle)^2/2\sigma_E^2\right)$. Next, we expand the logarithm of the multiplicity of the macrostates about the mean energy:

$$\ln \Omega(E_{\alpha}) \approx \ln \Omega(\langle E \rangle) + (E_{\alpha} - \langle E \rangle) \frac{\partial \ln \Omega(\langle E \rangle)}{\partial \langle E \rangle} = \ln \Omega(\langle E \rangle) + \frac{E_{\alpha} - \langle E \rangle}{kT}, \tag{8.25}$$

making use of the equality between the $\beta$ parameter of the reservoir $(1/kT)$ and that of the most likely macrostate (for which $E_{\alpha} = \langle E \rangle$), as discussed in Section 5.4. This implies that $\langle \ln \Omega(E_{\alpha}) \rangle \approx \ln \Omega(\langle E \rangle)$, which allows us to write

$$\begin{aligned} S_{\mathrm{G}} = k \sum_{\alpha} P(E_{\alpha}) \ln \Omega(E_{\alpha}) - k \sum_{\alpha} P(E_{\alpha}) \ln P(E_{\alpha}) &\approx k \ln \Omega(\langle E \rangle) \\ - k \sum_{\alpha} P(E_{\alpha}) \ln P(E_{\alpha}). \end{aligned} \tag{8.26}$$

The final term may be written

$$-k \sum_{\alpha} P(E_{\alpha}) \ln P(E_{\alpha}) = \frac{k}{2\sigma_E^2} \langle (E_{\alpha} - \langle E \rangle)^2 \rangle + k \ln \mathcal{N} = k \left( \frac{1}{2} + \ln \mathcal{N} \right). \tag{8.27}$$

We now recognise that as the thermodynamic limit is approached, very few macrostates are actually visited, and in the limit, only the one with $E_{\alpha} = \langle E \rangle$. This implies that the normalisation constant $\mathcal{N}$ is of the order of one, while $\Omega(\langle E \rangle)$, the microstate multiplicity of the macrostate at the mean energy, is enormous. Thus from (8.26) and (8.27), we find that the Gibbs entropy takes the form

$$S_{\mathrm{G}} \approx k \ln \Omega(\langle E \rangle), \tag{8.28}$$

for large systems, taking the Boltzmann form in terms of the multiplicity of the macrostate at the mean energy appropriate to the prevailing reservoir temperature.

### 8.2.3 Third Law Revisited

Its correspondence with the Boltzmann entropy in the thermodynamic limit suggests that the Gibbs entropy is compatible with the third law of thermodynamics, but this can be demonstrated more directly.

The canonical probability of any microstate goes to zero as $T \to 0$ owing to the form of the Boltzmann factor $\exp(-E_i/kT)$, unless the microstate energy is zero. Since the limit of the quantity $x \ln x$ is zero as $x \to 0$, contributions to the sum $-k\sum_i P_i \ln P_i$ from nonzero energy microstates vanish as $T \to 0$. Therefore, if there is a unique ground state with zero energy, then as $T \to 0$ its probability of occupation approaches unity. Thus $S_{\mathrm{G}} \to -k(1)\ln(1) = 0$ as $T \to 0$. The third law is satisfied unless there are numerous ground states of the system.

## 8.3 Shannon Entropy

Shannon entropy is a further generalisation of Gibbs entropy for which various claims are made, including applicability away from equilibrium. It is also known as information entropy $S_I$ because of its links with information theory, a branch of mathematical logic with its roots in the optimisation of transmission rates in communication systems, and developed initially in 1948 by Claude Shannon (1916–2001).

The idea is that the combination of microstate probabilities used in the Gibbs entropy provides a measure of the uncertainty embodied by a probability distribution, and that the actual probabilities should be determined by maximising this uncertainty subject to any known constraints. These ideas are based on the information theoretic interpretation of probability as a best judgement of the likelihood of outcomes, rather than as a representation of their frequencies of occurrence in trials, as alluded to in Section 4.1.

The starting point is the expression

$$S_I = -k \sum_i P_i \ln P_i, \tag{8.29}$$

in terms of a set of microstate probabilities $P_i$. This looks the same as the Gibbs entropy, but the equilibrium probabilities $P_i$ are not this time assumed to be canonical or grand canonical, on the basis of the principle of equal a priori microstate probabilities applied to a microcanonical ensemble of system and environment. The Shannon expression is claimed to be universal. The first piece of evidence in favour of such a viewpoint is to notice that the form of the Shannon entropy is compatible not only with the Gibbs entropy, but also with the Boltzmann entropy for an isolated system. If the appropriate microstate probabilities $P_i = 1/\Omega$ are inserted, where $\Omega$ is the number of microstates, we find that

$$S_I = -k \sum_i P_i \ln P_i = k \sum_i \Omega^{-1} \ln \Omega = k \ln \Omega = S_{\mathrm{B}}, \tag{8.30}$$

As just mentioned, it is *axiomatic* that the probabilities $P_i$ are determined by maximising the Shannon entropy, taking due consideration of appropriate constraints. This has the appealing feature that the equilibrium probabilities in the microcanonical, canonical and grand canonical ensembles can be obtained using the same methods. The maximisation may be carried out using the Lagrange multiplier approach employed in Section 8.1.2.

For example, in the microcanonical case, the only constraint is the normalisation condition $\sum_i P_i = 1$, and the maximisation of $S_I - \lambda \sum_i P_i$ with respect to the $P_i$, where $\lambda$ is a Lagrange multiplier, proceeds as

$$\frac{\partial}{\partial P_j}\left(-k\sum_i P_i \ln P_i - \lambda \sum_i P_i\right) = -k\ln P_j - k - \lambda = 0, \tag{8.31}$$

which implies that all the $P_i$ are equal. This is clearly the most uncertain we can be about the microscopic state of a system, and the Shannon entropy would be $k \ln \Omega$ as we just showed. In contrast, if we knew that a system was definitely in a particular microstate, then the Shannon entropy would be zero (since either $P_i$ or $\ln P_i$ would be zero for all microstates), which would then be a representation of the minimum of uncertainty.

If the maximisation of $S_I$ were carried out under the additional constraint of a known mean energy $\langle E\rangle = \sum_i E_i P_i$, the resulting microstate probabilities would be determined from

$$\frac{\partial}{\partial P_j}\left(-k\sum_i P_i \ln P_i - \lambda \sum_i P_i - \lambda' \sum_i E_i P_i\right) = -k\ln P_j - k - \lambda - \lambda' E_j = 0, \tag{8.32}$$

implying that the $P_i$ would be canonical, namely $P_i = Z^{-1}\exp(-\beta E_i)$ with $Z = \sum_i \exp(-\beta E_i)$ and $\beta = \lambda'/k$. Notice that the parameter $\beta$ in this expression derives from a Lagrange multiplier, and is to be identified through the expression $\langle E\rangle = Z^{-1}\sum_i E_i \exp(-\beta E_i)$ that relates it to the known mean energy. In contrast, in the usual derivation of the canonical ensemble, $\beta$ is a property of the reservoir. In a similar way, by specifying a known mean particle number, we would be able to generate the grand canonical microstate probabilities.

This differs from Boltzmann's viewpoint, which emphasised that the appropriate probabilities should arise through an analysis of the microscopic dynamics. It also goes beyond the principle of equal a priori probabilities, the basis of Gibbs' development, by providing an underlying rationale for that principle. By maximising the uncertainty in the statistical description, as embodied by the Shannon entropy, while constraining the description to be consistent with whatever information we possess, we specify the least biassed probability distribution, or the logical best guess. On the other hand, Shannon entropy does require us to accept a particular view of probabilities, and opinions about this can differ.

Yet another definition of entropy, similar to Shannon entropy, and named after von Neumann, has been used to explore uncertainty in quantum mechanics, where the probabilities of state occupation are affected by intrinsic quantum uncertainties, as well as by classical statistical uncertainty. This is an important tool in the fields of quantum information processing and quantum computers.

## 8.4 Fine and Coarse Grained Entropy

Boltzmann's insight that entropy is a measure of the number of accessible microstates raises a rather fundamental question. How do we define a microstate? So far, we have discussed simple systems such as sets of oscillators where there is no doubt that a microstate is defined by an arrangement of quanta, but we intend to apply the same ideas to physical systems, notably ones involving atoms and molecules. Is it enough to specify positions and velocities at the level of atoms, or should we dig deeper and specify the coordinates of the nuclei and electrons, or even of the protons and neutrons inside the nucleus? Surely this affects the value of the entropy?

Of course, the absolute value of entropy does not matter if all we need is an entropy *difference*. For example, imagine that each of the 55 microstates in the phase space of $N = 3$ oscillators with $Q = 9$ quanta, illustrated in Figure 4.3, is actually a collection of $\tilde{\Omega}$ underlying microstates. Is the Boltzmann entropy equal to $k \ln 55$ or $k \ln(55\tilde{\Omega})$? To a certain extent, it does not matter. Evaluating the difference in entropy on releasing a constraint to move from the spikiness $\mathrm{Sp} = 9$ macrostate into the entire phase space gives an entropy change $\Delta S = k \ln(55/3)$ and the dependence on $\tilde{\Omega}$ has vanished. This assumes that the underlying multiplicity of each (apparent) microstate is the same. But if we did want to work with the absolute entropy, we simply do not know how many layers might lie beneath the subatomic scale. So we have to satisfy ourselves with an entropy that is defined with respect to some graining of the structure of matter.

A similar issue arises in the following circumstances. If a gas is discovered to have two isotopes, should this invalidate any previously published tables of entropy measurements? After all, from that moment on, when we perform an experiment on the gas we are aware of additional uncertainty with regard to the mass of every atom. But if the isotopic composition has no measurable effect on the thermodynamic processes of interest to us, then the answer is no: the additional microscopic degree of freedom is irrelevant and need not be taken into account in calculating the entropy.

The next matter that comes to mind is whether entropy depends on the choice by which we divide phase space into macrostates. This is apparent in the procedure where we take the entropy of the maximum entropy macrostate as an approximation to the total entropy. If we choose a finer scale macrostate patchwork across the phase space, does this not imply that the entropy is smaller than that obtained from a coarser scale patchwork? The answer is no: we recall that the entropy of the maximum entropy macrostate is only an approximation to the total entropy. The size of the accessible phase space is independent of the way we carve it up into macrostates, and this is the correct measure of the entropy.

The practice of using larger patches of phase space to provide a coarser scale description of the system configuration is known as *coarse graining*. While a probability distribution across such coarse macrostates would appear to embody less uncertainty, when properly interpreted through (8.23), for example, it nevertheless provides a measure of the number of underlying microstates. Much confusion can arise from failing to appreciate these points.

## 8.5 Entropy at the Nanoscale

Statistical thermodynamics can be applied to equilibrium systems of any size, as long as we are prepared to impose the principle of equal a priori probabilities, and this implies that entropy is a property of systems large and small. This is a departure from its classical roots in macroscopic thermodynamic behaviour, but a perfectly acceptable extension. In recent years, the feasibility of performing well-controlled heat and mass transfer experiments on systems at the nanoscale (which often means at submicron length scales) has created a requirement for a formalism to match. The increase in the Gibbs or Shannon entropy upon equilibration of a system still holds, but the Boltzmann concept of a time-dependent entropy estimate $k \ln \Omega_\alpha$ that tends to rise but sometimes falls is also useful. This is a matter for current development.

## 8.6 Disorder and Uncertainty

Finally, let us revisit the question raised in Chapter 1: is entropy a measure of disorder or of uncertainty? The development in this chapter so far would seem to support the latter: in fact disorder has not been mentioned at all. But the idea that entropy represents disorder is nevertheless extremely well embedded in the literature, and is how most introductory discussions are framed.

I have personally found the emphasis on disorder confusing in that it raises unnecessary questions. For example, the simple point can be made that the evolution of the universe does not always involve a transition from order into disorder. Galaxies and planets form from the cosmic chaos, and intelligent life itself has developed from the chemical soup here on Earth. This sounds like order emerging from disorder, and the suggestion is sometimes even made that life, with its ability to adapt to its environment, and intelligence, with its appreciation of order and desire to impose it on its surroundings, somehow does not obey the second law. And then, just what is meant by order? Is it spatial repetition? Why is this special? Nature simply evolves one arrangement of the world into another. A snapshot of a system is just an configuration of its atoms and any perceived order within it can be simply a matter of taste.

Furthermore, it confuses me to suggest that energy exists in forms that are ordered or disordered and that the imperative is to evolve from one to the other. The concept of the reduction in the 'quality', or concentration of energy, alluded to in Section 2.16, does help since it is a more abstract formulation, but applying the concept of order to energy, or to the particles that possess it, for me carries too great an implication of regularity in spatial pattern. We need to unpick such arguments with care.

We could perhaps rephrase the explanation by noting that systems of particles often have a tendency towards *disorderly* or chaotic evolution, in certain circumstances. The word disorderly suggests uncertainty in future behaviour while disordered suggests a lack of spatial pattern in the present: a subtle distinction. A disorderly crowd is less easy to describe, and less predictable, than an orderly one.

On the other hand, the idea of disorder does capture something of Boltzmann's insight that entropy represents uncertainty in microscopic configuration. Declaring that a set of particles is spatially ordered means we impose demanding geometrical criteria that few configurations can satisfy, and relaxing the requirement of order allows many more to meet the specification: ordered systems therefore come in fewer possible arrangements than disordered ones, and the connection with uncertainty follows. The mistake is to claim that low uncertainty is only ever associated with spatial order: we could have considered a demanding set of criteria that had nothing to do with spatial order: that the particles should lie within a short distance of some randomly selected points, for example. We are then referring to certainty, not order.

The growth of disorder is a useful shorthand for the increase in entropy, but it is not the whole story. Nevertheless, it does convey the idea of decay and decline in natural processes, and of dissipation of energy and matter, the ultimate destination being the 'heat death' of the universe, when all distinctions have been lost leaving uniformity everywhere. It is a good story, though clearly there are exciting events that take place alongside the decline. And we should not spend time wondering whether we should regard the heat death universe as ordered (in that there is uniformity) or disordered (it must have lots of entropy). Its microscopic state is simply maximally uncertain.

## Exercises

**8.1** Consider two ideal gases, one coloured red and one coloured green. A container is divided in half by a partition that is permeable to red gas particles only. Initially, the right hand subvolume of the container holds red gas at a certain pressure, while the left hand subvolume holds red and green gas in equal proportions such that the total pressure on each side of the partition is the same. The container is in thermal contact with a heat bath at temperature $T$. Determine the chemical potential of red gas in the two subvolumes and deduce whether the partition, if it is free to move, is likely to slide to the left, to the right, or remain in place. A colour-blind professor observes the system and believes a law of thermodynamics is being violated: which one and why?

**8.2** Show that the microstate multiplicity of a macrostate of $N$ quantum harmonic oscillators, labelled by the set of populations $\{n_k\} \equiv \{n_0, n_1, \cdots, n_Q\}$, where $n_k$ is the number of oscillators that possess $k$ quanta, is given by

$$\Omega(N, Q, \{n_k\}) = \frac{N!}{n_0! \cdots n_Q!},$$

such that $\sum_{k=0}^{Q} n_k = N$ and $\sum_{k=0}^{Q} k n_k = Q$, where $Q$ is the fixed total number of quanta in the system. For the case $N = 3$ and $Q = 4$, identify the four macrostates and their microstate multiplicities. A measurement is made of the 'spikiness' of the oscillator system, defined as the difference in number of quanta possessed by the highest occupied and lowest occupied oscillator at a given instant of time. Show that the four population macrostates labelled by the $\{n_k\}$ are each characterised by a unique spikiness value, and list those values. Determine the probability distribution of the system over the macrostates, assuming that the statistics are governed by the principle of equal a priori probabilities. For a system where $N$ and the $n_k$ are very large, show that $\ln \Omega \approx -\sum_{k=0}^{Q} n_k \ln(n_k/N)$. Hence show that the maximum entropy macrostate of this system is characterised by the populations $n_k^* = N \exp(-k\beta)/\sum_{m=0}^{Q} \exp(-m\beta)$, where $\beta$ is a constant.

**8.3** Derive the grand canonical microstate probabilities by a constrained maximisation of the Shannon entropy.

# 9

# Statistical Thermodynamics of the Classical Ideal Gas

In this chapter, we develop the statistical thermodynamics of a system of noninteracting particles in order to model the classical properties of ideal gases at high temperatures or low densities. Some of the strangest phenomena in science are to be found when we cool a gas towards absolute zero, or increase its density far enough until deviations from the classical gas laws emerge. Such systems are known as quantum gases, and will be treated in Chapters 10–12. Nevertheless, our treatment of a classical gas is based on a quantum mechanical treatment of the particles, where we establish the system microstates and their energies, and then construct canonical averages through the partition function. Our aim will be to recover the entropy function describing the monatomic classical ideal gas, which was the focus of attention in Chapter 2. In order to do so, we shall find it necessary to employ another ingredient of quantum mechanics: the indistinguishability of particles and the requirement that the eigenstates of a system of many particles should satisfy certain symmetry requirements.

A system of noninteracting particles confined by a common external potential, such as a box, is relatively straightforward to analyse quantum mechanically. We can construct the description from the solution to the problem of a *single* particle in the system. The eigenstates of the particle are known as 'single particle states'. The procedure is similar to determining the shells available to an electron orbiting a nucleus, and then filling them with as many electrons as are available, subject to various rules.

## 9.1 Quantum Mechanics of a Particle in a Box

The problem of $N$ noninteracting particles in a cubic box of volume $V$ is reducible to one of a single particle because the Hamiltonian operator in the Schrödinger equation separates into kinetic energy terms for each particle. This means that the wavefunction of the $N$ particles is a product of wavefunctions of individual particles, or indeed a sum

*Statistical Physics: An Entropic Approach*, First Edition. Ian Ford.

of such products, as we shall see. Therefore, we focus our attention on determining the wavefunctions and energies of a single particle in the box. We start with

$$-\frac{\hbar^2}{2m}\nabla^2\psi(\mathbf{r}) = \epsilon\psi(\mathbf{r}), \tag{9.1}$$

where $m$ is the particle mass and $\psi(\mathbf{r})$ is the single particle wavefunction. Writing $\psi(\mathbf{r}) = \psi_{n_x}(x)\psi_{n_y}(y)\psi_{n_z}(z)$, this separates further into equations for 1-d wavefunctions of the type

$$-\frac{\hbar^2}{2m}\frac{\mathrm{d}^2\psi_{n_x}(x)}{\mathrm{d}x^2} = \epsilon_{n_x}\psi_{n_x}(x), \tag{9.2}$$

where the wavefunctions and energies are labelled with the index $n_x$, and where $\epsilon = \epsilon_{n_x} + \epsilon_{n_y} + \epsilon_{n_z}$. The boundary conditions are $\psi_{n_x}(x) = 0$ at $x = 0$ and $x = l$, where $l$ is the length of a side of the cubic box. The solutions are

$$\psi_{n_x}(x) \propto \sin(k_x x), \tag{9.3}$$

in terms of an $x$-component of the wavevector $k_x = \pi n_x/l$, with $n_x$ a positive integer. The associated energy is

$$\epsilon_{n_x} = \frac{\hbar^2 k_x^2}{2m} = \frac{h^2 n_x^2}{8ml^2}. \tag{9.4}$$

Notice that these are not evenly spaced, in contrast to the quantised energies of a 1-d harmonic oscillator.

## 9.2 Densities of States

The single particle states may be visualised as a set of points arranged as a 3-d cubic lattice, each specified by a wavevector $\mathbf{k} = (k_x, k_y, k_z)$ with magnitude $k$. The points lie in what is called **k**-space, or sometimes reciprocal space, since in a manner of speaking it is an inverse of the space corresponding to particle position **r**, in the sense that functions of **r** can be represented as Fourier transforms of functions of **k**, and vice versa. A point in this **k**-space represents a standing wave in **r**-space.

From the quantisation conditions, the nearest neighbour distance between **k**-space lattice points is $\Delta k = \pi/l$. This is illustrated in Figure 9.1. The energy of each state is

$$\epsilon_{\mathbf{k}} = \frac{h^2}{8ml^2}(n_x^2 + n_y^2 + n_z^2), \tag{9.5}$$

and this is proportional to the square of the distance from the origin to the lattice point **k** representing the state. All states lying at the same distance from the origin in **k**-space have the same energy. Each state has a degeneracy of $(2s + 1)$ for particles of spin $s$, corresponding to different values of the quantum number $m_s$. Thus, the points in Figure 9.1 represent the phase space available to a particle with a specific value of $m_s$ and the whole picture would need to be duplicated $(2s + 1)$ times to include all orientations of the spin.

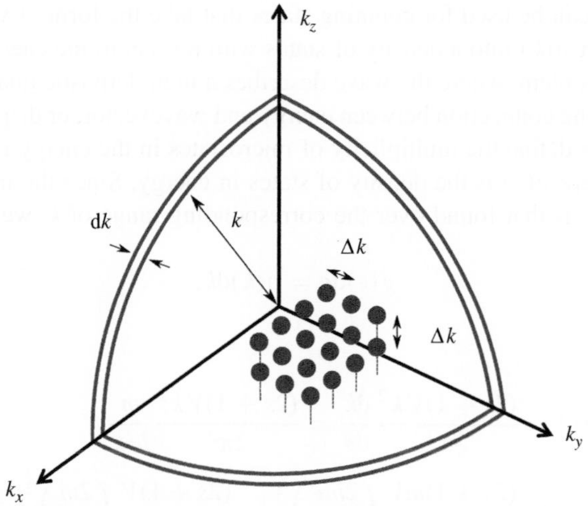


**Figure 9.1** Wavefunctions describing a single particle confined to a cubic volume $V = l^3$ take the form of standing waves specified by wavevectors that form a cubic array in what is called **k**-space, with allowed components $k_j = n_j \Delta k$ defined by positive integers $n_j$, some of which is shown. We divide this phase space of a single particle into macrostates with a magnitude of wavevector in the range $k \to k + dk$ and estimate their microstate multiplicity by counting the number of single particle microstates within the octant shell indicated.

The canonical partition function is a sum of Boltzmann factors over all available microstates:

$$Z_1 = \sum_{n_x, n_y, n_z} (2s+1) \exp\left(-\beta \epsilon_{\mathbf{k}}\right), \tag{9.6}$$

where the spin degeneracy factor is seen to play the role of a multiplicity of microstates for the macrostate labelled by the wavevector **k**. It is more convenient, however, to group the microstates into macrostates labelled by a range of the *magnitude* of the wavevector. We write

$$Z_1 = \int_0^\infty \rho(k) \exp\left(-\beta \epsilon(k)\right) \mathrm{d}k, \tag{9.7}$$

where $\rho(k)\mathrm{d}k$ is the multiplicity of microstates in the range $k \to k + \mathrm{d}k$. $\rho(k)$ is called the density of states in wavevector.

We calculate $\rho(k)$ by counting the number of wavevector lattice points lying between spherical shells of radius $k$ and $k + \mathrm{d}k$ in the sector of **k**-space with positive wavevector components. The volume of this region, multiplied by the spin degeneracy, and divided by the volume per lattice point $(\Delta k)^3$, gives the microstate multiplicity:

$$\rho(k)\mathrm{d}k = (2s+1)\frac{4\pi k^2}{8}\mathrm{d}k \frac{1}{(\Delta k)^3} = \frac{(2s+1)k^2 l^3}{2\pi^2}\mathrm{d}k, \tag{9.8}$$

and so we write

$$\rho(k) = \frac{(2s+1)Vk^2}{2\pi^2}, \tag{9.9}$$

since $l^3$ is equal to the volume of the box $V$.

This approach can be used for counting states that take the form of waves in boxes. It is usual to convert $\rho(k)$ into a density of states with respect to the energy $\epsilon$ of the wave. For the present problem, where the wave describes a nonrelativistic quantum mechanical particle in a box, the connection between energy and wavevector, or dispersion relation, is $\epsilon = \hbar^2 k^2/2m$. We define the multiplicity of microstates in the energy range $\epsilon \to \epsilon + \mathrm{d}\epsilon$ to be $g(\epsilon)\mathrm{d}\epsilon$, where $g(\epsilon)$ is the density of states in energy. Since the multiplicity in this range is the same as that found over the corresponding range of $k$, we have

$$g(\epsilon)\mathrm{d}\epsilon = \rho(k)\mathrm{d}k, \tag{9.10}$$

in which case

$$\begin{aligned} g(\epsilon) &= \frac{(2s+1)Vk^2}{2\pi^2}\frac{\mathrm{d}k}{\mathrm{d}\epsilon} = \frac{(2s+1)Vk^2}{2\pi^2}\frac{m}{\hbar^2 k} \\ &= \frac{(2s+1)mV}{2\pi^2\hbar^2}\left(\frac{2m\epsilon}{\hbar^2}\right)^{\frac{1}{2}} = \frac{(2s+1)V}{(2\pi)^2}\left(\frac{2m}{\hbar^2}\right)^{\frac{3}{2}}\epsilon^{\frac{1}{2}}. \end{aligned} \tag{9.11}$$

This density is illustrated in Figure 9.2. Its nonlinearity is a consequence of the uneven spacing of the single particle energy levels alluded to in Section 9.1.

## 9.3 Partition Function of a One-Particle Gas

The partition function representing the one-particle system is now given by

$$Z_1 = \int_0^\infty g(\epsilon)\exp(-\beta\epsilon)\,\mathrm{d}\epsilon = \frac{(2s+1)V}{(2\pi)^2}\left(\frac{2m}{\hbar^2}\right)^{\frac{3}{2}}\int_0^\infty \epsilon^{\frac{1}{2}}\exp(-\beta\epsilon)\,\mathrm{d}\epsilon. \tag{9.12}$$

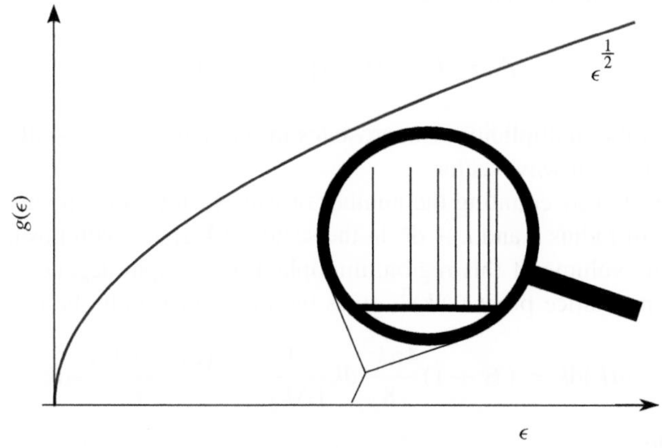


**Figure 9.2** The density of states $g(\epsilon)$, with respect to energy $\epsilon$, of a single particle in a box. The magnification of the energy axis indicates explicitly that the spectrum of states steadily becomes denser as $\epsilon$ increases.

The integral may be evaluated by the transformation $\beta\epsilon = z^2$ and employing $\int_0^\infty z^2 \exp(-z^2)dz = \pi^{1/2}/4$, giving

$$Z_1 = (2s+1)\frac{V}{\lambda_{\text{th}}^3}, \tag{9.13}$$

where we define the thermal de Broglie wavelength as

$$\lambda_{\text{th}} = \left(\frac{\beta h^2}{2\pi m}\right)^{\frac{1}{2}} = \left(\frac{h^2}{2\pi mkT}\right)^{\frac{1}{2}}. \tag{9.14}$$

We shall show that this is the same quantity as was introduced in (2.52).

It is tempting now to write the partition function of an $N$-particle gas as the product of $N$ single particle partition functions. A microstate would be identified by the set of wavevectors $\{\mathbf{k}_j\}$ labelling the single particle states occupied by particles $j = 1 \ldots N$, together with a specification of the particle spin orientations. The total energy would be $E = \sum_j \epsilon_{\mathbf{k}_j}$. After all, we did precisely this for the system of vacancies in (6.31). We would get

$$\begin{aligned} Z_N &= (2s+1)^N \sum_{\{\mathbf{k}_j\}} \exp\left(-\sum \beta\epsilon_{\mathbf{k}_j}\right) \\ &= \left[(2s+1)\sum_{\mathbf{k}_1} \exp(-\beta\epsilon_{\mathbf{k}_1})\right] \ldots \left[(2s+1)\sum_{\mathbf{k}_N} \exp(-\beta\epsilon_{\mathbf{k}_N})\right], \end{aligned} \tag{9.15}$$

so that $Z_N = Z_1^N$, since each factor in brackets is the partition function of a single particle in the box. But this would be **wrong**, and would lead to the so-called *Gibbs paradox*. We have failed to take account of particle indistinguishability.

## 9.4 Distinguishable and Indistinguishable Particles

We now need to discuss the effect of the indistinguishability of particles in statistical thermodynamics. The issues can be explored using the system of harmonic oscillators discussed extensively in Chapter 5. A microstate of this system is a specification of the energy possessed by each harmonically bound particle. It was not emphasised in Chapter 5, but we do so now, that the particles were distinguished by their interaction with distinct, spatially separated, harmonic potentials, as illustrated in Figure 5.1. Equivalently, all the particles could be imagined to be moving in a single potential, but could be distinguished by each having a different colour, for example. The enumeration of microstates was performed on this basis.

But this changes when a system is composed of particles that cannot be distinguished. For example, consider the case of two harmonically bound particles viewed from an angle that makes them indistinguishable, as illustrated in Figure 9.3. Our perception would be that there is just one microstate of the system with the particle oscillation amplitudes shown, not two. For the case of two colourless particles interacting with the same potential, this is more than just a perception. According to the rules of quantum

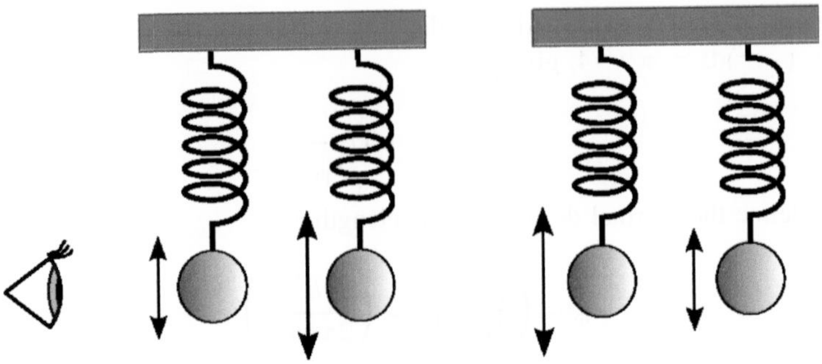

**Figure 9.3** Two harmonically bound particles are viewed from the side. The two situations shown, where the right hand particle and then the left hand particle has the higher energy, cannot be distinguished. By analogy, when two indistinguishable particles are held in the same potential, we must take care not to overcount the microstates when constructing thermodynamic ensembles.

mechanics, the two arrangements simply merge into one. Because of this, we have to be careful not to overcount the microstates in systems involving particles confined to the same spatial region.

Taking an explicit example, the indistinguishability of particles means we cannot label microstates of the oscillator system as we did in Section 4.2.2 using the numbers of quanta in each oscillator $\{q_j\}$. For three particles in the same harmonic potential, the microstates previously denoted (1,0,0), (0,1,0) and (0,0,1) in notation $(q_1, q_2, q_3)$ are in fact identical if we cannot distinguish the particles. To repeat: this is not just a matter of failing to perceive the difference. There *is* no difference.

The approach described in Section 9.1 might suggest that three states exist with wavefunctions $\psi_1(\mathbf{r}_1)\psi_0(\mathbf{r}_2)\psi_0(\mathbf{r}_3)$, $\psi_0(\mathbf{r}_1)\psi_1(\mathbf{r}_2)\psi_0(\mathbf{r}_3)$ and $\psi_0(\mathbf{r}_1)\psi_0(\mathbf{r}_2)\psi_1(\mathbf{r}_3)$, where $\psi_0(\mathbf{r}_j)$ and $\psi_1(\mathbf{r}_j)$ are the ground state and first excited state of the $j$th harmonically bound particle. But this is incorrect. The point is that quantum mechanics ought to be built without particle labelling. If we insist on using labels as an intuitive device, then all physical results need to be independent of the way in which the labels are assigned to individual particles. So the three states must be fused into one, and the ways in which this can be done are explored in greater detail in Chapter 10.

We need to find another description of a microstate of the oscillators, and the natural one to turn to is to specify the population distribution of particles in the so-called single particle states of the potential. This is intrinsically free of particle labelling. For a harmonic potential, the single particle states have energies $\hbar\omega/2$, $3\hbar\omega/2$ and so on. So the three oscillating particles with one quantum of energy between them just considered are represented as a single microstate corresponding to a population distribution of two particles in the ground state and one in the first excited single particle state. Such a description avoids specifying which particle is in which state, a situation that is then quite acceptable.

The labelling of the states of this system according to populations of oscillators with specified numbers of quanta $(n_0, n_{1,\ldots})$ was explored in Section 8.1.2. We introduced this scheme to describe *macrostates* of a system of distinguishable particles. Now we employ it to provide a way to specify *microstates* when the particles are indistinguishable. There are fewer microstates than was the case when we studied distinguishable particles in separate potentials. For example, the $N = 3$, $Q = 4$ system has 15 microstates if the particles are distinguishable, but only four if they are indistinguishable. In the notation $(n_0, n_1, n_2, n_3, n_4)$ the microstates are labelled (2,0,0,0,1), (1,1,0,1,0), (1,0,2,0,0), (0,2,1,0,0), as we found in Section 8.1.3.

Indistinguishability has a profound effect on statistical thermodynamics. According to Boltzmann's principle, the entropy for three indistinguishable particles that together possess four quanta is $k\ln 4$ when they are held in the same potential, in contrast to $k\ln 15$ if the particles are held in spatially separated potentials. The number of possible microstates changes and ensemble averages of system properties differ as a consequence.

As another example, consider the $N = 3$, $Q = 9$ oscillator system. The phase space for distinguishable oscillators, or particles, was given in Figure 4.3, divided into spikiness macrostates with multiplicities shown in Figure 4.4. If the particles are indistinguishable, however, the phase space is cut down in size as illustrated in Figure 9.4. Microstates in Figure 4.3 that are distinguishable from one another only through the swapping of particle labels are merged. Thus the three red microstates at the vertices of Figure 4.3, denoted as $(q_1, q_2, q_3) = (9, 0, 0)$, $(0, 9, 0)$, and $(0, 0, 9)$, and considered to be different if the particles are distinguishable, are replaced by a single red microstate in Figure 9.4 that could be labelled $(n_0, n_1, n_2, n_3, n_4, n_5, n_6, n_7, n_8, n_9) = (2, 0, 0, 0, 0, 0, 0, 0, 0, 1)$. We can regard the new phase space as a folded down version of the old phase space, in this case approximately one-sixth of the size.

The revised number of microstates in each spikiness macrostate is illustrated in Figure 9.5. Notice that the statistics of spikiness, such as the mean and standard deviation, differ considerably with respect to the situation in Figure 4.4 for distinguishable particles.

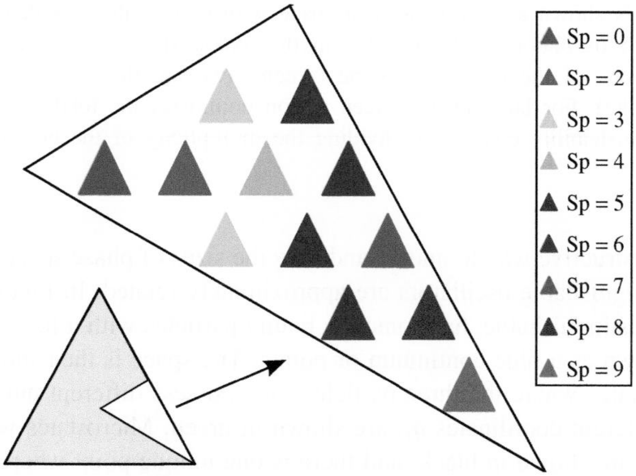


**Figure 9.4** An illustration of the phase space available to $N = 3$ indistinguishable particles (or oscillators) possessing $Q = 9$ quanta. The reduced phase space of twelve microstates comprises approximately 1/6, or $1/N!$ of the original phase space given in Figure 4.3, shown in outline in the lower left. Spikiness macrostates are shown in different colours.

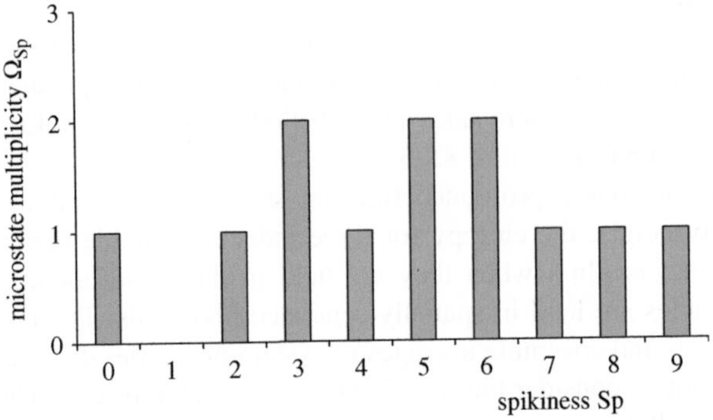


**Figure 9.5** Microstate multiplicity of spikiness for three indistinguishable oscillators holding nine quanta between them. This should be contrasted with Figure 4.4 for distinguishable oscillators.

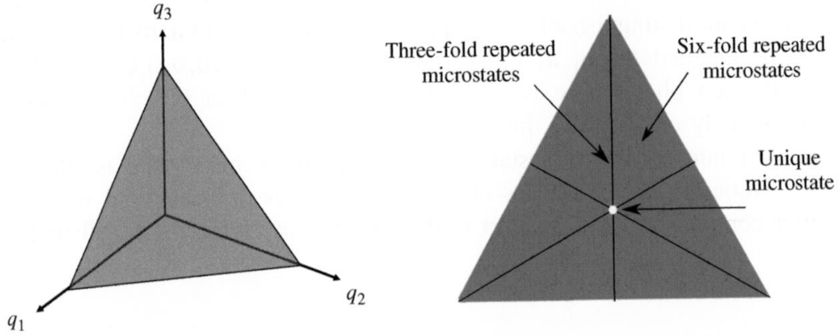


**Figure 9.6** The phase space of three distinguishable oscillators with a constant total number of quanta $Q$ is shown as a triangle in a space with coordinate axes denoting the number of quanta in each oscillator $q_j$. Microstates in this phase space can be categorised according to whether none of the $q_j$ are the same (green); two are the same (black); or all three are the same (white). For large $Q$, the green region dominates the total area, and the correction for indistinguishability consists of dividing the multiplicity of the blue phase space by a factor of 6.

There is an instructive way to understand how the sizes of phase spaces of distinguishable and indistinguishable oscillators are approximately related. In Figure 9.6 the phase space of three distinguishable, harmonically bound particles with a fixed total number of quanta $Q$ is shown as a blue continuum of points. The space is then shown divided into regions. Microstates where the three particles each possess different numbers of quanta, indicated by different coordinates $q_j$, are shown in green. Microstates where two of the $q_j$ are the same are shown in black, and there is one unique point where all three $q_j$ are the same.

We have built a phase space for distinguishable oscillators, but now we seek to understand how it can be reduced in size to correct for indistinguishability. The number of microstates $\Omega_{\text{indist}}$ available to indistinguishable oscillators comes from merging microstates in the six-fold repeated regions coloured green, and the three-fold repeated regions coloured black, such that

$$\Omega_{\text{indist}} = \tfrac{1}{6}\Omega_{\text{green}} + \tfrac{1}{3}\Omega_{\text{black}} + \Omega_{\text{white}}. \tag{9.16}$$

For example, with $Q = 9$ we can deduce from Figure 4.3 that $\Omega_{\text{green}} = 42$, $\Omega_{\text{black}} = 12$ and $\Omega_{\text{white}} = 1$ such that $\Omega_{\text{indist}} = 12$ as shown in Figure 9.4.

Large $Q$ at a given $N$ suggests a system at high temperature. In this regime, it is visually clear that phase space is dominated by the *green* region, and so we can make the approximation $\Omega_{\text{indist}} \approx (1/6)\Omega_{\text{green}}$ for a system consisting of three oscillators with a large number of quanta. For high temperatures, or equivalently classical conditions, it becomes a very good approximation to relate the sizes of the phase spaces of distinguishable and indistinguishable oscillators by

$$\Omega_{\text{indist}} \approx \tfrac{1}{6}\Omega_{\text{dist}}. \tag{9.17}$$

We now generalise the argument and make the claim that for the case of $N$ oscillators, the correction factor is approximately $1/N!$ for high temperature, classical conditions.

## 9.5 Partition Function of an *N*-Particle Gas

We return to the system of particles in a box and attempt to take account of particle indistinguishability in computing the partition function $Z_N$. For a set of $N$ distinguishable particles in the box, the partition function $Z_N^{\text{dist}}$ is indeed given by (9.15), but from the arguments given in Section 9.4, this involves an overcounting of microstates if the particles are indistinguishable. The particles are held in the same potential (the box); they are not in separate potentials such as atoms attached to specific lattice sites.

The partition function in (9.15) is a sum of contributions where each of the $N$ distinguishable particles assumes one of the wavevectors in the single particle phase space. Let us consider instead a sum where all the particles are in *different* single particle states, which is clearly a smaller number:

$$Z_N^{\text{diff}} = (2s+1)^N \sum_{\{\mathbf{k}_j \neq \mathbf{k}_i\}} \exp\left(-\sum \beta\epsilon_{\mathbf{k}_j}\right) < Z_N^{\text{dist}} = Z_1^N. \tag{9.18}$$

For the conditions of classical gas behaviour, namely, high temperature and low density, this sum makes the dominant contribution to the partition function. The argument is the same as the one we employed for the oscillators in the previous section: the sum $Z_N^{\text{diff}}$ is analogous to the green region of the three oscillator phase space, and as long as the phase space is large, the partition function of *distinguishable* particles $Z_N^{\text{dist}}$ is approximately given by $Z_N^{\text{diff}}$. Now, a specific set of $N$ different wavevectors $\{\mathbf{k}_j\}$ will appear $N!$ times in $Z_N^{\text{diff}}$, corresponding to the ways in which $N$ particle labels can be assigned to $N$ wavevectors, and so $Z_N^{\text{diff}}$ is $N!$ times too large if the particles are in fact indistinguishable. Therefore, in order to take into account the indistinguishability, we divide this contribution by $N!$.

But since $Z_N^{\text{diff}} \approx Z_N^{\text{dist}} = Z_1^N$, the partition function of a classical $N$-particle gas is then to be written as

$$Z_N^{\text{indist}} \equiv Z_N \approx \frac{Z_N^{\text{diff}}}{N!} = \frac{Z_1^N}{N!}, \tag{9.19}$$

which replaces (9.15) and is now inclusive of an approximate treatment of indistinguishability. For low temperatures, we would need to find another way to construct a partition function to account for indistinguishability, and we shall return to this in Chapters 11 and 12. But for now we investigate the properties of $Z_N$ for indistinguishable particles and try to establish connections with the treatment of the ideal gas in classical thermodynamics.

## 9.6 Thermal Properties and Consistency with Classical Thermodynamics

From the partition function we are able to calculate canonical averages of system properties. First we evaluate the mean energy using (6.13):

$$\langle E \rangle = -\frac{\mathrm{d} \ln Z_N}{\mathrm{d}\beta}. \tag{9.20}$$

Using $Z_N = Z_1^N/N!$ together with (9.13) and (9.14), we find

$$\langle E \rangle = -N\frac{\mathrm{d} \ln Z_1}{\mathrm{d}\beta} = N\frac{\mathrm{d} \ln \lambda_{\mathrm{th}}^3}{\mathrm{d}\beta} = \frac{3}{2}\frac{N}{\beta} = \frac{3}{2}NkT, \tag{9.21}$$

consistent with the calculation made in Section 6.1.1, and with the equipartition theorem as well.

Next we evaluate the Helmholtz free energy. Assuming $N$ is large, we can write

$$\ln Z_N \approx N\ln Z_1 - N\ln N + N = N\ln(\mathrm{e}Z_1/N) = N\ln\left(\frac{(2s+1)\mathrm{e}V}{N\lambda_{\mathrm{th}}^3}\right), \tag{9.22}$$

where $\mathrm{e} \approx 2.7183$ is the base for natural logarithms, and so

$$F = -NkT\ln\left(\frac{(2s+1)\mathrm{e}V}{N\lambda_{\mathrm{th}}^3}\right) = -NkT\ln\left(\frac{(2s+1)\mathrm{e}(2\pi mkT)^{\frac{3}{2}}V}{Nh^3}\right). \tag{9.23}$$

Therefore, from (8.11) the Gibbs entropy is

$$\begin{aligned} S_{\mathrm{G}} &= \frac{\langle E \rangle}{T} - \frac{F}{T} = \frac{\langle E \rangle}{T} + k\ln Z_N = \frac{3}{2}Nk + Nk\ln\left(\frac{(2s+1)\mathrm{e}V}{N\lambda_{\mathrm{th}}^3}\right) \\ &= Nk\ln\left(\frac{(2s+1)\mathrm{e}^{\frac{5}{2}}(2\pi mkT)^{\frac{3}{2}}}{(N/V)h^3}\right), \end{aligned} \tag{9.24}$$

This should be compared with the result (2.22) from classical thermodynamics. We have worked out the entropy of the monatomic ideal classical gas, according to the principles of statistical thermodynamics, and shown that it is compatible with the form derived on the basis of classical thermodynamics. Equation (9.24) is known as the Sackur–Tetrode expression for the entropy of a monatomic ideal gas in the classical regime of high temperature and low density. This is a key result, to demonstrate that statistical thermodynamics accounts quantitatively for a result derived in classical thermodynamics!

In (2.54), we showed that the chemical potential of a monatomic ideal classical gas was

$$\mu = kT\ln\left(\frac{N\lambda_{\text{th}}^3}{(2s+1)V}\right), \tag{9.25}$$

and comparing this with (2.53) implies that the thermal de Broglie wavelength defined in (9.14), with $s = 0$, corresponds to the quantity introduced in our discussion of classical thermodynamics (where naturally spin was taken to be zero) in (2.52). Moreover, we can identify the constant $\hat{c}$ that appears in the classical thermodynamic expression for entropy in (2.22): we have

$$\hat{c} = \frac{e^{-\frac{5}{2}}}{(2s+1)}\left(\frac{h^2}{2\pi m}\right)^{\frac{3}{2}}. \tag{9.26}$$

This shows that $\hat{c}$ is related to Planck's constant, a quantum concept. We could not have obtained this value within the framework of classical thermodynamics.

Finally, we comment on the Gibbs paradox referred to in Section 9.3. If we had not divided by the $N!$ factor in (9.19), we would not have obtained the factor of $N$ inside the logarithm in the Sackur–Tetrode expression (9.24) for the entropy. We could not have matched the form of the ideal gas entropy derived from classical thermodynamics. Even more seriously, the entropy would then not have been extensive, that is, it would not double when $N$ and $V$ are both doubled. Historically, this was a puzzle that was resolved only by recognising that the particles of a gas are *indistinguishable*, and that as a result, there are fewer microstates available than might otherwise be expected.

## 9.7 Condition for Classical Behaviour

We have stated several times that we expect intuitively that the classical regime of gas behaviour should correspond to high temperatures and low densities, but now we obtain a more precise condition.

Recall that the key requirement for deriving the approximate form of the $N$-particle partition function was that there was a dominant contribution to the partition function arising from microstates where single particle states in the box are occupied by at most one particle. We need a criterion to specify conditions where the probability of occupation of a single particle state by *more* than one particle is negligible.

First, we use the Gibbs entropy to determine a rough estimate of the size of the phase space at the average system energy, using the argument in Section 8.2.2. From (9.24) we have

$$S_{\text{G}} = Nk\ln\left(\frac{(2s+1)e^{\frac{5}{2}}V}{N\lambda_{\text{th}}^3}\right) \approx k\ln\Omega(\langle E\rangle), \tag{9.27}$$

and so $\Omega(\langle E\rangle) \sim (n_q V/N)^N$, where we have defined $n_q = 1/\lambda_{\text{th}}^3$. $\Omega(\langle E\rangle)$ is analogous to the area of the blue triangle in Figure 9.6. Now, the argument that single particle states are rarely occupied by more than one particle rests on the idea that the system phase space is very large, that is, $\Omega(\langle E\rangle) \gg 1$. This is analogous to arguing that the green

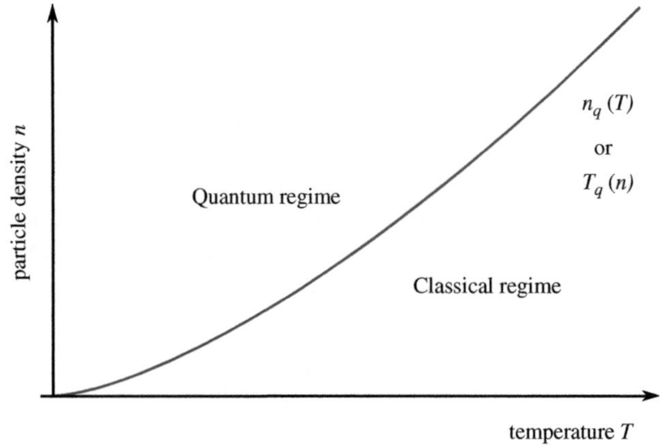


**Figure 9.7** The quantum concentration $n_q(T)$ or its inverted form $T_q(n)$ separates the density–temperature plot of gas conditions into regions where classical and quantum behaviour is to be expected.

regions of the triangle in Figure 9.6 are dominant. The condition holds when $n_q V \gg N$, or when the particle density is much less than $n_q$, namely,

$$n = \frac{N}{V} \ll n_q = \frac{1}{\lambda_{\text{th}}^3} = \left(\frac{2\pi mkT}{h^2}\right)^{\frac{3}{2}}. \tag{9.28}$$

The temperature dependence of $n_q$ is sketched in Figure 9.7.

The classical regime corresponds to densities $n$ well below $n_q$ for a given temperature, or temperatures well above a threshold $T_q = n^{2/3}h^2/(2\pi mk)$ for a given density. We interpret this to mean that the volume per particle should be much larger than the volume $\lambda_{\text{th}}^3$ in order that the gas should behave classically, in the sense that its thermodynamic properties should match those we obtained in Chapter 2. The particles need to be many thermal de Broglie wavelengths apart, on average. For a proton at room temperature, $\lambda_{\text{th}}$ is about 0.1 nm; so such a gas would need to be quite dense to depart from classical behaviour.

The root mean square velocity of a gas particle in the classical regime is $v_{\text{rms}} = \langle v^2 \rangle^{1/2} = (3kT/m)^{1/2}$, and the quantum de Broglie wavelength of a particle at such a speed is $h/(mv_{\text{rms}}) = h/(3mkT)^{1/2}$, which is similar in form to $\lambda_{\text{th}}$ in (9.14), and this accounts for the name. The criterion (9.28) involving $\lambda_{\text{th}}$ therefore suggests that departure from classical thermodynamic gas behaviour is a quantum effect. It is partly so because at such densities the discrete nature of energy levels in the system becomes apparent, making the indistinguishability correction more complicated than the simple factor of $1/N!$, but also because the non-negligible multiple occupancy of single particle states at high gas densities brings into the discussion some fundamental rules of quantum mechanics that we meet in Chapter 10. In any case, nonclassical behaviour emerges for particle densities greater than $n_q = \lambda_{\text{th}}^{-3}$ and $n_q$ is known as the *quantum concentration* for this reason.

## Exercises

**9.1** Show that the canonical partition function of a single atom with spin zero in a box of volume $V$ is $Z_1 = V\lambda_{th}^{-3}$, where $\lambda_{th} = [h^2/(2\pi mkT)]^{1/2}$ is the thermal de Broglie wavelength. Estimate the thermal de Broglie wavelength of a sodium atom at room temperature. Calculate the partition function for a gas of $N \gg 1$ indistinguishable noninteracting atoms in the volume $V$, such that $N/V \ll \lambda_{th}^{-3}$, and demonstrate that the Helmholtz free energy is extensive. Would this model apply to sodium vapour at room temperature?

**9.2** For a system of three oscillators holding four quanta, determine the number of microstates for cases where they are distinguishable or indistinguishable.

**9.3** Calculate the mean and standard deviation of the spikiness of a system of three indistinguishable oscillators that hold nine quanta, and compare the outcome with the case of distinguishable oscillators discussed in Section 4.3.

**9.4** Write down the partition function of $N$ particles confined to a 1-d harmonic potential and in contact with a heat bath at temperature $T$, in the classical limit, given that the classical partition function of one particle in an oscillator of frequency $\omega$ is $Z_1 = kT/\hbar\omega$. Show that the chemical potential of the system is given by $kT\ln(N\hbar\omega/kT)$.

**9.5** Two identical particles are tethered harmonically to points far apart and coupled to a heat bath such that the partition function might be written $Z_1^2$ and the free energy $-2kT\ln Z_1$. On the other hand, if the tether points were brought together, then we should take account of indistinguishability and write the partition function as $(1/2)Z_1^2$ and the free energy as $-2kT\ln Z_1 + kT\ln 2$, such that the system free energy increases by $kT\ln 2$. Is this an energy change or an entropy change? [Hint: consider (6.13) or (6.15)]. Is an external force required to bring the particles together? [Hint: consider Section 3.2.2].

**9.6** Under what circumstances might the atoms in a system be distinguishable and when might they be indistinguishable?

# 10

# Quantum Gases

We saw in Chapter 9 that a statistical thermodynamic treatment of a set of indistinguishable noninteracting particles in a box gave us thermodynamic properties corresponding to those of a classical ideal gas as long as the particle density did not exceed a temperature-dependent quantum concentration $n_q$, defined in (9.28). When this condition is violated, the thermodynamic properties begin to depart from classical behaviour because the probability of occupation of at least some single particle states in the box becomes significant. When this happens, the correction for the overcounting of microstates due to indistinguishability has to be revised, and we also have to implement certain quantum mechanical rules regarding the multiple occupations of states. We describe these systems as quantum gases, and the determination of their properties, while demonstrating that they correspond to real behaviour, provides considerable experimental support for statistical thermodynamic methods.

## 10.1 Spin and Wavefunction Symmetry

We resolved the Gibbs paradox in Section 9.5 by regarding particles as indistinguishable. It seems that it is meaningless to attach a label to a particular particle. Yet this is naturally the way both classical and quantum mechanics have been developed. For example, if we wish to describe the behaviour of a system of two particles (electrons in the helium atom, for example), then the standard approach is to solve a Schrödinger equation involving a Hamiltonian $H(\mathbf{r}_1, \mathbf{r}_2)$, to obtain a wavefunction $\psi(\mathbf{r}_1, \mathbf{r}_2)$ where the first position $\mathbf{r}_1$ is the location of particle A and the second is the location of particle B. We take $|\psi(\mathbf{r}_1, \mathbf{r}_2)|^2$ to be proportional to the probability that particle A is located at position $\mathbf{r}_1$ and particle B at $\mathbf{r}_2$.

But the particles are indistinguishable, so the event that we have just mentioned cannot be meaningful. Instead, we must ask for the probability that an *unspecified* particle is found at position $\mathbf{r}_1$ while the other is at $\mathbf{r}_2$. We ought to develop quantum mechanics without individual particle labels at all. But if we do introduce particle labelling into a description of indistinguishable particles, it must be in such a way that the choice of

*Statistical Physics: An Entropic Approach*, First Edition. Ian Ford.

labelling doesn't matter to the physics. The probability that A is at $\mathbf{r}_1$ and B at $\mathbf{r}_2$ must be equal to the probability that B is at $\mathbf{r}_1$ and A at $\mathbf{r}_2$.

This invariance of the square modulus of the wavefunction under the swapping of particle *labels* is equivalent to an invariance under the swapping of particle *positions*, namely

$$|\psi(\mathbf{r}_1, \mathbf{r}_2)|^2 = |\psi(\mathbf{r}_2, \mathbf{r}_1)|^2, \tag{10.1}$$

or more generally, for a wavefunction describing more than two particles:

$$|\psi(\mathbf{r}_1, \mathbf{r}_2, \cdots)|^2 = |\psi(\mathbf{r}_2, \mathbf{r}_1, \cdots)|^2, \tag{10.2}$$

where the dots denote additional particle positions. This means that

$$\psi(\mathbf{r}_1, \mathbf{r}_2, \cdots) = e^{i\theta}\psi(\mathbf{r}_2, \mathbf{r}_1, \cdots), \tag{10.3}$$

and a second swapping implies that $\psi(\mathbf{r}_1, \mathbf{r}_2, \cdots) = e^{2i\theta}\psi(\mathbf{r}_1, \mathbf{r}_2, \cdots)$ indicating that $\exp(2i\theta) = 1$, or $\theta = 0$ or $\pi$. The so-called *exchange* symmetry therefore leads to two possibilities: the wavefunction can be either even or odd under the swapping of labels:

$$\psi(\mathbf{r}_1, \mathbf{r}_2, \cdots) = \pm\psi(\mathbf{r}_2, \mathbf{r}_1, \cdots). \tag{10.4}$$

It is a remarkable fact that the choice of symmetry (+ sign) or antisymmetry (– sign) of such a wavefunction under the exchange of labels is connected to the spin of the particle. The *spin-statistics theorem* states that the wavefunction of particles with integer spin must be symmetric under exchange of particle labels, and that of particles with half-integer spin must be antisymmetric. Integer spin particles are known as *bosons*, and half-integer spin particles are called *fermions*, in recognition of this fundamental difference. The spin-statistics theorem arises from considering the very meaning of particle spin in quantum mechanics, and is beyond the scope of this book.

## 10.2 Pauli Exclusion Principle

The antisymmetry requirement for fermions means that only one particle can occupy each single particle state. This is the familiar Pauli exclusion principle in atomic physics that limits the occupation of atomic orbitals to one electron (a spin half fermion) of each spin orientation. Let us see how this arises.

Consider two electrons in a helium atom, each with the same orientation of spin. The spatial part of their wavefunction $\psi(\mathbf{r}_1, \mathbf{r}_2)$ possesses an antisymmetry under label exchange, such that $\psi(\mathbf{r}_1, \mathbf{r}_2) = -\psi(\mathbf{r}_2, \mathbf{r}_1)$. It is immediately apparent that the probability $|\psi(\mathbf{r}_1, \mathbf{r}_1)|^2$ that the two electrons should be found at the same position is zero. The electrons appear to avoid each other, for a reason that has nothing to do with their electrostatic repulsion, but instead is due to the antisymmetry of their joint wavefunction.

Now imagine constructing a wavefunction of two noninteracting fermions in a box by forming a product of two single particle states, as we claimed was possible in Section 9.1. We consider single particle states with wavevectors $\mathbf{k}$ and $\mathbf{k}'$ such that $\psi_{\mathbf{k}\mathbf{k}'}(\mathbf{r}_1, \mathbf{r}_2) = \psi_{\mathbf{k}}(\mathbf{r}_1)\psi_{\mathbf{k}'}(\mathbf{r}_2)$ where $\psi_{\mathbf{k}}(\mathbf{r}_1) \propto \sin(\mathbf{k} \cdot \mathbf{r}_1)$ and so on. However, this expression is not antisymmetric in particle label. For particles with the same spin orientation, a wavefunction with acceptable exchange symmetry has the spatial part

$$\psi_{\mathbf{kk}'}(\mathbf{r}_1, \mathbf{r}_2) \propto \psi_{\mathbf{k}}(\mathbf{r}_1)\psi_{\mathbf{k}'}(\mathbf{r}_2) - \psi_{\mathbf{k}}(\mathbf{r}_2)\psi_{\mathbf{k}'}(\mathbf{r}_1). \tag{10.5}$$

Clearly, the wavefunction for a situation where both particles have the same wavevector vanishes. A second fermion is excluded from a single particle state that is already occupied. Furthermore, $\psi_{\mathbf{kk}'}(\mathbf{r}_1, \mathbf{r}_1) = 0$ and the two particles avoid each other spatially, as in helium. Our understanding of atomic structure rests upon this principle, and the same ideas apply to the statistical thermodynamics of the fermion gas. Equation (10.5) demonstrates explicitly that what we might have thought were two microstates actually corresponds to just one.

We shall pursue the consequences of this in Chapter 12, but first we should consider exchange symmetry in the boson gas. The wavefunction of two particles needs to be symmetric in label exchange, and so $\psi_{\mathbf{kk}'}(\mathbf{r}_1, \mathbf{r}_2) \propto \psi_{\mathbf{k}}(\mathbf{r}_1)\psi_{\mathbf{k}'}(\mathbf{r}_2) + \psi_{\mathbf{k}}(\mathbf{r}_2)\psi_{\mathbf{k}'}(\mathbf{r}_1)$ is the admissible combination. The wavefunction does not vanish when both particles are assigned the same wavevector. Bosons do not avoid each other: $\psi_{\mathbf{kk}'}(\mathbf{r}_1, \mathbf{r}_1) \neq 0$. In fact, a single particle state can be occupied by any number of particles, leading to interesting nonclassical behaviour when the density exceeds the quantum concentration, as we shall see in Chapter 11.

## 10.3 Phenomenology of Quantum Gases

We have mentioned that nonclassical effects should emerge for gases that are denser than the quantum concentration, and it is important to estimate the necessary conditions for various example systems. The quantum concentration $n_q$ is given by

$$n_q = \left(\frac{2\pi mkT}{h^2}\right)^{\frac{3}{2}}, \tag{10.6}$$

and its magnitude can be reduced, and therefore made more attainable, if we consider a low temperature or a particle with a low mass. If we insert the proton mass, we get $n_q \approx 2 \times 10^{26} T^{3/2}\ \mathrm{m}^{-3}$, where $T$ is in kelvin. Atomic or molecular gases at room temperature and pressure have a particle density of about $p/kT$, which is of order $10^{26}\ \mathrm{m}^{-3}$. So, if a gas in a closed box at such a density were cooled, nonclassical effects might be expected at temperatures below about 1 K. By then, however, it will most likely no longer be a gas, having condensed into a liquid or a solid.

Nevertheless, the boson liquid $^4$He, with a particle density of $2 \times 10^{28}\ \mathrm{m}^{-3}$, exhibits some rather amazing phenomena at a temperature of around 2.17 K. The most striking effect is that the liquid appears to lose most of its viscosity, giving rise to strange flow behaviour, such as an ability to pass through tiny pores, or to creep up surfaces drawn by capillary action, unimpeded by viscous drag. The phenomenon is known as superfluidity. The liquid also appears to have an anomalously high thermal conductivity, and has a peak in its heat capacity in this temperature region. Might this be an effect that can be explained by the statistical thermodynamic properties of a quantum gas?

More recently, similar peculiar phenomena have been detected in atomic gases, rather than in liquids, trapped in small quantities and cooled to very low temperatures in various ways. The signature, in this case, is a departure from the Maxwell–Boltzmann distribution of atomic speeds. The statistical behaviour of the particles differs radically from that expected of a classical gas. Again, might this be a quantum effect?

It is tempting to ascribe other unexpected phenomena of macroscopic systems at low temperature to quantum gas properties. One of the most dramatic of these effects is the loss of electrical resistance of many materials below a transition temperature: the phenomenon of superconductivity. This is accompanied by peculiar magnetic effects whereby the material appears to be unable to accommodate magnetic flux, leading to repulsion from magnetic fields, the so-called Meissner effect. The temperature for the superconducting transition varies over a wide range for different materials, and great technological gains would be available if a material with room temperature superconductivity could be discovered. Is this something to do with a gas exceeding its quantum concentration?

The final example might come as a surprise, but evidence would suggest that we interact every day with a quantum gas at room temperature. The Drude model of the behaviour of conduction electrons in a metal is based on the idea that they act like a gas, in that they are free to move around their container (the metal) and suffer relatively few collisions with each other or with the atoms in the sample. When driven by a potential gradient, they flow like a gas to create an electric current. It has been proposed that this freedom of transport can also account for the high thermal conductivity of metals. So, what might be the quantum concentration of an electron gas in a metal? From (10.6), and using the electron mass, we arrive at an estimate of $n_q \approx 2 \times 10^{21} T^{3/2}$ m$^{-3}$. In a metal, the concentration of conduction electrons is about one per atom, or about $10^{28}$ m$^{-3}$. The threshold temperature below which we might expect nonclassical thermodynamic behaviour is therefore about $10^4$ K! At room temperature, the electron gas in a metal is nonclassical. So might this explain anything about them? Well, the Drude model, introduced in 1900 before the advent of quantum mechanics, had a significant problem. If the electron gas in a metal were classical, then it would contribute a heat capacity of $(3/2)k$ per electron, and this is simply not observed. Might this be a quantum effect? We shall address these questions in the next two chapters.

## Exercises

**10.1** What physical property determines whether a particle is a fermion or a boson? What values of this property are distinctive for bosons? State an important symmetry property that must be satisfied by the wavefunction of a system of many bosons. What is the corresponding property that must be possessed by the wavefunction of a system of many fermions?

**10.2** The exchange symmetry of a wavefunction of two fermions suggests that the probability of their being found at the same spatial point is zero. Interpret this in terms of an effective interaction force between the particles.

**10.3** The muon is an unstable elementary particle analogous in many ways to the electron but 200 times heavier. If we could replace all the free electrons in a metal with muons, would the result be a classical or a quantum gas? What (brief!) difference would it make to the heat capacity of the metal?

# 11 Boson Gas

Bosons are particles with integer spin, named in honour of Satyendra Nath Bose (1894–1974), who, along with Einstein, developed the main ideas behind the statistical thermodynamics of a boson gas. We saw in Chapter 10 that quantum mechanics imposes no limit on the number of bosons that can occupy the same single particle energy level in a system of noninteracting particles. The population in each energy level is controlled by the prevailing temperature through the Boltzmann factor and so at low temperature, multiple occupancy of some low-lying energy levels is expected to be significant. In these circumstances, we cannot follow the procedure employed in Section 9.5 to construct the $N$-particle canonical partition function $Z_N$. We used the one-particle partition function $Z_1$ to build a partition function $Z_N^{\rm dist} = Z_1^N$ for $N$ distinguishable particles and then took approximate account of the indistinguishability by dividing by $N!$, a step that presumes that multiple occupancy of single particle states can be neglected. If we cannot rely on this assumption, the correction factor is not available in a simple form.

Instead, the best way to explore the properties of the boson gas is through the grand canonical ensemble. We consider first the statistical thermodynamic properties of a system consisting of a single quantised standing wave state in **k**-space, and then deduce the properties of an entire collection of such single particle states.

## 11.1 Grand Partition Function for Bosons in a Single Particle State

The grand canonical ensemble that we developed in Chapter 7 is a treatment of a system able to exchange energy and particles with a reservoir at temperature $T$ and chemical potential $\mu$. All the statistical properties are to be derived from the grand partition function, which from (7.14) is given by

$$Z_G(\mu, V, T) = \sum_{\text{microstates } i} \exp\left(-\frac{(E_i - \mu N_i)}{kT}\right). \tag{11.1}$$

The sum is over all microstates of the system in a volume $V$, each characterised by an energy $E_i$ and a number of particles $N_i$.

*Statistical Physics: An Entropic Approach*, First Edition. Ian Ford.

We shall apply this to a system consisting of a single particle state at energy $\epsilon$. The energy of a microstate is therefore $E_i = N_i\epsilon$, such that each microstate is fully specified by the number of particles held, in which case we get

$$Z_G^\epsilon(\mu, V, T) = \sum_{N_i=0}^{\infty} \exp\left(\frac{(\mu - \epsilon)N_i}{kT}\right) = \frac{1}{1 - \exp\left(\frac{\mu-\epsilon}{kT}\right)}, \tag{11.2}$$

having evaluating the sum as a geometric series. The superscript indicates that this is the grand partition function of the single particle state at energy $\epsilon$. The reservoir that provides energy and particles is essentially the remainder of the boson gas, distributed across all the other single particle states in the container, and presumed to act as a heat and particle bath with fixed temperature and chemical potential. Equivalently, we can instead imagine a more abstract external source of energy and particles.

## 11.2 Bose–Einstein Statistics

From (7.16), the mean population in the single particle state at energy $\epsilon$, also known as the *occupation number*, is

$$\begin{aligned} \langle N\rangle_\epsilon = kT\left(\frac{\partial \ln Z_G^\epsilon}{\partial \mu}\right)_{T,V} &= \left[1 - \exp\left(\frac{\mu - \epsilon}{kT}\right)\right] \frac{\exp\left(\frac{\mu-\epsilon}{kT}\right)}{\left[1 - \exp\left(\frac{\mu-\epsilon}{kT}\right)\right]^2} \\ &= \frac{1}{\exp\left(\frac{\epsilon-\mu}{kT}\right) - 1}, \end{aligned} \tag{11.3}$$

using $Z_G^\epsilon$ from (11.2). This expression is known as Bose–Einstein statistics. It has several interesting features.

If the chemical potential of the reservoir increases, while its temperature remains constant, the mean population in the single particle state rises; explicitly

$$\left(\frac{\partial \langle N\rangle_\epsilon}{\partial \mu}\right)_{T,V} = \frac{1}{kT} \frac{\exp\left(\frac{\epsilon-\mu}{kT}\right)}{\left(\exp\left(\frac{\epsilon-\mu}{kT}\right) - 1\right)^2} > 0, \tag{11.4}$$

which makes physical sense. However, the reservoir chemical potential cannot be raised above the energy of the state $\epsilon$, in order that the denominator in (11.3) should never be negative (mean populations must be positive). In the limit that $\mu \to \epsilon$ from below, the mean population can become very large, which of course is allowed for bosons.

Now recall that our strategy is to regard the standing wave single particle states as independent systems in contact with the same particle reservoir. These have energies $E_{\mathbf{k}}$ given by (9.5) varying from $3h^2/(8mV^{2/3})$ (which is very small when $V$ is large) up to infinity. We conclude that the chemical potential of the particle reservoir has to be less than $3h^2/(8mV^{2/3})$, or roughly speaking never *positive*. This is slightly strange, and we'll return to the implications later.

Considering (11.3) with $\mu < \epsilon$, the mean populations in the single particle states decrease as the state energy $\epsilon$ increases, for a given $\mu$ and $T$, as illustrated in Figure 11.1. The singularity in the Bose–Einstein statistics expression lies safely in the unphysical region of negative $\epsilon$.

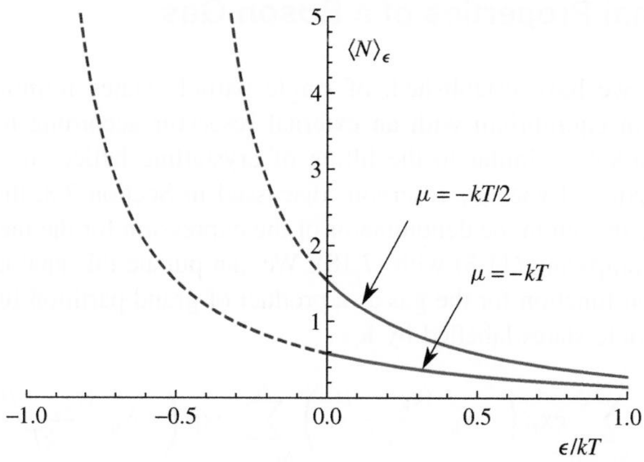


**Figure 11.1** Bose–Einstein occupation number of states with (positive) energy $\epsilon$, illustrated for two (negative) values of reservoir chemical potential $\mu$ and a fixed temperature. The dashed lines are continuations of $\langle N \rangle_\epsilon$ for unphysical negative values of $\epsilon$, to illustrate the singularity at $\epsilon = \mu$.

A further feature of interest is that as the temperature of the reservoir is increased, at constant chemical potential, the mean population in the state again increases; explicitly

$$\left(\frac{\partial \langle N \rangle_\epsilon}{\partial T}\right)_{\mu,V} = \frac{(\epsilon - \mu)}{kT^2} \frac{\exp\left(\frac{\epsilon-\mu}{kT}\right)}{\left(\exp\left(\frac{\epsilon-\mu}{kT}\right) - 1\right)^2} > 0. \tag{11.5}$$

The picture we have, therefore, is that the mean populations in a set of single particle states in a box can be increased or decreased by varying the external reservoir parameters $\mu$ and $T$. The sum of the occupation numbers of all the single particle states, divided by the volume of the box, then corresponds to the mean particle density $n = \sum_\epsilon \langle N \rangle_\epsilon / V$, and we can presume that the chemical potential of such a gas will match that of the reservoir since the two will be in equilibrium.

It is quite illuminating to insert the expression in (9.25) for the reservoir chemical potential appropriate to the classical gas regime (with spin zero particles), namely $\mu = kT \ln(n/n_q)$ with $n \ll n_q$, into (11.3). For $\mu/kT \ll 0 < \epsilon/kT$ such that $\exp[(\epsilon - \mu)/kT] \gg 1$ we get

$$\langle N \rangle_\epsilon \approx \exp\left(-\frac{(\epsilon - \mu)}{kT}\right) = \frac{n}{n_q} \exp\left(-\frac{\epsilon}{kT}\right) \ll 1, \tag{11.6}$$

demonstrating that the likelihood of multiple occupancy of any of the single particle states is very small if the density is much less than the quantum concentration $n_q$. Quantum gas thermodynamic properties emerge when the reservoir parameters are such that the mean density of particles in the system exceeds $n_q$. Equation (9.25) and Figure 11.1 suggest that this would involve the approach of $\mu$ towards zero from below, such that the single particle states with lowest energy become multiply occupied. But the relationship between $\mu$ and $n$ will then no longer be given by the classical expression, and the gas will deviate from classical behaviour, which we explore next.

## 11.3 Thermal Properties of a Boson Gas

The picture that we have established, of single particle states forming a cubic array in **k**-space, and in equilibrium with an external reservoir according to Bose–Einstein statistics, is remarkably similar to the filling of crystalline lattice sites with vacancies drawn from an external vacancy reservoir, discussed in Section 7.3, though there is an important change in sign in the denominator of the expression for the mean population at each site when comparing (11.3) with (7.18). We can pursue this analogy and construct the grand partition function for the gas as a product of grand partition functions for each of the single particle states labelled by $\mathbf{k}_i$:

$$Z_G(\mu, V, T) = \sum_{N_{\mathbf{k}_1}=0}^{\infty} \exp\left(-N_{\mathbf{k}_1}\frac{(\epsilon_{\mathbf{k}_1} - \mu)}{kT}\right) \sum_{N_{\mathbf{k}_2}=0}^{\infty} \exp\left(-N_{\mathbf{k}_2}\frac{(\epsilon_{k_2} - \mu)}{kT}\right) \cdots . \quad (11.7)$$

The microstates are labelled by the set of populations $\{N_{\mathbf{k}_i}\}$ across the array. Notice that this avoids assigning labels to any of the particles such that we can be sure that the result will be appropriate for indistinguishable particles.

The grand partition function of the gas is therefore

$$Z_G = \prod_{\mathbf{k}_i} Z_G^{\epsilon_{\mathbf{k}_i}}, \quad (11.8)$$

with $Z_G^{\epsilon_{\mathbf{k}_i}}$ given by $[1 - \exp[(\mu - \epsilon_{\mathbf{k}_i})/kT]]^{-1}$ according to (11.2). This is equivalent to $\ln Z_G = \sum_{\mathbf{k}_i} \ln Z_G^{\epsilon_{\mathbf{k}_i}}$, which can be written in the form $\ln Z_G = \sum_\epsilon \Omega(\epsilon) \ln Z_G^\epsilon$ where the sum is over the energies of single particle states and $\Omega(\epsilon)$ is the multiplicity of such states at a given energy $\epsilon$. As the single particle energies lie very close together, we might consider representing the sum as an integral over a continuous single particle energy $\epsilon$, denoting the multiplicity of single particle states in the range $\epsilon \to \epsilon + d\epsilon$ as $g(\epsilon)d\epsilon$, where $g(\epsilon)$ is a density of states as discussed in Section 9.2. We write

$$\ln Z_G \approx \int_0^\infty g(\epsilon) \ln Z_G^\epsilon d\epsilon = -\int_0^\infty g(\epsilon) \ln\left[1 - \exp\left(\frac{\mu - \epsilon}{kT}\right)\right] d\epsilon. \quad (11.9)$$

The average number of particles in the $g(\epsilon)d\epsilon$ single particle states within the energy range $\epsilon \to \epsilon + d\epsilon$ is equal to $\langle N \rangle_\epsilon g(\epsilon)d\epsilon$, with $\langle N \rangle_\epsilon = (\exp[(\epsilon - \mu)/kT] - 1)^{-1}$ according to (11.3). Thus the mean number of particles in the system, across the entire set of single particle energy states, is

$$\langle N \rangle = \sum_{\mathbf{k}_i} \langle N \rangle_{\epsilon_{\mathbf{k}_i}} = \sum_\epsilon \Omega(\epsilon)\langle N \rangle_\epsilon \approx \int_0^\infty g(\epsilon)\langle N \rangle_\epsilon d\epsilon. \quad (11.10)$$

This can be obtained more formally from (11.9) using

$$\langle N \rangle = kT\left(\frac{\partial \ln Z_G}{\partial \mu}\right)_{T,V} = kT\int_0^\infty g(\epsilon)\left(\frac{\partial \ln Z_G^\epsilon}{\partial \mu}\right)_{T,V} d\epsilon = \int_0^\infty g(\epsilon)\langle N \rangle_\epsilon d\epsilon. \quad (11.11)$$

Similarly, from the definition of $Z_G$ in (11.1), we can write

$$\langle E - \mu N \rangle = kT^2 \left( \frac{\partial \ln Z_G}{\partial T} \right)_{\mu,V} = kT^2 \int_0^\infty g(\epsilon) \left( \frac{\partial \ln Z_G^\epsilon}{\partial T} \right)_{\mu,V} d\epsilon$$

$$= \int_0^\infty g(\epsilon)(\epsilon - \mu)\langle N \rangle_\epsilon d\epsilon, \tag{11.12}$$

where we have used

$$kT^2 \left( \frac{\partial \ln Z_G^\epsilon}{\partial T} \right)_{\mu,V} = -\frac{kT^2}{Z_G^\epsilon} \frac{\left[ -\exp\left( -\frac{(\epsilon-\mu)}{kT} \right) \right]}{\left[ 1 - \exp\left( -\frac{(\epsilon-\mu)}{kT} \right) \right]^2} \frac{(\epsilon - \mu)}{kT^2}$$

$$= \frac{\epsilon - \mu}{\exp\left( \frac{\epsilon-\mu}{kT} \right) - 1} = (\epsilon - \mu)\langle N \rangle_\epsilon. \tag{11.13}$$

Comparing (11.11) and (11.12), this is consistent with our requirement that

$$\langle E \rangle = \sum_{\mathbf{k}_i} \epsilon_{\mathbf{k}_i} \langle N \rangle_{\epsilon_{\mathbf{k}_i}} = \sum_\epsilon \Omega(\epsilon)\epsilon \langle N \rangle_\epsilon \approx \int_0^\infty g(\epsilon)\epsilon \langle N \rangle_\epsilon d\epsilon, \tag{11.14}$$

on the basis of the intuition that the total mean energy is the sum of mean energies $\epsilon\langle N \rangle_\epsilon$ for each single particle state.

Inserting (9.11) and (11.3) into (11.10) we then obtain

$$n(\mu, T) = \frac{\langle N \rangle}{V} = \frac{(2s+1)}{(2\pi)^2} \left( \frac{2m}{\hbar^2} \right)^{\frac{3}{2}} \int_0^\infty \frac{\epsilon^{\frac{1}{2}} d\epsilon}{\exp\left( \frac{\epsilon-\mu}{kT} \right) - 1}, \tag{11.15}$$

for the mean density of the gas at a given chemical potential and temperature. If this could be inverted, it would give us the chemical potential $\mu(n, T)$ of the gas in terms of the mean number of particles in the system at a given temperature, namely a replacement for the classical expression $\mu = kT \ln[n/n_q(T)]$ appropriate for spin zero particles. However, this is not straightforward. Nevertheless the classical limit may be recovered under the appropriate assumption $\epsilon - \mu \gg kT$, or very negative $\mu$, and $s = 0$, such that

$$n \approx \frac{1}{(2\pi)^2} \left( \frac{2m}{\hbar^2} \right)^{\frac{3}{2}} \int_0^\infty \epsilon^{\frac{1}{2}} \exp\left( -\frac{(\epsilon - \mu)}{kT} \right) d\epsilon$$

$$= \frac{1}{(2\pi)^2} \left( \frac{2mkT}{\hbar^2} \right)^{\frac{3}{2}} \left[ 2 \int_0^\infty z^2 \exp(-z^2) dz \right] \exp\left( \frac{\mu}{kT} \right)$$

$$= 2\pi \left( \frac{2mkT}{h^2} \right)^{\frac{3}{2}} \frac{\pi^{\frac{1}{2}}}{2} \exp\left( \frac{\mu}{kT} \right) = n_q \exp\left( \frac{\mu}{kT} \right), \tag{11.16}$$

having used the substitution $z = (\epsilon/kT)^{1/2}$, such that $\mu \approx kT \ln(n/n_q)$ as before.

The mean energy of the gas per unit volume, for a given $\mu$ and $T$, is written

$$e(\mu, T) = \frac{\langle E \rangle}{V} = \frac{(2s+1)}{(2\pi)^2}\left(\frac{2m}{\hbar^2}\right)^{\frac{3}{2}} \int_0^\infty \frac{\epsilon^{\frac{3}{2}} d\epsilon}{\exp\left(\frac{\epsilon-\mu}{kT}\right) - 1}, \tag{11.17}$$

and in the classical limit with $s = 0$ this specific energy becomes

$$\begin{aligned} e(\mu, T) &\approx \frac{1}{(2\pi)^2}\left(\frac{2m}{\hbar^2}\right)^{\frac{3}{2}} \int_0^\infty \epsilon^{\frac{3}{2}} \exp\left(-\frac{(\epsilon-\mu)}{kT}\right) d\epsilon \\ &= \frac{kT}{(2\pi)^2}\left(\frac{2mkT}{\hbar^2}\right)^{\frac{3}{2}} \left[2\int_0^\infty z^4 \exp\left(-z^2\right) dz\right] \exp\left(\frac{\mu}{kT}\right) \\ &= 2\pi kT \left(\frac{2mkT}{h^2}\right)^{\frac{3}{2}} \frac{3\pi^{\frac{1}{2}}}{4} \exp\left(\frac{\mu}{kT}\right) = \frac{3}{2} nkT, \end{aligned} \tag{11.18}$$

just as we would expect for an ideal gas, having inserted $\mu = kT \ln(n/n_q)$.

The Gibbs entropy of the gas is obtained from the grand partition function through the relation (8.17): we write

$$TS_G = \langle E \rangle - \mu \langle N \rangle + kT \ln Z_G, \tag{11.19}$$

such that according to (11.12) and (11.9) we get

$$\begin{aligned} S_G &= \frac{1}{T}\int_0^\infty g(\epsilon)(\epsilon-\mu)\langle N \rangle_\epsilon d\epsilon + k\int_0^\infty g(\epsilon) \ln Z_G^\epsilon d\epsilon \\ &= \frac{1}{T}\int_0^\infty g(\epsilon)\left[\frac{\epsilon-\mu}{\exp\left(\frac{\epsilon-\mu}{kT}\right)-1} - kT\ln\left(1-\exp\left(\frac{\mu-\epsilon}{kT}\right)\right)\right] d\epsilon, \end{aligned} \tag{11.20}$$

which can be shown to reduce to (2.54) in the classical limit where $\epsilon - \mu \gg kT$, namely an entropy per unit volume $s = S_G/V$ (not to be confused with spin, nor with the specific entropy *S/N* introduced in Section 3.9.2) that goes to $s_{cl} = nk(5/2 - \mu/kT)$, or

$$s_{cl} = nk \ln\left(\frac{e^{\frac{5}{2}} n_q(T)}{n}\right) = \frac{3}{2} nk \ln\left(\frac{e^{5/3} T}{T_q(n)}\right), \tag{11.21}$$

returning it to the form derived in (2.22) or the Sackur–Tetrode version (9.24), with spin zero particles, and having inserted

$$T_q(n) = \frac{h^2 n^{\frac{2}{3}}}{2\pi mk}. \tag{11.22}$$

Finally, we consider the pressure of the gas. This is straightforward to obtain from the grand partition function when we write

$$p = kT\left(\frac{\partial \ln Z_G}{\partial V}\right)_{T,\mu}, \tag{11.23}$$

which is an analogue of (8.22) for the grand canonical instead of the canonical ensemble. From (11.9) the only $V$-dependence in $\ln Z_G$ arises from the factor of $V$ in the density of single particle states $g(\epsilon)$ defined in (9.11), so

$$p = \frac{kT}{V}\ln Z_G = \frac{kT}{V}\int_0^\infty g(\epsilon)\ln Z_G^\epsilon d\epsilon$$
$$= -\frac{(2s+1)}{(2\pi)^2}\left(\frac{2m}{\hbar^2}\right)^{\frac{3}{2}} kT\int_0^\infty \epsilon^{\frac{1}{2}}\ln\left(1-\exp\left(\frac{\mu-\epsilon}{kT}\right)\right)d\epsilon, \quad (11.24)$$

which has the classical limit $p \to nkT$.

These expressions for the principal thermodynamic properties of the gas are slightly forbidding, but they yield understanding when we choose a chemical potential and temperature and evaluate the associated particle, energy and entropy densities $n$, $e$, $s$ and the pressure $p$. An example is given in Figure 11.2, where the temperature is fixed and $\mu$ is varied. Classical behaviour is recovered when $\epsilon - \mu \gg kT$ such that $e \to (3/2)nkT$ and $p \to nkT$ while the entropy density tends towards the form $s_{\text{cl}} = nk(5/2 - \mu/kT)$. All these quantities start to deviate from classical behaviour when the particle density approaches and exceeds the quantum concentration.

Nonclassical behaviour can also be exposed by eliminating $\mu$ between $e(\mu, V, T)$ and $n(\mu, V, T)$, for example, to give the energy density $e(n, V, T)$ in terms of more familiar variables. As an illustration, we plot $e$ and $s$ against $T$ at constant $n$ in Figure 11.3 for spin zero particles, and normalise the results by the classical expressions $e_{\text{cl}} = (3/2)nkT$ and $s_{\text{cl}} = nk\ln(e^{5/2}n_q/n)$. To a certain extent, the behaviour looks familiar. The classical mean energy is not appropriate for temperatures below $T_q = h^2n^{2/3}/(2\pi mk)$. This is the threshold temperature for quantum effects that was discussed in connection with Figure 9.7. The entropy is also reduced, which brings to mind the freezing out of degrees of freedom in the low temperature thermal behaviour of a diatomic gas, or of the Einstein solid, though the reasons for the deviation here are rather different.

However, these expressions harbour a very serious problem, and simply cannot be entirely correct. This is partly illustrated by the fact that the energy and entropy are not reported below a temperature $T \approx 0.5T_q$ in Figure 11.3. We investigate this problem in the next section.

## 11.4 Bose–Einstein Condensation

The nonclassical modifications to the thermal properties of a boson gas that we saw in the previous section as the temperature is reduced or chemical potential increased are perhaps not dramatic enough to explain the strange low temperature phenomena that were discussed in Section 10.3. But there is a flaw hidden within (11.15), in particular, and by investigating it we shall discover a rationale for such behaviour.

The density of particles in the gas derived in (11.15) may be rewritten as

$$n = \frac{2(2s+1)}{\pi^{\frac{1}{2}}}n_q\int_0^\infty \frac{x^{\frac{1}{2}}dx}{\exp\left(-\frac{\mu}{kT}\right)\exp(x)-1}, \quad (11.25)$$

showing that the density increases as the (negative) chemical potential increases; explicitly

$$\left(\frac{\partial n}{\partial \mu}\right)_{T,V} = \frac{2(2s+1)}{kT\pi^{1/2}} n_q \int_0^\infty \frac{\exp\left(-\frac{\mu}{kT}\right)\exp(x)\, x^{\frac{1}{2}}}{\left(\exp\left(-\frac{\mu}{kT}\right)\exp(x) - 1\right)^2} \mathrm{d}x > 0. \qquad (11.26)$$

The problem is that we concluded in Section 11.2 that the chemical potential has an upper limit of (approximately) zero, such that the integral in (11.25) has an upper limit of $\int_0^\infty x^{1/2}(\exp(x) - 1)^{-1}\mathrm{d}x \approx 1.306\pi^{1/2}$. For lower (i.e. negative) $\mu$, the denominator in the integrand is larger than $(\exp(x) - 1)$ for all $x \geq 0$ and the integral is therefore smaller. This suggests that the mean particle density of the system must satisfy

$$n \leq n_{\mathrm{max}}(T) = 2.612(2s+1)n_q(T), \qquad (11.27)$$

and this upper limit is apparent in the plot of $n/n_q$ for spin zero particles in Figure 11.2. Does this mean that there is a temperature-dependent maximum number of particles that can be accommodated in the box? What happens if we try to pump in some more? Or more tellingly, what happens if we have a certain density of particles $n$, then shut off their exchange with the reservoir and reduce the temperature until $n_{\mathrm{max}}(T)$ falls below $n$, or equivalently when $T \lesssim 0.5T_q$ in Figure 11.3? The particles cannot disappear!

Fortunately, the flaw in the development is easily identified. The approximation (11.10) is the root of the problem. An integral representation of the sum of mean populations over all the single particle states is an inadequate treatment for the very lowest states. The square root density of states (9.11) is a poor approximation at low $\epsilon$. Recall that it was derived in Section 9.2 by counting standing wave states in a thin shell in **k**-space, on the assumption that there are many such states to count and the granularity of the lattice of allowed states did not matter. But as the gas approaches the quantum regime,

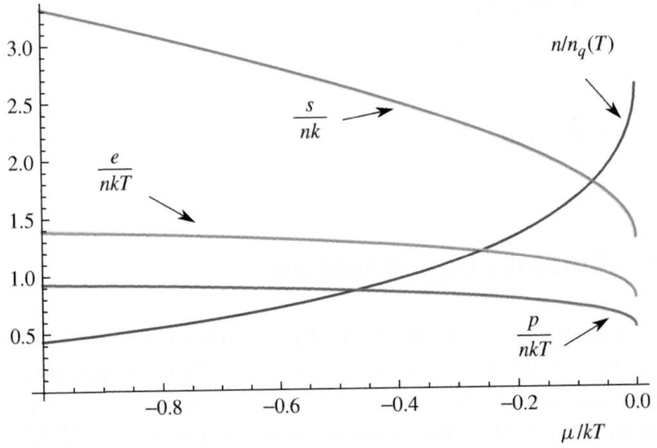


**Figure 11.2** Properties of a gas of spin zero bosons as a function of the chemical potential at constant temperature. As $\mu$ approaches zero from below, the particle density increases, and it exceeds the quantum concentration for $\mu \gtrsim -0.4kT$. The energy per particle then begins to deviate from the classical value $(3/2)kT$, the pressure falls below $nkT$ and the entropy per particle deviates from the classical result $k(5/2 - \mu/kT)$.

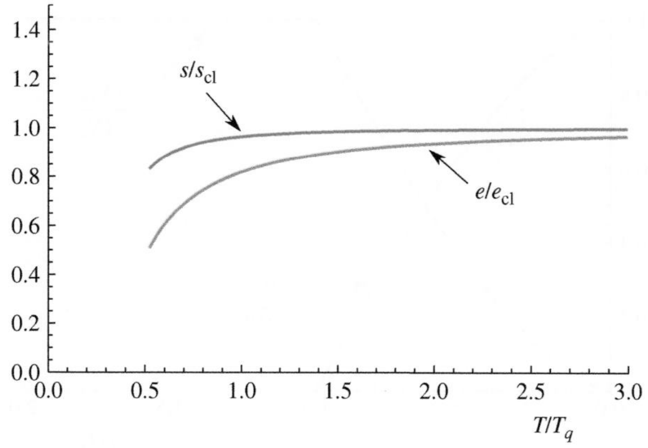


**Figure 11.3** Energy and entropy of a gas of spin zero bosons, per unit volume, divided by the classical expressions, as a function of temperature, illustrating quantum gas behaviour when $T \lesssim T_q$.

the mean populations of particles in the low energy states become significant and an approximate treatment will not do.

In order to remove this deficiency, we recognise that the integral of populations over the model density of states accounts for only some of the particles. The populations in the lowest energy states should be represented explicitly. In the simplest treatment, we consider just the population in the lowest energy state $\epsilon = 3h^2/8mV^{2/3} \approx 0$. This solves the problem of the apparent disappearance of particles when $T$ falls below $0.5T_q$: the particles that seemed not to find a home are actually accommodated in this state. The mean population in the single particle ground state at $\epsilon \approx 0$ is therefore given by

$$\langle N\rangle_0 = \langle N\rangle - V n_{\max}(T), \tag{11.28}$$

for temperatures where $n_{\max}(T) < n$, a condition that corresponds to

$$T < \left(\frac{n}{2.612(2s+1)}\right)^{\frac{2}{3}} \frac{h^2}{2\pi mk} = \left(\frac{1}{2.612(2s+1)}\right)^{\frac{2}{3}} T_q = T_C, \tag{11.29}$$

defining a temperature $T_C$, approximately equal to $0.5T_q$ for a spin of zero. Since $n_{\max} \propto n_q \propto T^{3/2}$, the relative proportion of particles occupying the ground state is given by

$$\frac{\langle N\rangle_0}{\langle N\rangle} = 1 - \frac{n_{\max}(T)}{n} = 1 - \left(\frac{T}{T_C}\right)^{\frac{3}{2}}, \tag{11.30}$$

for $T \le T_C$. The ground state appears to accumulate enormous numbers of particles when the temperature is decreased below $T_C$, and as $T \to 0$, all the particles fall into that state. The fractional occupation of the ground state as a function of temperature is illustrated in Figure 11.4, together with the proportion of particles that are *not* in the ground state, written $\langle N\rangle_{\neq 0}/\langle N\rangle$. This is multiple occupancy taken to an extraordinary degree.

The model now makes perfect sense. At a temperature of absolute zero, it satisfies our expectation that the particles should take the lowest available energy level. This would

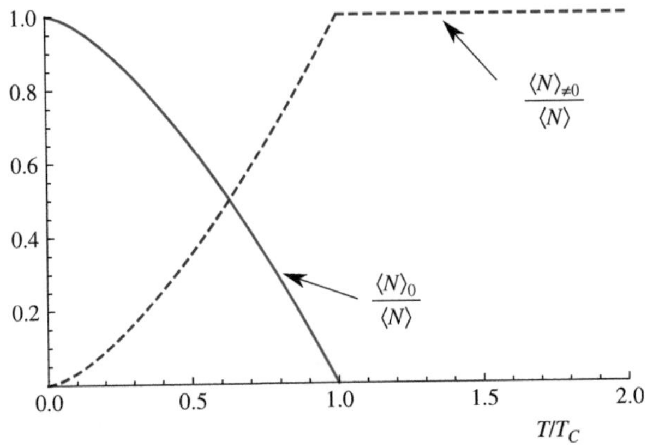


**Figure 11.4** The mean number of particles of a boson gas occupying the ground state $\langle N \rangle_0$ and occupying all other states $\langle N \rangle_{\neq 0}$, as a proportion of the whole population as a function of temperature. $T_C$ is the Bose–Einstein condensation temperature.

also be consistent with the third law: because there is just a single microstate available in this limit, the entropy of the gas will be zero. However, what is perhaps not expected is that there is a distinct temperature below which a gross 'condensation' of particles into the lowest energy single particle state begins. The word condensation suggests a phase transition of some kind, and indeed it is rather like that: the gas switches rapidly from a phase with properties that are a distorted version of those of a classical gas, to a phase that is altogether different. The phenomenon of the mass occupation of the lowest energy single particle state of the system is known as Bose–Einstein condensation and $T_C$ is called the Bose–Einstein condensation temperature.

We can illustrate the change in behaviour by sketching the entropy of the gas as a function of temperature at constant density in Figure 11.5. The entropy takes its classical logarithmic dependence on $T$ for $T \gg T_C$, but this cannot remain valid as the temperature is reduced since it would imply that the entropy eventually became negative. Instead, from Figure 11.2 we see that the entropy falls to approximately $1.3k$ per particle at $\mu \approx 0$, or equivalently at $T = T_C$, and then the entropy continues towards zero as the temperature is reduced below this threshold in proportion to the number of particles $\langle N \rangle_{\neq 0}$ that do not lie in the single particle ground state. The population in the ground state $\langle N \rangle_0$ contributes zero entropy. It is as though the gas separates into two components with very different thermodynamic properties.

Figure 11.6 illustrates how the chemical potential is logarithmic in temperature in the classical regime $T \gg T_C$, taking the form $\mu \approx -(3/2)kT \ln(T/T_q)$, but approaches zero asymptotically from below for $T < T_C$. Further implications are that the energy of the gas is proportional to $T$ in the classical regime, but is considerably suppressed for temperatures below $T_C$ as a result of particle condensation into the $\epsilon \approx 0$ ground state. The gas pressure is also suppressed.

So what can we say about a gas in which a considerable fraction of the particles reside in the lowest standing wave quantum mechanical state? The first thing we must abandon

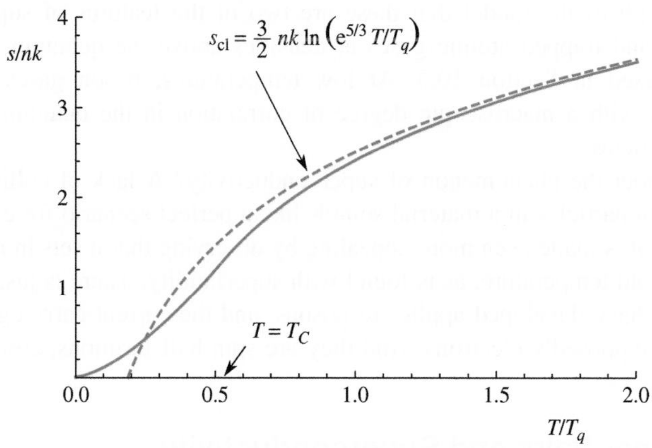


**Figure 11.5** Entropy per particle $s/n$ for a gas of spin zero bosons as a function of temperature at constant density, illustrating its deviation from the classical Sackur–Tetrode expression as the temperature falls below the Bose–Einstein condensation temperature $T_C$.

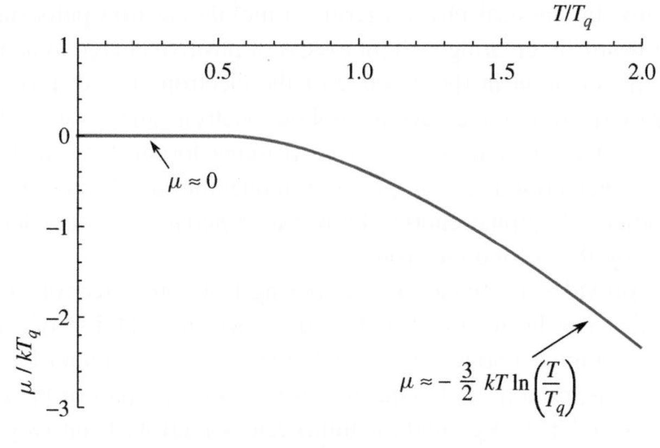


**Figure 11.6** Temperature dependence of the chemical potential of a gas of spin zero bosons, at a fixed density, showing the change in behaviour at the Bose–Einstein condensation temperature $T_C \approx 0.5T_q$.

is the classical picture of particles moving around in a chaotic fashion and colliding with one another. The picture provided by quantum mechanics is one where particles are moving (slowly!) backwards and forwards between the walls and never seeming to interact with each other. It is hard to get a feel for such a gas, but there ought to be at least two clear signatures: firstly that the distribution of particle speed is very narrow, as most of the particles are occupying the same energy level, and secondly that the viscosity of such a fluid ought to be very low, as the particles rarely transfer momentum to each other. Resistance to the shearing of a fluid fundamentally arises from collision processes on a microscopic scale that spread out the effects of an external impulse.

It is a triumph of the model that these are two of the features of superfluids, such as liquid $^4$He and trapped atomic gases at densities above the quantum concentration, that we discussed in Section 10.3. At low temperatures, boson gases adopt a new state of matter with a macroscopic degree of correlation in the quantum states of the component particles.

But what about the phenomenon of superconductivity? A lack of collisions between current-carrying particles in a material sounds like a perfect scenario for explaining zero resistivity, and it is made even more appealing by observing that it sets in rather abruptly below a threshold temperature, as is found with superfluidity. There is just one problem. The model we have developed applies to bosons, and the current-carrying particles in a conductor are supposedly electrons. And they are spin half fermions, aren't they?

## 11.5 Cooper Pairs and Superconductivity

When might a gas of fermions behave like a gas of bosons? Constructing an argument to allow this to happen might allow us to regard superconductivity as a Bose–Einstein condensation phenomenon. It has been shown that electrons in a material feel a weak attractive force towards one other, caused by their distortion of the arrangement of the surrounding atoms. In classical physics terms, a mobile electron pulls surrounding ions of the material towards it, creating a slight excess of positive charge in its neighbourhood compared with the situation in the absence of the electron. The distortion produces an attractive electrostatic well for a second mobile electron, and in spite of their mutual repulsion, the two electrons tend to associate, if rather loosely, into a state known as a *Cooper pair*. The interaction is nicely pictured in quantum mechanics as the emission of a quantum of lattice vibrational energy, known as a *phonon*, by one electron, followed by its absorption by the second electron.

The attraction provides a rationale for supposing that some electrons pair up as composite particles that can be regarded as bosons, since the addition of two half integer spins would make an integer spin. This is analogous to the formation of a helium atom, a boson, from electrons, protons and neutrons, all fermions. So part of the electron gas (in the presence of a suitable background medium) can convert itself into a gas of bosons at sufficiently low temperature. Then the condensation of the bosons into the lowest energy single particle state can proceed, giving rise to a quantum gas of Cooper pairs, each of which can carry an electric current with little or no collisional resistance, and we have a scheme for explaining superconductivity. Furthermore, the ease with which electric currents flow in superconductors means that exposure to a magnetic field elicits an extreme response that has the effect of cancelling the magnetic field within the material. The effective repulsion of magnetic flux is the Meissner effect.

The details of this so-called BCS mechanism of superconductivity, named after John Bardeen (1908–1981), Leon Cooper (1930–) and John Schrieffer (1931–), are much more elaborate than suggested by the above sketch, and understanding the mechanism relies to a considerable extent on an appreciation of the properties of the underlying fermion gas, which we discuss in Chapter 12. The main message to absorb here is that bosons can be composites of fermions (but not vice versa) and that a theory based on this idea does appear to explain the behaviour of at least some types of superconductor. In short, it is another triumph of statistical physics!

## Exercises

**11.1** Four bosons are held in a harmonic potential that has single particle states of energy $\epsilon = n\hbar\omega$, where $n = 0$, 1, 2 and so on. (a) Indicate the pattern of particle occupation of states, and total energy, of the first five microstates of the system (ordered by energy) if the bosons are distinguishable. (b) Indicate the particle occupation and total energy of the first seven microstates of the system if the bosons are indistinguishable.

**11.2** Derive the classical limits of the entropy and pressure of a boson gas starting with the expressions given in (11.20) and (11.24).

**11.3** Sketch the energy per particle of a boson gas as a function of temperature.

**11.4** Considering the temperature dependence of the entropy of a boson gas shown in Figure 11.5, or that of the energy per particle considered in question 11.3, sketch the heat capacity of the gas as a function of temperature. What feature appears in the vicinity of $T = T_C$?

**11.5** At what temperature would you expect a trapped gas of rubidium atoms with density $10^{19}\ m^{-3}$ to show signs of Bose–Einstein condensation?

**11.6** If the mass of a Cooper pair is twice that of an electron, estimate their density in a material for which superconductivity sets in at 10 K.

# 12

# Fermion Gas

Half integer spin particles are known as fermions in honour of Enrico Fermi (1901–1954). At low temperature or high density, we might naturally expect the properties of a gas of noninteracting fermions to differ from those of a classical gas, and we explore these quantum effects in this chapter.

We have made the point already that the classical gas model is founded on the assumption that the vast majority of contributions to the partition function arise from microstates where there is no multiple occupancy of the same single particle states. But fermions satisfy the Pauli Exclusion Principle, and we might wonder why the classical model developed in Section 9.5 does not apply to a fermion gas in the quantum regime. The problem is that the partition function for distinguishable particles $Z_N^{\rm dist} = Z_1^N$, upon which the classical model of a gas is based, where $Z_1$ is the partition function of a single particle in the box, implicitly contains terms where the particles occupy the same single particle state; so it will not do. We need to develop a more appropriate model, along lines similar to those developed for bosons in Chapter 11.

## 12.1 Grand Partition Function for Fermions in a Single Particle State

Once again, we construct a grand canonical ensemble for a system consisting of a single particle state at a specified energy. The grand partition function is in general

$$Z_G(\mu, T) = \sum_{\text{microstates } i} \exp\left(-\frac{(E_i - \mu N_i)}{kT}\right), \tag{12.1}$$

where $\mu$ and $T$ are parameters of the reservoir. For a single particle state at energy $\epsilon$, the microstate energy is given by $E_i = N_i\epsilon$ as for bosons. The occupancy of the state $N_i$, however, is now strictly zero or unity, according to the Pauli Exclusion Principle, and so

$$Z_G^\epsilon = \sum_{N_i=0}^{1} \exp\left(\frac{(\mu - \epsilon)N_i}{kT}\right) = 1 + \exp\left(\frac{\mu - \epsilon}{kT}\right). \tag{12.2}$$

*Statistical Physics: An Entropic Approach*, First Edition. Ian Ford.

This allows us to calculate the statistical properties of the single particle state, and by extension, the entire fermion gas.

## 12.2 Fermi–Dirac Statistics

We proceed as we did for bosons in Section 11.2. We calculate the mean population, or occupation number, for the single particle state as a function of reservoir chemical potential and temperature:

$$\langle N \rangle_\epsilon = kT\left(\frac{\partial \ln Z_G^\epsilon}{\partial \mu}\right)_{T,V} = \frac{\exp\left(\frac{\mu-\epsilon}{kT}\right)}{1+\exp\left(\frac{\mu-\epsilon}{kT}\right)} = \frac{1}{\exp\left(\frac{\epsilon-\mu}{kT}\right)+1}. \quad (12.3)$$

This result is known as Fermi–Dirac statistics or the Fermi–Dirac function, in recognition of the seminal contributions of Paul Dirac (1902–1984) to the theory of fermions. The difference in sign in the denominator with respect to the Bose–Einstein version in (11.3) is extremely important.

The occupation number as a function of state energy is illustrated in Figure 12.1. On the left, we also see the variation in mean population as $\mu$ is increased. The situation for $\mu = -kT$ is not too dissimilar to the Bose–Einstein occupation number for the same conditions in Figure 11.1. However, there is now no reason why $\mu$ cannot exceed zero, and for cases $\mu = 5kT$ and $10kT$, we see the progressive filling of the single particle states up to the capacity of unity set by the Pauli Exclusion Principle. The single particle state that has a mean population of 1/2 has an energy equal to the imposed chemical potential.

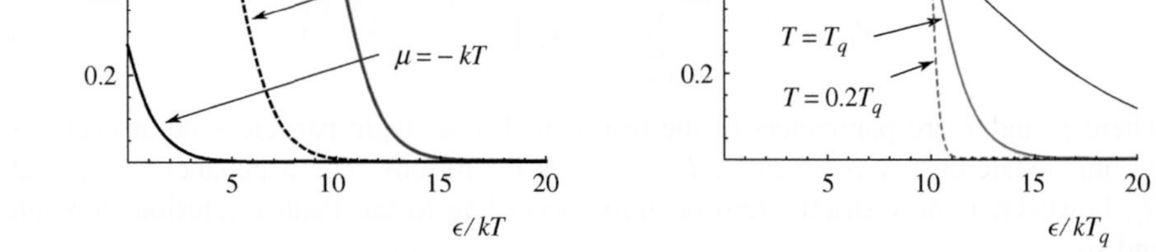

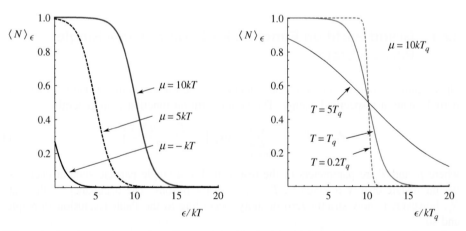


**Figure 12.1** Fermi–Dirac occupation numbers for single particle states at energies $\epsilon$ in contact with a reservoir at chemical potential $\mu$ and temperature $T$. The variation with $\mu$ at constant $T$ is shown on the left, and the variation with $T$ at constant $\mu$ on the right. A mean occupation number of $\frac{1}{2}$ prevails at an energy equal to the reservoir chemical potential. The occupation number varies between zero and unity over a range of a few $kT$ about $\epsilon = \mu$.

On the right in Figure 12.1, we explore the dependence of $\langle N\rangle_\epsilon$ on temperature at a constant chemical potential of $\mu = 10kT_q$, where $T_q$ is given in (11.22). The mean populations at $T = 5T_q$, $T_q$ and $(1/5)T_q$ illustrate that as the temperature is reduced, the state energy range over which the variation from unity to zero takes place becomes narrower. In the limit $T \to 0$, the occupation number takes the form of a step function, equal to unity for $\epsilon < \mu$ and zero for $\epsilon > \mu$. This behaviour is analogous to the manner in which electrons occupy atomic energy levels up to a certain point, leaving higher energy levels empty. The electrons are prevented from congregating in the lowest energy atomic state by the Pauli Exclusion Principle, and the same rule applies for fermions in a box. The microscopic state of the fermion gas at $T = 0$ is thus quite different from the case of a boson gas, and this brings about distinctive nonclassical thermodynamic behaviour that we now investigate.

## 12.3 Thermal Properties of a Fermion Gas

The properties of the fermion gas are deduced by placing all the single particle states, described by the density of states in energy $g(\epsilon)$ as given in (9.11), in contact with the same heat and particle reservoir. As in (11.9) the grand partition function is written as an integral:

$$\ln Z_G(\mu, T) \approx \int_0^\infty g(\epsilon) \ln Z_G^\epsilon \, \mathrm{d}\epsilon = \int_0^\infty g(\epsilon) \ln\left(1 + \exp\left(\frac{\mu - \epsilon}{kT}\right)\right) \mathrm{d}\epsilon, \tag{12.4}$$

using $Z_G^\epsilon$ from (12.2), and the statistical properties of the gas then follow by differentiation. The mean total number of fermions in the system is

$$\langle N\rangle = kT\left(\frac{\partial \ln Z_G}{\partial \mu}\right)_{T,V} = \int_0^\infty g(\epsilon)\left(\frac{\partial \ln Z_G^\epsilon}{\partial \mu}\right)_{T,V} \mathrm{d}\epsilon = \int_0^\infty g(\epsilon)\langle N\rangle_\epsilon \mathrm{d}\epsilon, \tag{12.5}$$

where $\langle N\rangle_\epsilon$ is given by (12.3), corresponding to the intuitive requirement that the mean number of particles in the gas is a sum of mean populations in all the single particle states. The particle density for spin $s$ and mass $m$ is then written as

$$n(\mu, T) = \frac{\langle N\rangle}{V} = \frac{(2s+1)}{(2\pi)^2}\left(\frac{2m}{\hbar^2}\right)^{\frac{3}{2}} \int_0^\infty \frac{\epsilon^{\frac{1}{2}} \mathrm{d}\epsilon}{\exp\left(\frac{\epsilon-\mu}{kT}\right) + 1}, \tag{12.6}$$

which differs from (11.15) only by the replacement of a minus sign by a plus sign in the denominator of the integrand. A nice way to remember this is that the minus sign is cut in *half* to make a plus sign, when the particles in question have a *half* integer spin.

The thermodynamic properties of the fermion gas can be derived in just the same way that we employed for bosons in Section 11.3. In particular, we find that the mean energy may be written as

$$\langle E\rangle = \int_0^\infty g(\epsilon)\epsilon\langle N\rangle_\epsilon \mathrm{d}\epsilon, \tag{12.7}$$

such that the energy density is

$$e(\mu, T) = \frac{\langle E\rangle}{V} = \frac{(2s+1)}{(2\pi)^2}\left(\frac{2m}{\hbar^2}\right)^{\frac{3}{2}} \int_0^\infty \frac{\epsilon^{\frac{3}{2}} \mathrm{d}\epsilon}{\exp\left(\frac{\epsilon-\mu}{kT}\right) + 1}, \tag{12.8}$$

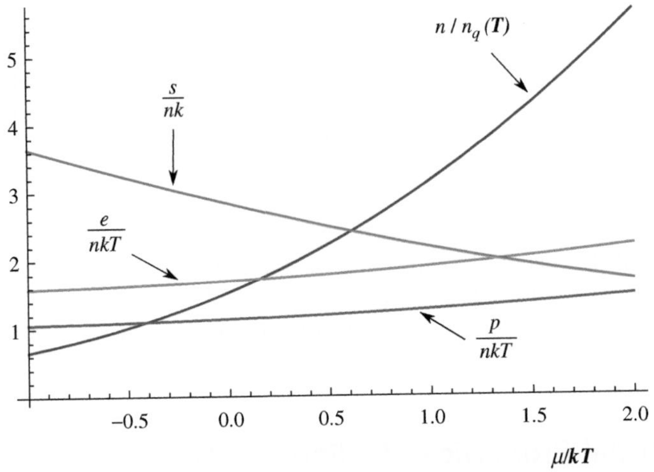


**Figure 12.2** Properties of a gas of spin half fermions as a function of chemical potential at a given temperature. Quantum deviations from classical properties $e_{\rm cl} = (3/2)nkT$ and $p_{\rm cl} = nkT$ appear when $n \gtrsim n_q(T)$.

which again differs only slightly with respect to (11.17). The entropy of the gas is

$$S_{\rm G} = \frac{1}{T}\int_0^\infty g(\epsilon)\left(\frac{\epsilon-\mu}{\exp\left(\frac{\epsilon-\mu}{kT}\right)+1} + kT\ln\left[1+\exp\left(\frac{\mu-\epsilon}{kT}\right)\right]\right)\mathrm{d}\epsilon, \tag{12.9}$$

in contrast to (11.20), and finally the pressure is given by

$$p = \frac{kT}{V}\int_0^\infty g(\epsilon)\ln Z_G^\epsilon\,\mathrm{d}\epsilon = \frac{(2s+1)}{(2\pi)^2}\left(\frac{2m}{\hbar^2}\right)^{\frac{3}{2}} kT\int_0^\infty \epsilon^{\frac{1}{2}}\ln\left[1+\exp\left(\frac{\mu-\epsilon}{kT}\right)\right]\mathrm{d}\epsilon, \tag{12.10}$$

which is to be compared with (11.24). We can develop (12.10) to establish a link between the pressure and the energy density. We write

$$\begin{aligned} p &= \frac{(2s+1)kT}{(2\pi)^2}\left(\frac{2m}{\hbar^2}\right)^{\frac{3}{2}}\left(\left[\frac{2}{3}\epsilon^{\frac{3}{2}}\ln\left(1+\exp\left(\frac{\mu-\epsilon}{kT}\right)\right)\right]_0^\infty \right.\\ &\quad\left. + \frac{2}{3kT}\int_0^\infty \frac{\epsilon^{\frac{3}{2}}\mathrm{d}\epsilon}{\exp\left(\frac{\epsilon-\mu}{kT}\right)+1}\right)\\ &= \frac{(2s+1)kT}{(2\pi)^2}\left(\frac{2m}{\hbar^2}\right)^{\frac{3}{2}}\frac{2}{3kT}\int_0^\infty \frac{\epsilon^{\frac{3}{2}}\mathrm{d}\epsilon}{\exp\left(\frac{\epsilon-\mu}{kT}\right)+1} = \frac{2}{3}\frac{\langle E\rangle}{V}, \end{aligned} \tag{12.11}$$

showing that relation (2.5) applies to the gas in both classical and quantum regimes.

The classical limits of these expressions are the same as those of the boson gas obtained in Section 11.3. However, in the nonclassical regime, the deviations from classical behaviour are rather different. We reiterate that unlike the boson gas, there is no

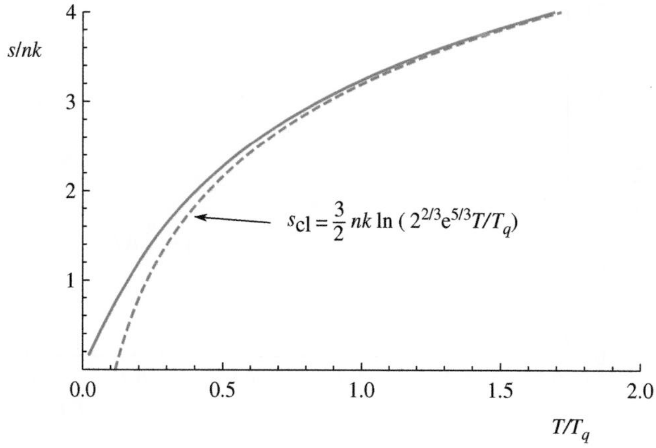


**Figure 12.3** Entropy per particle $s/n$ of a gas of spin half fermions against temperature, compared with the classical Sackur–Tetrode expression.

apparent upper limit to the chemical potential, and therefore no reason for an abrupt change in behaviour analogous to Bose–Einstein condensation. As a function of particle density at a constant temperature, the properties of the gas are illustrated in Figure 12.2 for spin half particles, which should be compared with those of the boson gas in Figure 11.2. The key conclusions are that the mean energy and pressure *exceed* classical expectations when $n \gtrsim n_q(T)$. In Figure 12.3 we see that the entropy per particle tends to zero as the temperature is increased at constant density. In contrast, the Sackur–Tetrode expression $s_{\mathrm{cl}} = nk \ln((2s+1)e^{5/2}n_q/n) = nk \ln((2s+1)e^{5/2}(T/T_q)^{3/2})$ becomes negative for sufficiently low temperature, which is obviously unphysical.

By combining $e(\mu, V, T)$and $n(\mu, V, T)$ to give the energy density in terms of particle density, $e(n, V, T)$, we can contrast the low temperature gas behaviour at constant density with the classical result $e_{\mathrm{cl}} = (3/2)nkT$ in Figure 12.4. The gradient $(\partial e/\partial T)_{n,V}$ provides the heat capacity per unit volume, and clearly this is suppressed with respect to classical gas behaviour when $T \lesssim T_q$. This result provides crucial experimental support for the approach, as it explains the missing heat capacity of electrons in metals.

## 12.4 Maxwell–Boltzmann Statistics

It will have been noticed that the wavefunction symmetry rules that apply to gases of bosons and fermions, and which are responsible for their distinctive nonclassical physical behaviour, give rise to expressions for the mean particle population in the single particle states that differ only slightly. These are the key results of Bose–Einstein and Fermi–Dirac statistics, respectively. The mean population is

$$\langle N \rangle_\epsilon = \frac{1}{\exp\left(\frac{\epsilon-\mu}{kT}\right) \pm 1}, \tag{12.12}$$

where the minus sign applies to bosons and the plus sign to fermions. We have already explored the classical limit of the properties of these gases, starting with (11.16) for

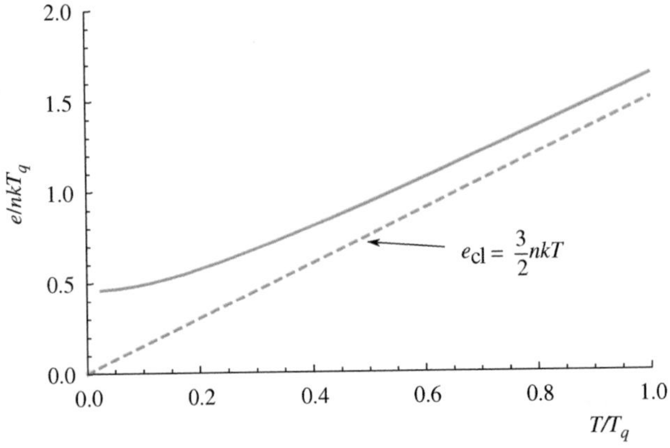


**Figure 12.4** Energy per particle $e/n$ of a gas of spin half fermions against temperature, compared with the classical behaviour. As $T \to 0$, the gradient of the curve, and hence the heat capacity at constant volume, goes to zero.

bosons in Section 11.3, and continuing with fermions in Section 12.3. Both forms of (12.12) coalesce into a classical version that goes by the name of Maxwell–Boltzmann statistics, and we shall use this to derive a connection between the exact formulation of the grand partition function of ideal quantum gases, presented here and in Chapter 11, and the approximate canonical partition function constructed for classical conditions in Section 9.5.

Recall that the classical limit corresponds to conditions where the probability of occupancy of single particle states is very small, namely $\langle N \rangle_\epsilon \ll 1$. This is equivalent to $(\epsilon - \mu)/kT \gg 1$, implying that the chemical potential of the reservoir should lie below the lowest energy single particle state by a considerable multiple of $kT$. There are indications in Figures 11.1 and 12.1 that when $\mu$ is very negative, the occupation number of all states is small. Figure 12.5 is a sketch of the chemical potential of boson and fermion

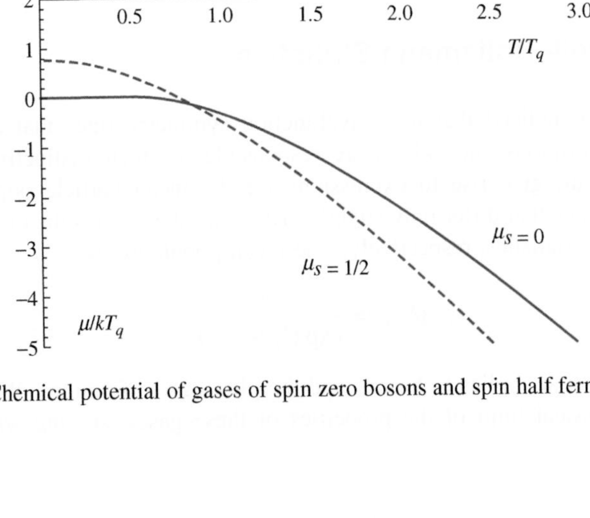


**Figure 12.5** Chemical potential of gases of spin zero bosons and spin half fermions as a function of temperature.

gases as a function of temperature, at constant particle density, that illustrates the same point. It is a generalisation of Figure 2.9, but with low temperature limits included on the basis of (11.15) and (12.6). Clearly a very negative chemical potential corresponds to high temperatures.

In these circumstances, both forms of the occupation number go over to

$$\langle N \rangle_\epsilon \approx \exp\left(-\frac{(\epsilon-\mu)}{kT}\right), \tag{12.13}$$

which is known as Maxwell–Boltzmann statistics. The distinction between boson and fermion is lost, because if there is a low probability of occupancy of single particle states, the implications of wavefunction symmetry with regard to *multiple* occupancy are not important. Maxwell–Boltzmann statistics as a function of single particle state energy $\epsilon$ are illustrated in Figure 12.6, alongside Bose–Einstein and Fermi–Dirac statistics for the same chemical potential and temperature.

The convergence of the expressions for the grand partition function of boson and fermion gases in the classical limit can also be explored. We have

$$Z_G^\epsilon = \left(\frac{1}{1 \mp \exp\left(\frac{\mu-\epsilon}{kT}\right)}\right)^{\pm 1} \approx 1 + \exp\left(-\frac{(\epsilon-\mu)}{kT}\right), \tag{12.14}$$

when $(\epsilon-\mu)/kT \gg 1$ or $\exp\left[(\mu-\epsilon)/kT\right] \ll 1$, and so

$$\ln Z_G^\epsilon \approx \ln\left[1 + \exp\left(-\frac{(\epsilon-\mu)}{kT}\right)\right] \approx \exp\left(-\frac{(\epsilon-\mu)}{kT}\right), \tag{12.15}$$

and

$$\ln Z_G = \int_0^\infty g(\epsilon)\exp\left(-\frac{(\epsilon-\mu)}{kT}\right) d\epsilon. \tag{12.16}$$

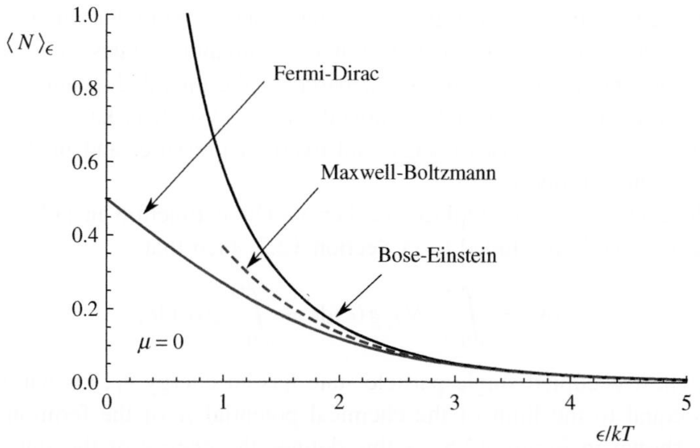


**Figure 12.6** Maxwell–Boltzmann statistics, indicating the mean occupation number of states of energy $\epsilon$ in contact with a reservoir at chemical potential $\mu$ (chosen in this example to be zero) and temperature $T$, valid for particles of any spin when the occupancy is small. The more general Fermi–Dirac and Bose–Einstein statistics are shown for comparison.

We note that the right hand side is proportional to the one-particle canonical partition function for the gas derived in Section 9.3. Thus $\ln Z_G = \exp(\mu/kT)Z_1$ or equivalently

$$Z_G = \exp\left(\exp\left(\frac{\mu}{kT}\right)Z_1\right) = \sum_{N=0}^{\infty} \exp\left(\frac{N\mu}{kT}\right)\frac{Z_1^N}{N!}. \tag{12.17}$$

But recalling from (7.9) that $Z_G = \sum_0^\infty Z_N \exp(N\mu/kT)$, we can read off the canonical partition function for the classical gas of $N$ particles, namely

$$Z_N = \frac{Z_1^N}{N!}, \tag{12.18}$$

just as we constructed in (9.19).

This result underlines the point that the development of gas properties using the grand partition function automatically takes indistinguishability into account: it has reproduced the $1/N!$ correction factor for a classical gas. This is due to its foundation upon populations of single particle states and not on the circumstances of individually labelled particles.

## 12.5 The Degenerate Fermion Gas

As noted in Section 10.3, we have ready access to fermion gases that exceed their quantum concentration, and so it is to the properties of such gases in the condition $n \gg n_q$ that we turn next.

In the extremes of high density or low temperature, the single particle states of the fermion gas system become occupied with probability unity up to a maximum single particle state energy, and states above this level are empty. In this condition, the fermion gas is described as *degenerate*. The meaning of the word in this context is that the particles have sunk into the lowest available energy states. Note that this is distinct from the use of 'degeneracy' to indicate the number of quantum states of a system at the same energy. Degenerate in the sense of 'debased' is the intended meaning. The entropy of the gas is zero, as the microstate assumed by the system is known with certainty. However, the gas still possesses energy and exerts a pressure, and in this section we shall determine these properties.

For the degenerate gas, we replace the Fermi–Dirac function in (12.5) with a step function, the $T \to 0$ limit alluded to in Section 12.2, such that

$$\langle N \rangle = \int_0^\infty \langle N \rangle_\epsilon g(\epsilon) d\epsilon = \int_0^{\epsilon_F} g(\epsilon) d\epsilon, \tag{12.19}$$

where the highest occupied single particle state has an energy $\epsilon_F$, known as the *Fermi energy*. It is equal to the limit of the chemical potential $\mu$ of the fermion gas at zero temperature, shown in Figure 12.5, as this defines the energy of the state that is half filled. Inserting the usual density of states, we get

$$\langle N \rangle = \frac{(2s+1)V}{(2\pi)^2}\left(\frac{2m}{\hbar^2}\right)^{\frac{3}{2}} \int_0^{\epsilon_F} \epsilon^{\frac{1}{2}} d\epsilon, \tag{12.20}$$

and if we specify a spin of one half, we obtain a mean particle density

$$n = \frac{\langle N \rangle}{V} = \frac{2}{(2\pi)^2}\left(\frac{2m}{\hbar^2}\right)^{\frac{3}{2}} \frac{2\epsilon_F^{3/2}}{3}, \tag{12.21}$$

or

$$\epsilon_F = \frac{\hbar^2}{2m}(3\pi^2 n)^{\frac{2}{3}}. \tag{12.22}$$

The Fermi energy is related to the temperature below which quantum effects become substantial. We define the Fermi temperature $T_F = \epsilon_F/k$ and then write

$$T_F n_q^{2/3} = \frac{\hbar^2}{2mk}(3\pi^2 n)^{\frac{2}{3}} \frac{2\pi mkT}{h^2} = \left(\frac{3\pi^{\frac{1}{2}}}{8}\right)^{\frac{2}{3}} T n^{\frac{2}{3}} \quad \Rightarrow \frac{T_F}{T} = \left(\frac{3\pi^{\frac{1}{2}} n}{8n_q}\right)^{\frac{2}{3}}. \tag{12.23}$$

Hence the condition for degeneracy $n \gg n_q$ is equivalent to $T \ll T_F$, and indeed the expression for the Fermi energy has a form very similar to that of the Bose–Einstein condensation temperature $T_C$ defined in (11.29) and the temperature $T_q$ given in (11.22); in fact $T_F = (3\pi^{1/2}/8)^{2/3} T_q \approx 0.76 T_q$.

Similarly, we can determine the total energy of the degenerate fermion gas as

$$\langle E \rangle = \frac{(2s+1)V}{(2\pi)^2}\left(\frac{2m}{\hbar^2}\right)^{\frac{3}{2}} \int_0^{\epsilon_F} \epsilon^{\frac{3}{2}} d\epsilon, \tag{12.24}$$

and together with (12.21) this implies

$$\frac{\langle E \rangle}{\langle N \rangle} = \frac{2\epsilon_F^{5/2}/5}{2\epsilon_F^{3/2}/3} = \frac{3\epsilon_F}{5}, \tag{12.25}$$

such that the mean energy per fermion is $3\epsilon_F/5$. The pressure of the degenerate gas is $p = 2\langle E \rangle/(3V)$ according to (12.11), so

$$p = \frac{2}{3}\frac{\langle E \rangle}{\langle N \rangle} n = \frac{2}{5} n \epsilon_F, \tag{12.26}$$

and a sketch of its temperature dependence is given in Figure 12.7. The pressure does not go to zero as $T \to 0$. Let us now apply these results to electrons in metals, the prime example of a degenerate fermion gas.

## 12.6 Electron Gas in Metals

In a metal, the electrons in the outermost shell of the constituent atoms are only loosely bound, and are able to wander around the material leaving behind a positively charged ion. These electrons can be roughly considered to form a gas of noninteracting particles of spin 1/2: the free electron or Drude model mentioned in Section 10.3. We ignore the fact that charged electrons ought to interact strongly with one another and with the ions; these forces can be neglected to a first approximation because of electrostatic screening

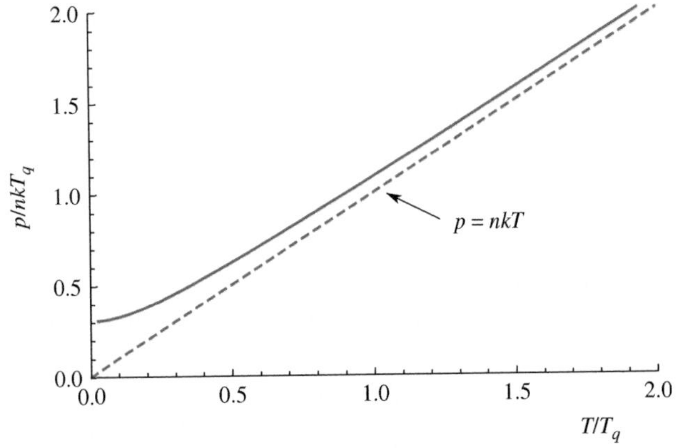


**Figure 12.7** Pressure of a gas of spin half fermions at constant density as a function of temperature, compared with the classical ideal gas law. As $T \to 0$ the pressure tends towards the degeneracy pressure $p = (2/5)n\epsilon_F$.

effects. The ability of the electrons to move around explains why metals are able to conduct electricity and heat so well.

The concentration of conduction electrons $n$ is about $10^{28}\,\mathrm{m}^{-3}$, and so from (12.22) the Fermi temperature $T_F = \epsilon_F/k$ is of order $10^4$ K. Thus at room temperature, $T \ll T_F$ and the electron gas is highly degenerate. Almost all electrons are in states from which no thermally driven transitions into empty states can easily be made: they are locked into their lattice points in **k**-space. Nevertheless, the electrons are free to drift in a collective manner under the influence of an electric field, and hence are able to conduct electricity.

The mean energy per electron is $(3/5)\epsilon_F$, equal to about 10 eV for the conditions given. This should be compared with the expected classical energy $(3/2)kT$ of around 25 meV at room temperature, and so the electrons are typically moving much more rapidly than particles of a classical gas at the same temperature. The pressure of the gas is $(2/5)n\epsilon_F$ which is approximately equal to the classical pressure of the gas at the Fermi temperature, namely $10^9$ Pa or $10^4$ atmospheres. This enormous electron pressure makes an important repulsive contribution to the cohesion of metals, balancing the electrostatic attraction between ions. While the energy and pressure of the gas are considerably greater than values expected of a classical gas at the same density, the entropy can be taken to be approximately zero. All these features are apparent from Figures 12.3, 12.4 and 12.7.

The typical electronic energy is nevertheless well below its rest mass energy of about 511 keV. This confirms that there is no inconsistency in using a nonrelativistic expression for the energy of an electron $\epsilon = \hbar^2k^2/(2m_e)$, where $m_e$ is the mass of the electron, in the derivation of the density of states $g(\epsilon)$ on which the model is built. However, if we imagined increasing the density of an electron gas, such an inconsistency would ultimately emerge since the Fermi energy and the mean energy per particle increase in proportion to $n^{2/3}$. In the next section we consider a situation where a relativistic treatment of the electrons would be more appropriate, and we shall see that such an extreme situation plays a role in the loss of stability of stars.

## 12.7 White Dwarfs and the Chandrasekhar Limit

The properties of the fermion gas allow us to understand the mechanical stability, or lack of it, of elderly stars. Leaving aside much of the complex physics of such systems, the issue can be understood as a balance between gravitational collapse and the resistance to compression of the gaseous core of the star.

Let us approach the question of stability in the following simplified way. The pressure at a depth $z$ below the surface of the *sea* is given by $\rho_w g z$, where $\rho_w$ is the mass density of water and $g$ is the gravitational acceleration, corresponding to an increase of about one atmosphere for every ten metres descent and where we have neglected atmospheric pressure. Let us estimate the pressure at the centre of a star of mass $M_s$ in a similar way. Replacing $g$ by the local acceleration towards the stellar centre according to Newtonian gravitation, namely $GM(r)/r^2$, where $G$ is the gravitational constant, $r$ is the distance to the centre of the star and $M(r)$ is the mass of the star contained within the radius $r$, we readily find that the difference in pressure between the centre of the star at $r = 0$ and its surface at $r = R$ is

$$p(0) - p(R) = \Delta p = \int_0^R \frac{GM(r)}{r^2} \rho_s(r) \mathrm{d}r, \tag{12.27}$$

where $\rho_s$ is the mass density of stellar material. We wish to obtain a rough estimate of $\Delta p$ and therefore assume the density to be a constant; then we can write $M(r) = M_s(r/R)^3$. Thus $\Delta p \sim G\rho_s M_s R^{-3} \int_0^R r\mathrm{d}r = G\rho_s M_s/(2R)$. Inserting $\rho_s = 3M_s/(4\pi R^3)$, we find the pressure at the centre of a star of mass $M_s$ and radius $R$ to be

$$p_G \sim \frac{3GM_s^2}{8\pi R^4}, \tag{12.28}$$

since $p(R) \sim 0$. Applying this to the sun with $R \sim 7 \times 10^8$ m and $M_s \sim 2 \times 10^{30}$ kg, we estimate the core pressure due to gravitational attraction to be $10^{14}$ Pa, using $G \approx 6.7 \times 10^{-11}\,\mathrm{m^3\,kg^{-1}s^{-2}}$. Better models of the sun might be more accurate but we are interested only in the rough dependence of the core pressure on the mass and radius of the star.

We assume that the gravitational pressure is balanced by the pressure of the hot gases of electrons and nuclei that are to be found in the centre of a star. Presuming that most of the mass of the star is carried by hydrogen nuclei, and continuing to assume that the stellar density is constant throughout the star, we estimate that the particle density in the star is $n_s \sim 3(M_s/m_p)/(4\pi R^3)$, where $m_p$ is the proton mass. For the sun, this is about $10^{30}\,\mathrm{m^{-3}}$, and there will be roughly equal numbers of electrons and protons. Assuming each species has the classical thermodynamic properties of an ideal gas, the pressure at the core would be $p_s = n_s k T_s \sim 10^7 T_s$ Pa. In order for this to balance the gravitational pressure, the core temperature $T_s$ needs to be of order $10^7$ K, and this is maintained by the process of nuclear fusion. The mean energy per particle is $(3/2)kT \sim 1$ keV, which is nonrelativistic, as it is much less than either the electron mass (511 keV) or the proton mass (938 MeV). Furthermore, the quantum concentration for the lightest available particle species, the electron, is

$$n_q = \left(\frac{2\pi m_e k T_s}{h^2}\right)^{\frac{3}{2}} \sim 10^{32}\,\mathrm{m^{-3}}, \tag{12.29}$$

and this is safely higher than the estimated particle density $n_s$. The mechanical balance at the centre of the star is quite adequately explained using the properties of classical ideal gases.

But when a star begins to run out of fusable nuclei, the supply of heat to maintain the core temperature and pressure fails, and the star begins to change. The sequence of events is rather complex, but we shall address a particular issue: what pressure might ultimately resist gravitational collapse if the core begins to cool? The quantum properties of the electron gas play a crucial role. As the core contracts, the particle density increases in proportion to $R^{-3}$. Without estimating how the electronic quantum concentration $n_q$ might change as the temperature falls, it is clear that a reduction in stellar radius will raise the density of electrons in the star until it enters the regime where degenerate gas properties take over from classical behaviour. Under these circumstances, the gas pressure is given by $p_s \approx (2/5)n_s\epsilon_F \propto n_s^{5/3}$ using (12.22). Thus the resistance to compression increases, until it can balance gravitational forces, in which case an equilibrium is established at a radius $R_d$ given by

$$p_s = \frac{2}{5}\frac{\hbar^2}{2m_e}(3\pi^2)^{\frac{2}{3}}n_s^{5/3} = \frac{\hbar^2}{5m_e}(3\pi^2)^{\frac{2}{3}}\left[\frac{3\left(M_s/m_p\right)}{4\pi R_d^3}\right]^{\frac{5}{3}} = \frac{3GM_s^2}{8\pi R_d^4}, \quad (12.30)$$

which leads to

$$R_d = \frac{2}{5}\left(\frac{9\pi}{4}\right)^{\frac{2}{3}}\frac{\hbar^2}{GM_s^{\frac{1}{3}}m_p^{\frac{5}{3}}m_e}. \quad (12.31)$$

If we insert values of the parameters, then the equilibrium radius for a stellar mass equal to that of the sun is $7 \times 10^6$ m. This is about 1% of the present radius of the sun! Consequently, the particle density $n_s$ in this star is about six orders of magnitude higher than that of the sun, or about $10^{36}$ m$^{-3}$.

Considering its quantum concentration estimated in (12.29), the electron gas in such a star is degenerate, as required. The gases of nuclei, on the other hand, have a quantum concentration that is approximately $(m_p/m_e)^{3/2} \sim 10^5$ higher, or about $10^{37}$ m$^{-3}$, and they remain approximately classical in behaviour. The burden of resisting gravity is carried by the electron gas alone. From a consideration of (12.28) the pressure at the centre of the star is eight orders of magnitude higher than at the centre of the sun, reaching the extraordinary value of $10^{22}$ Pa or $10^{17}$ atmospheres. The density of the star $\rho_s$ is $3M_s/(4\pi R_d^3)$ or about $10^9$ kg m$^{-3}$, about a million times denser than ordinary terrestrial matter.

Such a star is stable since the resistance to compression $p_s \propto R^{-5}$ increases more rapidly than the gravitational pressure $p_G \propto R^{-4}$ if the radius should fall slightly below $R_d$. The resistance derives from the Pauli Exclusion Principle and the requirement that electrons should avoid one another for reasons of wavefunction asymmetry, a matter alluded to in question 10.2. It is a mighty consequence of quantum statistical thermodynamics.

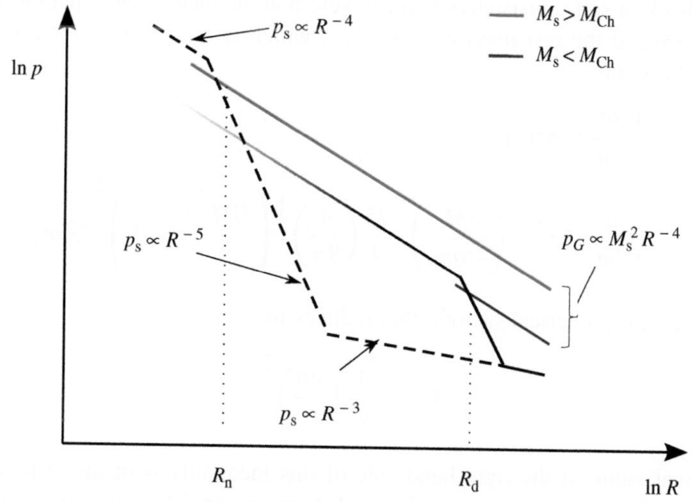


**Figure 12.8** A star of mass $M_s$ and radius $R$ that has exhausted its supply of fuel is stabilised by a balance between compressive pressure $p_G$ brought about by gravity, and resistive pressure $p_s$ of components of the stellar material. The electron gas becomes degenerate before the nucleon gas (formed from protons and ultimately neutrons) since the quantum concentration is proportional to particle mass. The electron pressure (black line) then rises rapidly on compression until it becomes equal to $p_G$ (blue line) at the radius $R_d$, stabilising a white dwarf. A star that exceeds the Chandrasekhar mass $M_{Ch}$, however, exerts too great a gravitational pressure to be balanced by this mechanism, since the electron gas softens at great compressions because of relativistic effects. For such cases, a balance between gravitational pressure (red line) and the pressure of the neutron gas (dashed line) in a nonrelativistic degenerate condition is possible at a radius $R_n$, producing an extremely highly compressed neutron star.

The star remains hot enough to radiate for some considerable time, indeed of the order of the age of the universe, but with little fusion going on at its centre, its final fate is to become a cold, stable stellar remnant. The balance between pressures is illustrated in Figure 12.8.

This sequence of events explains very well the existence of a peculiar class of stars, the *white dwarfs*. They are extraordinarily small in size, and emit dimly but with a white colouration. These features match very well the properties of the stellar remnant we have just described. It is estimated that almost all stars will eventually become white dwarfs after they exhaust their fuel supply. As they cool down and emit less strongly, they would evolve into remnants called black dwarfs, for obvious reasons.

But there is a flaw in this story, and it involves the assumption that the electrons in the degenerate core of the star are characterised by nonrelativistic energies. This gave us the density of single particle states $g(\epsilon) \propto \epsilon^{1/2}$ in (9.11) and then the degeneracy pressure (12.26) that we employed in the balance relation (12.30). So it would be an important check of self-consistency to make sure that the mean energy per electron in the star did not exceed the rest mass energy $m_e c^2$, which is the criterion for nonrelativistic behaviour. We write

$$\frac{3}{5}\epsilon_{\mathrm{F}} = \frac{3}{5}\frac{\hbar^2}{2m_e}(3\pi^2 n_{\mathrm{s}})^{\frac{2}{3}}$$

$$= \frac{3}{5}\frac{\hbar^2}{2m_e}(3\pi^2)^{\frac{2}{3}}\left(\frac{3M_{\mathrm{s}}}{4\pi m_p}\right)^{\frac{2}{3}}\frac{25}{4}\left(\frac{4}{9\pi}\right)^{\frac{4}{3}}\left(\frac{GM_{\mathrm{s}}^{\frac{1}{3}} m_p^{\frac{5}{3}} m_e}{\hbar^2}\right)^2 \lesssim m_e c^2, \quad (12.32)$$

and neglecting all numerical factors, this reduces to

$$M_{\mathrm{s}} \lesssim \frac{1}{m_p^2}\left(\frac{c\hbar}{G}\right)^{\frac{3}{2}}. \quad (12.33)$$

The numerical value of the right hand side of this inequality is of the order of $10^{30}$ kg, in other words, around the mass of the sun! A more careful analysis suggests that the upper limit on the mass of a star that could support itself as a white dwarf according to the model described above is about $2.9 \times 10^{30}$ kg, or about 1.44 times the solar mass.

This limitation of the model is rather serious. If we were to repeat the analysis of the fermion gas but this time assume that the energies of the single particle states were highly relativistic, such that the relationship between single particle energy and wavevector were $\epsilon = \hbar ck$ instead of $\epsilon = \hbar^2 k^2/(2m)$, then we would obtain the density of states $g(\epsilon) \propto \epsilon^2$ instead of $g(\epsilon) \propto \epsilon^{1/2}$ and Fermi energy $\epsilon_{\mathrm{F}} \propto n_{\mathrm{s}}^{1/3}$ rather than $\epsilon_{\mathrm{F}} \propto n_{\mathrm{s}}^{2/3}$. The resistance to gravitational collapse would now depend on particle density according to $p_{\mathrm{s}} \propto n_{\mathrm{s}}^{4/3}$ rather than $p_{\mathrm{s}} \propto n_{\mathrm{s}}^{5/3}$, and on stellar radius according to $p_{\mathrm{s}} \propto R^{-4}$ instead of $p_{\mathrm{s}} \propto R^{-5}$. The degenerate electron gas pressure will increase more slowly as the star is crushed when the particles are relativistic in energy, and in fact it increases at the same rate as the gravitational burden $p_G \propto R^{-4}$. The softening of the electron gas is illustrated in Figure 12.8. This is a crucial point: there is then *no* stable radius at which the opposing pressures will balance.

We conclude that any star with $M_{\mathrm{s}} > M_{\mathrm{Ch}} \approx 1.44 M_{\odot}$, where $M_{\odot}$ is the mass of the sun, would have to look elsewhere for support against gravitational collapse after its nuclear fuel runs out. $M_{\mathrm{Ch}}$ is known as the Chandrasekhar limit, after Subrahmanyan Chandrasekhar (1910–1995). But there are other possibilities to fall back on. An alternative support mechanism is considered in the next section, and with it a new class of stellar remnant.

## 12.8 Neutron Stars

If the star collapses so far that the electron gas has turned relativistic and soft, there is still a possibility that a balance can be established between gravitational pressure and the degeneracy pressure of the nuclei inside the star. These are mostly protons, which are also fermions, and as they are 1800 times more massive than electrons, they remain nonrelativistic at stellar densities beyond that to be found at the Chandrasekhar limit.

We might imagine that a balance between forces is achieved at a stellar radius

$$R_n = \frac{2}{5}\left(\frac{9\pi}{4}\right)^{\frac{2}{3}} \frac{\hbar^2}{GM_s^{\frac{1}{3}} m_p^{\frac{8}{3}}}, \tag{12.34}$$

which is just the equation for the dwarf radius $R_d$ in (12.31) with the electron mass replaced by the proton mass. This is illustrated in Figure 12.8. The radius is 1800 times smaller than the dwarf radius we estimated earlier, or around 4 km. Remember that this contains $10^{30}$ kg of material! The mass density is ten orders of magnitude greater than that of the already crushed white dwarf, or about $10^{19}\,\mathrm{kg\,m^{-3}}$, and the particle density is $10^{47}\,\mathrm{m^{-3}}$, such that the mean particle separation is of order 0.1 fm. An estimate of the pressure would be $10^{35}$ Pa! The entropy, on the other hand, would be essentially zero.

At such an enormous pressure, it becomes favourable for protons to combine with electrons to become neutrons, which are also fermions, and emit a neutrino. The star is converted into a gravitationally bound gas of neutrons at a density similar to that of the atomic nucleus. Such a stellar remnant is seriously astonishing. But is there evidence that they exist? Extraordinarily, the answer is yes: pulsars, extremely compact astronomical objects that emit pulses of radio emissions, are thought to be examples of these so-called *neutron stars*. Possessing a mass beyond the Chandrasekhar limit, these objects can be stabilised only by the degeneracy pressure of neutrons.

But immediately we can see that there is an upper limit to the mass of a neutron star. If the neutron gas goes relativistic, it will in turn become too soft to carry the burden of the gravitational pressure. What happens then is not clear. Ideally, the star would look for a similar way out, and try to form a fermion gas made up of still heavier particles from the stupendously dense nuclear material. We simply do not know what could happen, but we have an idea of another fate that might befall the star if its collapse continues too far. It might disappear behind its event horizon; or more exactly its radius $R$ might fall below the Schwarzschild radius $R_S = 2GM_s/c^2$, which is about 1 km for the parameters considered. After this point gravity at the surface of the stellar remnant becomes so strong that even light is trapped, and the elderly star turns into a black hole.

## 12.9 Entropy of a Black Hole

This is not the place to discuss black holes in detail, as it would require the mathematics of general relativity, but their thermodynamics are very interesting. Black holes possess entropy, and a great deal of it.

Various studies have suggested that the entropy of a black hole can be written as

$$S_{BH} = \frac{kc^3 \mathcal{A}}{4G\hbar}, \tag{12.35}$$

where $\mathcal{A} = 4\pi R_S^2$ is the area of the event horizon of the black hole, defined as the surface of a sphere at the Schwarzschild radius. The label BH can stand for either black hole or for the surnames of Jacob Bekenstein (1947–) and Stephen Hawking (1942–), the main developers of the ideas. The entropy per neutron (presuming that this is the right particle to use) of the hole is $S_{BH}/(M_s/m_p) = 4\pi kGM_s m_p/(c\hbar) = 10^{20}k$. In common with many other aspects of the physics of stellar remnants, this is a huge number, but it is also seriously confusing.

One puzzle arises from the so-called no-hair theorem of black holes, which suggests that a black hole has only a handful of measurable properties, such as mass, angular momentum and charge. How can such an apparently simple physical object possess so much internal uncertainty? The answer is that these gross properties are the macroscopic state variables; all the microscopic detail that was uncertain but in principle measurable just before the star disappeared behind its event horizon is still there, but is now definitively hidden from view.

A more serious puzzle is the following. The neutron material just before the disappearance of the remnant behind the event horizon was an extremely degenerate fermion gas, and we might have been under the impression that its entropy per particle was approximately zero. Where has the black hole entropy come from?

This is where the topic becomes speculative. The Bekenstein–Hawking entropy, if it is indeed a fundamental property of the object behind the event horizon, must be a fine grain entropy related to the number of microstates of the components of whatever the object is made of. We discussed coarse and fine grain entropy in Section 8.4: at the coarse grain level of the neutron gas the entropy is zero. The significance of the black hole entropy is that it might be giving us a count of arrangements at the very lowest level of the universe; way below atoms and quarks. Somehow these features are implicit in having used general relativity to study black hole behaviour and deduce its thermodynamic properties. Models of black holes consistent with general relativity but based on current variants of string theory appear to offer such a vast multiplicity of microstates, with their associated entropy. It could be that the thermodynamics of black holes provides a perspective on the ultimate level of physical reality, although it is a little too soon to be absolutely sure!

## Exercises

**12.1** Three fermions, all with the same spin orientation, are held in a harmonic potential that, with the neglect of zero point energy, has single particle states at energies $\epsilon_n = n\hbar\omega$, where $n = 0$, 1, 2 and so on. Indicate the particle arrangement and energy of the four lowest energy microstates of the system if the particles are (a) distinguishable and (b) indistinguishable. Assuming the canonical partition functions for each case may be approximated at low temperature as a sum over these four microstates, calculate the Helmholtz free energies of each system in terms of $x = \exp(-\hbar\omega/kT)$. Also calculate the mean energies.

**12.2** Show that the chemical potential of a gas of nonrelativistic electrons at a temperature much less than the Fermi temperature is proportional to $n^{2/3}$, where $n$ is the particle density.

**12.3** For a degenerate gas of ultrarelativistic electrons in volume $V$, the density of states $g(E)$ is proportional to $VE^2$. Show that the Fermi energy is proportional to $n^{1/3}$. Hence show that $p \propto n^{4/3}$ for an ultrarelativistic degenerate electron gas.

**12.4** Derive the Fermi energy $\epsilon_F$ of a gas of $N$ electrons confined to a volume $V$ at zero temperature and show that the mean energy per electron in such conditions is $3\epsilon_F/5$. The relationship between the mean energy $E$ and the entropy $S$ of a low temperature electron gas is

$$E = \frac{3}{5}N\epsilon_F\left[1 + \frac{5}{3}\left(\frac{S}{\pi Nk}\right)^2\right].$$

Show that the relationship between temperature and entropy for this system is given by $T = 2\epsilon_F S/(\pi^2 k^2 N)$. Does the gas satisfy the third law of thermodynamics? Express $E$ and $S$ in terms of $T$ to show that the Helmholtz free energy is given by

$$F = \frac{3}{5}N\epsilon_F - \frac{N(\pi kT)^2}{4\epsilon_F}.$$

Hence show that the pressure of the gas is

$$p = \frac{2NkT_F}{5V}\left[1 + \frac{5\pi^2}{12}\left(\frac{T}{T_F}\right)^2\right],$$

where $T_F$ is the Fermi temperature. In a particular metal, the Fermi temperature is 57000 K. Calculate the entropy per electron at $T = 300$ K.

**12.5** A system consists of two levels with energies 0 and $\epsilon$, respectively. Calculate the canonical partition function $Z_1$ of the system when it accommodates one particle and is in contact with a heat bath at temperature $T$. Express your result in terms of the parameter $y = \exp(-\epsilon/kT)$. Calculate the mean energy of the particle. Calculate the canonical partition function $Z_2$ when the system accommodates two indistinguishable fermions, if each level can occupy one particle at most. The system is exposed to an environment that is a source of indistinguishable fermions at a chemical potential $\mu$, and heat at a temperature $T$. Determine the mean number of particles in the system $\langle N\rangle$, in terms of $y$ and $w = \exp(\mu/kT)$.

**12.6** An astronomical object is discovered with an apparent radius of 2.5 km. Assuming it is a stabilised stellar remnant with a mass of that of the sun, deduce the approximate mass of the fermions within it, and their likely nature, assuming there is only one species present.

# 13

# Photon Gas

In this chapter we shall consider the statistical thermodynamics of light. We consider the dynamics of electric and magnetic fields inside a box, and then couple their standing wave modes to a heat bath such that they acquire a canonical equilibrium distribution of energy. Then thermodynamic properties such as energy density, pressure and entropy are derived, and used to explain various aspects of black-body or cavity radiation. It was the resolution of some of these puzzles, particularly by Max Planck (1858–1947), that set off the quantum revolution at the beginning of the twentieth century.

We shall find that electromagnetic radiation behaves rather like a gas of particles, involving quanta of the electromagnetic field known as *photons*. They have zero mass, move at the speed of light and do not interact with each other. On the other hand, they interact strongly with electrically charged matter, to a degree that the walls of a container can be considered to act as a particle reservoir for photons, but at a rather unexpected chemical potential.

## 13.1 Electromagnetic Waves in a Box

We begin this ambitious programme by establishing the classical standing wave modes of electric and magnetic fields inside an evacuated box. The wave equation controls their evolution, for example

$$\frac{\partial^2 \mathbf{E}}{\partial t^2} = c^2 \nabla^2 \mathbf{E}, \tag{13.1}$$

where $\mathbf{E}$ is the electric field and $c$ is the speed of light. Assuming that the box is a cube of side length $l$ and the walls are good conductors, we impose boundary conditions such as $E_y(x=0,t) = E_y(x=l,t) = 0$ together with a similar constraint on the $z$-component. This means we expect a standing wave mode along the $x$-axis with transverse polarisation in the $y$ direction given by $E_y(x,t) = E_{y0}\sin(k_x x)\sin(\omega t)$ where $E_{y0}$ is the maximum electric field amplitude, $k_x = n_x \pi / l$ with $n_x$ a non-negative integer, and $\omega = c k_x$. There are standing wave modes in the other two Cartesian directions, involving the respective components of wavevector $k_y$ and $k_z$, and each has two directions of transverse polarisation.

*Statistical Physics: An Entropic Approach*, First Edition. Ian Ford.

This specification of the independent oscillatory modes of the electric field is very reminiscent of the wavefunction solutions to the Schrödinger equation describing a single particle in a box, considered in Section 9.1. It is important to bear in mind that these modes are classical field oscillations, and not wavefunctions, but nevertheless we shall carry over many of the concepts. Using an argument similar to that employed in Section 9.2, we can deduce that the number of standing wave modes with a magnitude of wavevector in the range $k \to k + \mathrm{d}k$ is

$$\rho(k)\mathrm{d}k = 2\frac{Vk^2}{2\pi^2}\mathrm{d}k, \tag{13.2}$$

in terms of a density of states $\rho(k)$. Notice that instead of the factor of $(2s + 1)$ in (9.9) that denotes the number of spin orientations of a spin $s$ particle, we have a factor of two that counts the number of transverse polarisations associated with a standing wave with a given wavevector.

The corresponding magnetic field must satisfy Maxwell's equations of electromagnetism, particularly

$$\frac{\partial \mathbf{B}}{\partial t} = -\nabla \times \mathbf{E}, \tag{13.3}$$

and so $\partial B_z/\partial t = -\partial E_y/\partial x = -k_x E_{y0} \cos(k_x x) \sin(\omega t)$ giving $B_z(x,t) = B_{z0} \cos(k_x x) \cos(\omega t)$ with $B_{z0} = k_x E_{y0}/\omega = E_{y0}/c$, with similar specifications of a magnetic mode to accompany each of the other five electric field modes. The pattern of the fields is illustrated in Figure 13.1.

It is worth making explicit the correspondence between this behaviour and that of a mechanical oscillator. The energy $E$ of the specified field mode, not to be confused with the amplitude of electric field $E_y$, is given by

$$\begin{aligned} E &= \int \left(\frac{\varepsilon_0}{2}E_y^2 + \frac{1}{2\mu_0}B_z^2\right)\mathrm{d}V \\ &= l^2 \int \mathrm{d}x \left(\frac{\varepsilon_0}{2}E_{y0}^2 \sin^2 k_x x \sin^2 \omega t + \frac{1}{2\mu_0}B_{z0}^2 \cos^2 k_x x \cos^2 \omega t\right) = l^3 \frac{\varepsilon_0}{2}E_{y0}^2, \end{aligned} \tag{13.4}$$

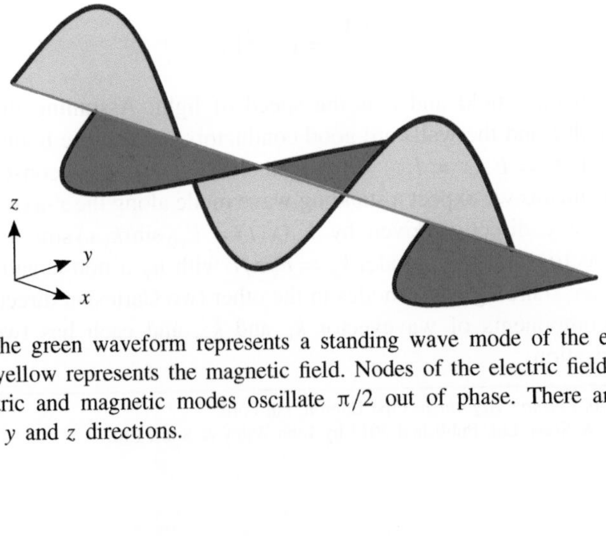


**Figure 13.1** The green waveform represents a standing wave mode of the electric field in a cavity, and the yellow represents the magnetic field. Nodes of the electric field lie at the cavity walls. The electric and magnetic modes oscillate $\pi/2$ out of phase. There are similar modes extending in the $y$ and $z$ directions.

using $c^2 = 1/(\varepsilon_0\mu_0)$, where $\varepsilon_0$ and $\mu_0$ are the permittivity and permeability of vacuum. A mechanical oscillator with mass $m$ and harmonic spring constant $\kappa$ evolves according to $\mathrm{d}p/\mathrm{d}t = -\kappa q$ and $\mathrm{d}q/\mathrm{d}t = p/m$, where $q$ is the displacement and $p$ the momentum, such that $p = p_0 \cos\omega t$ and $q = q_0 \sin\omega t$ where $\omega = (\kappa/m)^{1/2}$, and with amplitudes related by $p_0 = q_0\kappa/\omega$, giving an energy $E = (1/2)\kappa q^2 + (1/2m)p^2 = (1/2)\kappa q_0^2 \sin^2\omega t + (1/2m)p_0^2 \cos^2\omega t = \kappa q_0^2/2$, which is similar in form to (13.4). As a result of this correspondence, we conclude that the electric and magnetic fields *roughly* correspond to the displacement and momentum of the oscillator.

This strongly suggests that the standing wave mode should be treated in quantum mechanics in just the same way as a mechanical oscillator. Its energy cannot take arbitrary values, but only the quantised values $(n + 1/2)\hbar\omega$, where $n$ is a non-negative integer. The quantised electromagnetic field in the box is physically equivalent to a set of harmonic oscillators with specified frequencies. We can therefore use what we know about the statistical thermodynamics of oscillators to establish the thermal properties of electromagnetic radiation.

## 13.2 Partition Function of the Electromagnetic Field

We now construct the canonical partition function of the electromagnetic field. We have a set of oscillatory modes defined by a wavevector $k$ and angular frequency $\omega$. The density of states in $k$ allows us to obtain the corresponding density of states in $\omega$ using the *dispersion relation* $\omega = ck$. We write

$$g(\omega) = \rho(k)\frac{\mathrm{d}k}{\mathrm{d}\omega} = \frac{V\omega^2}{c^3\pi^2}. \tag{13.5}$$

Each standing wave mode of the electromagnetic field is in canonical equilibrium with a heat bath at temperature $T$. The analysis in Section 6.4 then gives us the canonical partition function of the mode:

$$Z_\omega = \frac{1}{2\sinh\left(\frac{\hbar\omega\beta}{2}\right)}. \tag{13.6}$$

The total partition function $Z$ is the product of $Z_\omega$ over all modes, and so

$$\ln Z = \sum_\omega \ln Z_\omega \approx \int_0^\infty g(\omega)\ln Z_\omega \mathrm{d}\omega, \tag{13.7}$$

since $g(\omega)\mathrm{d}\omega$ counts the number of modes in the frequency range $\omega$ to $\omega + \mathrm{d}\omega$. Inserting (11.30), we get

$$\ln Z = \int_0^\infty g(\omega)\ln\left(\frac{1}{2\sinh\left(\frac{\hbar\omega\beta}{2}\right)}\right)\mathrm{d}\omega = \frac{V}{c^3\pi^2}\int_0^\infty \omega^2 \ln\left(\frac{1}{2\sinh\left(\frac{\hbar\omega\beta}{2}\right)}\right)\mathrm{d}\omega, \tag{13.8}$$

and the mean energy of the system is

$$\langle E\rangle = -\frac{\partial \ln Z}{\partial \beta} = -\int_0^\infty g(\omega)\frac{\partial \ln Z_\omega}{\partial \beta}\mathrm{d}\omega = \int_0^\infty \frac{\hbar\omega g(\omega)\mathrm{d}\omega}{2\tanh\left(\frac{\hbar\omega\beta}{2}\right)}$$

$$= \int_0^\infty \hbar\omega\left(\langle N\rangle_\omega + \frac{1}{2}\right)g(\omega)\mathrm{d}\omega = \frac{\hbar V}{c^3\pi^2}\int_0^\infty \omega^3\left(\langle N\rangle_\omega + \frac{1}{2}\right)\mathrm{d}\omega, \quad (13.9)$$

where we have inserted the identity $\coth(x/2) = 2(\exp(x) - 1)^{-1} + 1$ and defined

$$\langle N\rangle_\omega = \frac{1}{\exp(\hbar\omega\beta) - 1}. \quad (13.10)$$

But it should be noticed that there is a serious problem with the mean energy in (13.9): it is infinite! The trouble comes from the term that is proportional to $\int_0^\infty \omega^3\mathrm{d}\omega$. We cannot work with this.

We can readily identify the source of the problem and find a way round. The expression for the mean energy in (13.9) clearly shows that the infinite term is the sum of zero point energies of the quantum oscillators. The problem can therefore be overcome if we set a limit on the number of oscillators that represent the fields in the box. In addition, this is physically sensible: it is unreasonable to suggest that field modes should exist for arbitrarily high frequency and wavevector, or equivalently for arbitrarily small wavelengths. We do not know whether Maxwell's equations apply for length scales that are arbitrarily small.

So we cut off the integral at some high but finite frequency, such that the mean energy is no longer infinite. Furthermore, we take the view that the large but constant zero point energy of the field is not of interest to us in statistical thermodynamics, and simply renormalise the energy scale by subtracting it away. We shall therefore work with

$$\langle E\rangle = \int_0^\infty \hbar\omega\langle N\rangle_\omega g(\omega)\mathrm{d}\omega = \frac{\hbar V}{c^3\pi^2}\int_0^\infty \frac{\omega^3}{\exp(\hbar\omega\beta) - 1}\mathrm{d}\omega, \quad (13.11)$$

or equivalently with

$$\ln Z = \int_0^\infty g(\omega)\ln Z_\omega^R \mathrm{d}\omega = \int_0^\infty g(\omega)\ln\left(\frac{1}{1 - \exp(-\hbar\omega\beta)}\right)\mathrm{d}\omega, \quad (13.12)$$

which is based on the partition function $Z_\omega^R$ of an oscillator for which the zero point energy has been ignored. Its energy levels are given by $n\hbar\omega$, and $Z_\omega^R = \sum_{n=0}^\infty \exp(-n\hbar\omega\beta) = (1 - \exp(-\hbar\omega\beta))^{-1}$, which is to be contrasted with the form derived in (6.23) and used in (13.8). These quantities satisfy $\langle E\rangle = -\partial \ln Z/\partial\beta$ as they should.

The quantity $\langle N\rangle_\omega$ is the mean number of quanta in the mode at frequency $\omega$. Now we arrive at a point where we place a physical interpretation on the model, and in particular on (13.11). The standing wave modes provide an analogue of the single particle states of particles in a box discussed extensively in the last two chapters. The mean energy of the field (13.11) is similar in form to the mean energies of boson and fermion gases in (11.14) and (12.7). It seems that the electromagnetic fields in the box, coupled to a heat bath, behave rather like a gas of particles that can occupy single particle states at energies $\hbar\omega$, with the density of mode frequencies $g(\omega)$ playing the role of the density of single particle states. The number of quanta in each mode is the analogue of the number of *particles* in the state, and we can define

$$\langle N\rangle = \int_0^\infty \langle N\rangle_\omega g(\omega)\mathrm{d}\omega = \int_0^\infty \frac{1}{\exp(\hbar\omega\beta)-1} g(\omega)\mathrm{d}\omega, \tag{13.13}$$

to be the total number of these particles in the box.

Furthermore, the dispersion relation $\omega = ck$ implies that these particles satisfy a linear relationship between energy and wavevector $\epsilon = c\hbar k$, suggesting that their rest mass $m_0$ is zero, since the general relationship is $\epsilon^2 = m_0^2c^4 + p^2c^2$ and the momentum of a freely propagating particle is $p = \hbar k$. A rest mass of zero implies that the particles travel at the speed of light, and the negative sign in the denominator of $\langle N\rangle_\omega$ suggests that they behave rather like bosons, with integer spin. These particles are the photons that were mentioned at the beginning of this chapter. This interpretation is an extremely important conceptual leap. We explore the idea of photons, and determine thermodynamic properties of the electromagnetic field, in the next section.

## 13.3 Thermal Properties of a Photon Gas

### 13.3.1 Planck Energy Spectrum of Black-Body Radiation

The properties of electromagnetic radiation inside cavities, determined by experiment in the later decades of the nineteenth century, provided the first indications that classical physics could not account for the workings of the world. Planck introduced the quantum view by suggesting that the energies of cavity fields were quantised, and obtained (13.11), which we write as

$$\langle E\rangle = V\int_0^\infty u(\omega)\mathrm{d}\omega, \tag{13.14}$$

in terms of $u(\omega)$, an energy density in frequency, per unit spatial volume.

This density could be determined by measuring a frequency spectrum of the intensity of radiation emerging from a small hole in a cavity, and regarding it as a sample of the equilibrium counter-propagating radiation to be found inside. The walls of the cavity are assumed to emit such radiation to balance the absorption of radiation from an opposing surface. The walls are regarded as the reservoir that supplies the modes with energy, through coupling of electromagnetic fields with charged matter. For ideal coupling, there is no reflection, such that all the incident radiation is absorbed before being re-emitted. Roughly speaking, absorbing walls are black, and the emission from them is called black-body radiation. It is a slight misnomer though, as a hot wall can emit very bright radiation, and not appear black in the least!

The point made by Planck was that the inferred experimental energy spectrum $u(\omega)$ matched the form suggested by (13.11), which we now know as the Planck spectrum

$$u(\omega) = \frac{\hbar}{c^3\pi^2}\frac{\omega^3}{\left(\exp\left(\frac{\hbar\omega}{kT}\right)-1\right)}, \tag{13.15}$$

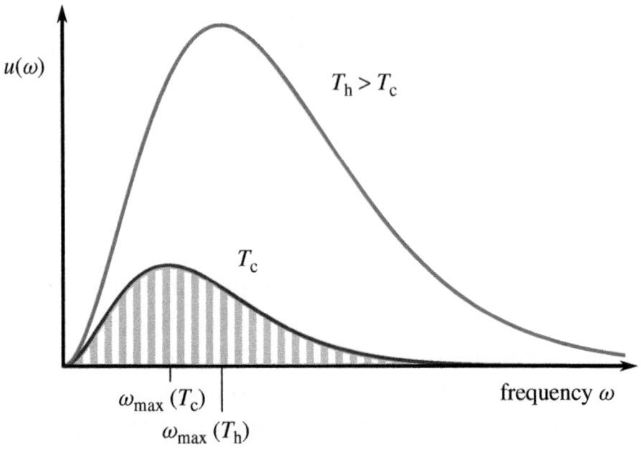


**Figure 13.2** Sketch of the Planck energy spectrum of black-body radiation at a low temperature $T_c$ (blue line) and a high temperature $T_h$ (red line). The frequency $\omega_{max}$ where the spectrum reaches a peak increases linearly with temperature (Wien's law) and the area under the curve, illustrated for the case $T = T_c$, is the total radiant energy and increases with the fourth power of the temperature (Stefan's law).

where the temperature of black-body radiation is that of the walls with which it couples. Examples at a high and low temperature are shown in Figure 13.2. The spectrum has a peak that corresponds to the condition

$$\frac{du}{d\omega} = \frac{\hbar}{c^3\pi^2}\left(\frac{3\omega^2}{\left(\exp\left(\frac{\hbar\omega}{kT}\right)-1\right)} - \frac{\omega^3}{\left(\exp\left(\frac{\hbar\omega}{kT}\right)-1\right)^2}\frac{\hbar}{kT}\exp\left(\frac{\hbar\omega}{kT}\right)\right) = 0, \tag{13.16}$$

and hence the peak frequency $\omega_{max}$ is given by $3(e^x - 1) = xe^x$, where $x = \hbar\omega_{max}/kT$. The numerical solution is

$$\omega_{max} \approx 2.821\frac{kT}{\hbar}, \tag{13.17}$$

and the peak frequency is therefore proportional to temperature, with the proportionality constant providing a value for the Planck constant. The experimental observation that the peak in the energy spectrum of black-body radiation scales linearly with temperature is known as Wien's law.

The Planck spectrum can be interpreted as a density of photons in frequency, per unit volume, with the form

$$n(\omega) = \frac{1}{V}\langle N\rangle_\omega g(\omega) = \frac{1}{c^3\pi^2}\frac{\omega^2}{\left(\exp\left(\frac{\hbar\omega}{kT}\right)-1\right)}. \tag{13.18}$$

The mean population of photons in a mode $\langle N\rangle_\omega = [\exp(\hbar\omega/kT) - 1]^{-1}$ has a form so similar to the Bose–Einstein statistics expression (11.3), when we regard $\hbar\omega$ as the analogue of the energy of a single particle state, that we can deduce two important properties of photons, namely that they are bosons, as already noted, and that they have zero chemical potential.

This might sound peculiar, but the fact is that photons differ from particles of matter in that their total population does not satisfy a conservation condition. There are no photons in the environment coupled to our system and their population in the box can be increased simply by increasing the temperature of the walls. The chemical potential was developed in our discussion of the grand canonical ensemble in Chapter 7 on the basis of particle conservation under exchange between the system and its environment. Photons do not fully correspond to the classical concept of particles, and one implication of this is that black-body radiation has zero chemical potential.

### 13.3.2 Photon Energy Density and Flux

The Planck spectrum allows us to calculate the total energy of black-body radiation per unit volume $e = \langle E \rangle / V$. This is

$$e = \int_0^\infty u(\omega)\mathrm{d}\omega = \frac{\hbar}{c^3\pi^2}\int_0^\infty \frac{\omega^3\mathrm{d}\omega}{\exp\left(\frac{\hbar\omega}{kT}\right)-1} = \frac{(kT)^4}{c^3\pi^2\hbar^3}\int_0^\infty \frac{x^3\mathrm{d}x}{(\mathrm{e}^x-1)}, \tag{13.19}$$

having inserted $x = \hbar\omega/kT$. The integral can be shown to be equal to $\pi^4/15$, and so the energy density is

$$e = \frac{4\sigma}{c}T^4, \tag{13.20}$$

where $\sigma = \pi^2 k^4/(60c^2\hbar^3) = 5.67\times 10^{-8}\,\mathrm{Wm^{-2}\,K^{-4}}$ is known as the Stefan–Boltzmann constant.

For a gas, the flux of particles through a plane is known from kinetic theory to be $\phi = nv/4$, where $v$ is the mean speed and $n$ is the particle density. In the photon picture, the mean speed of the particles is $c$, and so the flux of photons with frequencies in the range $\omega \to \omega + \mathrm{d}\omega$ is $cn(\omega)\mathrm{d}\omega/4$. The energy flux associated with this cohort is $\hbar\omega cn(\omega)\mathrm{d}\omega/4$ and therefore that associated with all frequencies is

$$\phi_E = \frac{c}{4}\int_0^\infty \hbar\omega n(\omega)\mathrm{d}\omega = \frac{c}{4}\int_0^\infty u(\omega)\mathrm{d}\omega = \frac{c}{4}e = \sigma T^4. \tag{13.21}$$

Black bodies therefore radiate energy according to the fourth power of their temperature, matching the experimental observation known as Stefan's law. The increase in area under the Planck spectrum $u(\omega)$ due to an increase in temperature is illustrated in Figure 13.2.

### 13.3.3 Photon Pressure

Black-body radiation exerts a pressure, further supporting an interpretation in terms of a photon gas. It is also known as radiation pressure. We employ (8.22):

$$p = kT\left(\frac{\partial \ln Z}{\partial V}\right)_T, \tag{13.22}$$

and (13.12), so that

$$p = kT\int_0^\infty \ln\left(\frac{1}{1-\exp\left(-\frac{\hbar\omega}{kT}\right)}\right)\frac{\omega^2}{c^3\pi^2}\mathrm{d}\omega. \tag{13.23}$$

We integrate by parts and refer to (13.19) to obtain

$$p = \frac{1}{3c^3\pi^2}\int_0^\infty \frac{\hbar\omega^3}{\exp\left(\frac{\hbar\omega}{kT}\right)-1}\mathrm{d}\omega = \frac{1}{3}e = \frac{4\sigma}{3c}T^4. \tag{13.24}$$

This expression may be combined with Stefan's law (13.20) such that the pressure and energy flux are related by $p = 4\phi_E/(3c)$. Apart from a difference in numerical factor, this is compatible with the idea that when radiation is reflected from a wall, the pressure exerted is proportional to the momentum flux carried by the photons, and this is proportional to the energy flux since the momentum of a photon with wavevector $k$ is $\hbar k = \hbar\omega/c$, while its energy is $\hbar\omega$.

Photon pressure acts on the walls of a cavity containing the radiation, or if it escapes from the cavity, on any surface upon which the radiation falls, in proportion to the incident intensity. Notice the contrast with the relation $p = 2e/3$ found for the gas of nonrelativistic particles, derived, for example in (2.5) and later in (12.11). The effect is typically very small, for example if the walls of a cavity are at $10^3$ K, the radiation pressure inside is only about $10^{-4}$ Pa, or one billionth of an atmosphere, but even the pressure of radiation from the sun can have an effect on the trajectories of spacecraft over long distances, and can even be exploited for propulsion, at least in principle.

### 13.3.4 Photon Entropy

Finally, we consider the entropy per unit volume of a photon gas. Using the Gibbs formulation from (8.11), we write

$$\begin{aligned} s = \frac{S_{\mathrm{G}}}{V} &= \frac{\langle E\rangle}{TV} + \frac{k}{V}\ln Z \\ &= \frac{e}{T} + \frac{k}{V}\int_0^\infty g(\omega)\ln\left(\frac{1}{1-\exp\left(-\frac{\hbar\omega}{kT}\right)}\right)\mathrm{d}\omega \\ &= \frac{4\sigma}{c}T^3 + \frac{4\sigma}{3c}T^3 = \frac{16\sigma}{3c}T^3 = \frac{4e}{3T}. \end{aligned} \tag{13.25}$$

Notice that the Gibbs free energy of the photon gas is zero:

$$\begin{aligned} G &= \langle E\rangle - TS_{\mathrm{G}} + pV \\ &= \langle E\rangle - \tfrac{4}{3}\langle E\rangle + \tfrac{1}{3}\langle E\rangle = 0, \end{aligned} \tag{13.26}$$

a result that is compatible with the earlier conclusion that the chemical potential for a photon gas is zero, since $G = \langle N\rangle\mu$. Photons can be added to the system by changing the temperature of the heat bath: there is no such thing as a photon bath with a controllable chemical potential because photons are not conserved particles.

Furthermore, the Helmholtz free energy $F = \langle E\rangle - TS$ is equal to $-\langle E\rangle/3$, and the Helmholtz free energy density $f$ is given by $-e/3$. Thermodynamic relations such as (3.35) as well as Maxwell relations such as (3.37) may be verified for a photon gas. The thermodynamic properties of photons are illustrated in Figure 13.3.

We can add to our photon gas interpretation by noting that photons carry an entropy flux $\phi_S = 4\phi_E/(3T)$ since the entropy density is proportional to the energy density divided by temperature. When radiation emitted from a high temperature source is absorbed by an object and re-emitted at a cooler temperature, the energy flows balance in a steady state, but the entropy flows do not because of the disparity in temperatures of the radiation. The emitted entropy flux is greater than the absorbed flux.

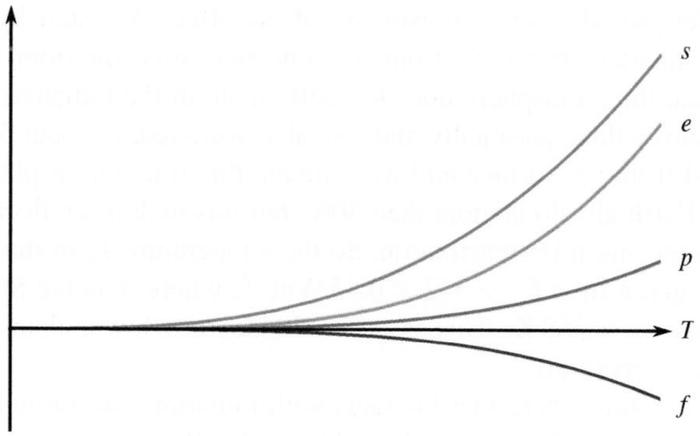


Figure 13.3 Dependence of photon gas thermodynamic properties on temperature.

The emitted photons have a lower mean frequency, according to Wien's law, and therefore a lower mean energy than the absorbed photons, such that in order to conserve energy there must be more photons emitted than absorbed. This creation of particles, with the implied increase in uncertainty of microscopic configuration, ties in with the conclusion that entropy is being generated by the absorber/emitter as it interacts with the radiation, which is what we would expect since the process is clearly out of equilibrium, and therefore should satisfy the second law.

## 13.4 The Global Radiation Budget and Climate Change

The sun is effectively a cavity that leaks black-body radiation at a temperature of about 5800 K. The intensity falls in proportion to the inverse square of the distance from the sun, but the escaping photons retain a Planck spectrum characterised by this temperature, as there are very few objects with which they can interact before they arrive at the top of the terrestrial atmosphere. However, the situation is different at ground level, owing to the absorption and re-radiation of photons from components of the atmosphere, including clouds, dust, aerosols and individual gas phase molecules, as well as reflection, particularly from clouds. As we noted in the previous section, these are in general entropy generating, nonequilibrium processes.

Gaps open in the Planck spectrum of the incident radiation caused by particular absorption mechanisms, for example in the ultraviolet part of the spectrum that interacts with stratospheric ozone. The Earth's surface reflects some of the remaining radiation, depending on the frequency, but absorbs the major part of it and re-radiates at a range of temperatures depending on the local climate, latitude as well as other geographical and meteorological features. The terrestrial emission is similarly reflected, absorbed and re-radiated by the atmosphere as it propagates outwards. The so-called global radiation budget, expressing the balance of radiant energy coming in and going out, involves much complex science, but it is easy to demonstrate that the transfers give rise to an important warming effect that has sustained life on this planet and should be interfered with at our peril. We refer here to the well-known mechanism of the greenhouse effect and the prospect of anthropogenic climate change.

A simple model provides a rough estimate of the effect. We start by considering the balance between incident radiation from the sun and emission from the Earth, under the assumption that the atmosphere does *not* participate in the radiative transfer process. The incident radiative flux, seasonally and globally averaged, is about 342 $\mathrm{Wm^{-2}}$. Some of this is reflected from the surface and we estimate this fraction, or planetary albedo, to be 15%: the true Earth albedo is more than 30%, but this includes reflection from clouds, and here we are ignoring this contribution. So the temperature $T_s$ of the terrestrial black-body radiation is given by $\sigma T_s^4 = 342 \times 0.85\,\mathrm{Wm^{-2}}$, where $\sigma$ is the Stefan–Boltzmann constant, and thus $T_s \approx 268$ K. This is some 20 degrees lower than the actual mean terrestrial surface temperature.

Now consider an atmosphere that interacts with radiation. To a rough approximation, the incident radiation is affected just by additional reflection, particularly off clouds; raising the planetary albedo to about 30%. Relatively little absorption and re-radiation takes place at the high frequencies of the incident Planck spectrum. However, matters are very different at lower frequencies characteristic of terrestrial emission. To first approximation we shall assume that all of it is intercepted and re-emitted as black-body radiation at an atmospheric temperature $T_a$. Half of this emission is directed back towards the Earth, and half out into space, as illustrated in Figure 13.4. It is the latter part that balances the incoming solar radiation, such that $\sigma T_a^4 = 342 \times 0.7\,\mathrm{Wm^{-2}}$, in which case $T_a \approx 255$ K. The surface temperature is then determined by a balance between the gain and loss of energy from the atmosphere:

$$\sigma T_s^4 = 2\sigma T_a^4, \tag{13.27}$$

such that $T_s = 2^{1/4} T_a \approx 303$ K.

In this simple case a warming of over 30 K has been provided by atmospheric participation in the radiative balance. Such a model cannot be expected to be fully accurate, of course, as it has neglected many important details. A significant flux that is absent

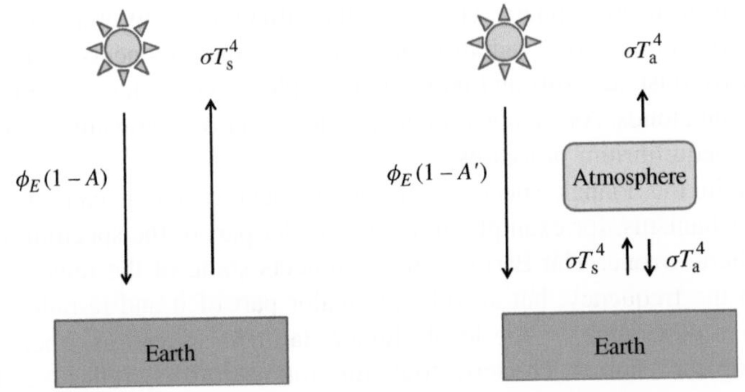


**Figure 13.4** A simple model of the greenhouse effect. Radiative transfer for the Earth system with no participation by the atmosphere, on the left, is contrasted with the more realistic but still simplified situation on the right, where greenhouse gases, aerosols and clouds interact with the terrestrial radiation.

in the analysis is the roughly 10% of terrestrial radiation that is not absorbed by the atmosphere and is emitted directly to space, according to present estimates. This is the component of radiative transfer that can be intercepted by anthropogenic radiatively active atmospheric components: additional greenhouse gases including carbon dioxide and methane, as well as extra aerosol or haze particles, or indeed additional cloud. The assessment of the overall effect on the global radiation budget is one of the most complex tasks currently being undertaken in science, but various studies have left little doubt that there is significant potential for disturbance of the terrestrial climate due to human activities.

## 13.5 Cosmic Background Radiation

Solar radiation is just an approximate version of the Planck spectrum, but the Earth is also illuminated by radiation that is extraordinarily close to the Planck shape, with a peak frequency in the microwave region. It is very nearly isotropic, meaning that it arrives from all directions in space at almost equal intensity, when certain biases such as the motion of the Earth is taken into account. This so-called cosmic background radiation is believed to consist of photons emitted in the early universe. It is a faint glow from the Big Bang, and one of the principal reasons why we are confident that such an event took place.

When the universe was extremely hot and dense, according to current cosmological models, photons coupled strongly to a plasma of charged particles and assumed a Planck spectrum at the evolving temperature. The ongoing expansion cooled the universe, in the normal way of an adiabatic, though possibly nonquasistatic, expansion of a system, and the density of matter fell until electrons found it thermodynamically favourable to combine with nuclei to form atoms, a process known as recombination. The reduction in charge density caused the photons to decouple from matter. The confining cavity walls melted away, at it were, leaving behind photons characterised by the temperature of the plasma at recombination, estimated to be about 3000 K.

These photons have since then propagated through the thinning vastness of space with very little change in the Planck spectrum, and these are what we detect today, except for one feature: the temperature of the radiation has cooled to about 2.725 K. The uniformity of the temperature of the radiation is very striking, with variations of just $10^{-4}$ K. The greyscale view of the entire sky shown in Figure 13.5 is a now-famous portrayal of the tiny fluctuations as speckles visible against a uniform temperature background, obtained by the Wilkinson Microwave Anisotropy Probe (WMAP). It is essentially an image of the largest and most featureless physical object we have ever seen.

The usual interpretation of the cooling is that it is space itself that is expanding. A picture where matter and photons are expanding into a fixed spatial arena cannot apply because, as with the escape of radiation from the sun, such a process would reduce the intensity but not change the temperature of the spectrum. We must imagine that the photons, classically regarded as spatially extended disturbances of the electromagnetic field, have distorted with the underlying space on which they propagate. Their wavelengths increase with time, but since $\omega = ck = 2\pi c/\lambda$, this means that their frequencies

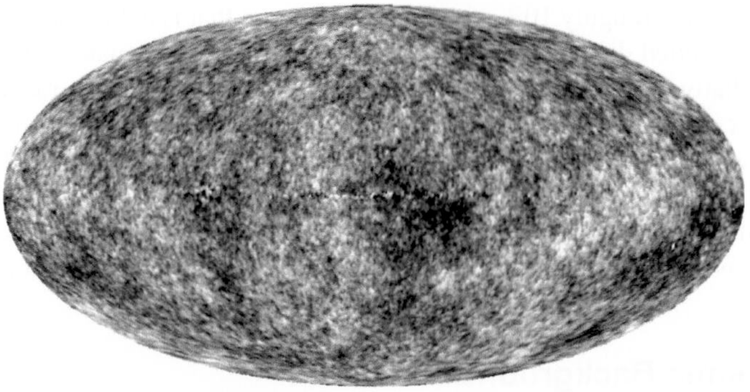

**Figure 13.5** Representation of spatial fluctuations in the temperature of cosmic background radiation. The lightest and darkest patches are hotter or cooler by just a few parts in $10^4$. Source: Reproduced from WMAP, http://map.gsfc.nasa.gov/.

decrease. The peak in the Planck spectrum has moved towards lower frequencies, and according to Wien's law, this corresponds to a reduction in the characteristic temperature of the radiation.

Another view to take is more akin to traditional thermodynamics. We consider the expansion of patches of space against a pressure exerted by other patches, with the pressure deriving from the electromagnetic radiation carried within each. Since the properties of each patch are identical, there is no heat transfer between them. Assuming that the expansion is quasistatic, the first law therefore takes the form $d\langle E\rangle = -p dV$; inserting the relation $p = \langle E\rangle/(3V)$ for radiation, this becomes $d\langle E\rangle/\langle E\rangle = -(1/3)dV/V$ or $\langle E\rangle \propto V^{-1/3}$, and as $\langle E\rangle \propto VT^4$, this means that $T \propto V^{-1/3}$.

The latter relationship may also be obtained through (13.25) by requiring that the photon entropy of the universe is constant as it undergoes a quasistatic adiabatic expansion. An increase in the linear dimension of the universe by three or four orders of magnitude over the period since recombination would account for the cooling in radiation from 3000 K to the present day value. This interpretation is slightly hard to sustain, though, since the radiation is supposed not to interact with matter during the expansion, and so it is difficult to argue that thermal equilibrium is continually re-established.

## Exercises

**13.1** Black-body radiation is contained within a box of volume $V$ and temperature $T$. Calculate the energy density of the radiation at 400 K. Calculate the photon density and hence the mean energy per photon at this temperature. You may assume that $\int_0^\infty x^3/(\exp(x)-1)dx = \pi^4/15$ and $\int_0^\infty x^2/(\exp(x)-1)dx \approx 2.404$.

**13.2** Above what temperature, approximately, would the photon density exceed the typical molecular density of air in a sealed room?

**13.3** Estimate the entropy production associated with the absorption and re-emission of radiation by the Earth.

**13.4** An evacuated and rigid container with volume $V$ at a temperature $T$ contains black-body radiation. The container is placed in thermal contact with a heat bath at temperature $T_\mathrm{r}$. If the heat capacity of the cavity material itself is negligible, show that the overall change in entropy of the universe after the system and heat bath have reached thermal equilibrium is $\Delta S_{\mathrm{tot}} = 4\sigma V T_\mathrm{r}^3[1 - t^3(4 - 3t)]/(3c)$ where $t = T/T_\mathrm{r}$. Comment on the sign of $\Delta S_{\mathrm{tot}}$ as a function of $t$.

# 14

# Statistical Thermodynamics of Interacting Particles

Having developed the statistical thermodynamics of ideal gases in both classical and quantum regimes of behaviour, we now discuss a broader perspective that includes systems of interacting particles. The areas of application are vast, but we must be brief.

## 14.1 Classical Phase Space

Modern presentations of statistical thermodynamics focus on the quantum point of view because it has enormous advantages over a classical treatment with regard to the enumeration of the multiplicities of microstates and the summation of partition functions, since the states are discrete. But this was not how statistical thermodynamics was initially developed by Boltzmann and Gibbs. In classical mechanics, microstates of a system form a continuum specified by the classical dynamical variables of the constituent particles. In what sense can we count them?

The classical procedure is to regard a summation over all microstates as an integral over the classical phase space of the system. This is a multidimensional space for which the positions and momenta of the constituent particles provide the coordinates. The instantaneous state of the system corresponds to a single point in this space that follows a continuous trajectory as time progresses, as determined by the equations of motion. We used such an integral in Section 6.1.2 when we considered the statistical properties of a classical oscillator.

For a system consisting of a particle described by position $x$ and momentum $p$ and with total energy $E$ in the range $E \to E + \mathrm{d}E$, the multiplicity of microstates may be written as $\mathrm{d}\Omega(E) = g(E)\mathrm{d}E$ with

$$g(E) = \int \mathrm{d}p\mathrm{d}x\,\rho(p,x)\delta(H(p,x) - E), \tag{14.1}$$

*Statistical Physics: An Entropic Approach*, First Edition. Ian Ford.

where we have included a density of microstates $\rho(p,x)$ over the phase space, such that $\rho(p,x)\mathrm{d}p\mathrm{d}x$ is the number of microstates in the range $x \to x+\mathrm{d}x$ and $p \to p + \mathrm{d}p$. The isolation of the system is represented by the delta function that constrains the Hamiltonian function of position and momentum to be equal to the energy. For example, the Hamiltonian of the 1-d harmonic oscillator is $H(p,x) = p^2/(2m) + \kappa x^2/2$. The canonical partition function of such a system in contact with a heat bath is written as

$$Z = \int_{-\infty}^{\infty} \mathrm{d}p\mathrm{d}x\, \rho(p,x)\, \exp\left(-\frac{H(p,x)}{kT}\right). \tag{14.2}$$

In earlier examples, we assumed $\rho$ to be a constant, on the grounds that quantum mechanics showed that this would be appropriate, but there is also a classical argument in support of this assumption, based on the idea that the density of states in phase space should not change if we decide to shift the origin of coordinates such that $x \to x + x_0$ and $p \to p + p_0$. Such an invariance rules out any variation in $\rho$, and we set it equal to a constant $h_0^{-1}$.

As an example, let us evaluate $Z$ for the 1-d oscillator Hamiltonian

$$\begin{aligned} Z &= \frac{1}{h_0}\int_{-\infty}^{\infty} \mathrm{d}p\ \exp\left(-\frac{p^2}{2mkT}\right)\int_{-\infty}^{\infty} \mathrm{d}x\ \exp\left(-\frac{\kappa x^2}{2kT}\right) \\ &= \frac{1}{h_0}(2\pi mkT)^{\frac{1}{2}}\left(\frac{2\pi kT}{\kappa}\right)^{\frac{1}{2}} = \frac{kT}{\left(\frac{h_0}{2\pi}\right)\omega}, \end{aligned} \tag{14.3}$$

where $\omega$ is the angular frequency of the oscillator, given by $(\kappa/m)^{1/2}$, and where we have used $\int_{-\infty}^{\infty} \mathrm{d}x\ \exp(-\alpha x^2) = (\pi/\alpha)^{1/2}$.

The classical (high temperature) limit of the canonical partition function of the quantum oscillator in (6.24) is

$$Z = \frac{\exp\left(-\frac{\hbar\omega}{2kT}\right)}{1-\exp\left(-\frac{\hbar\omega}{kT}\right)} \approx \frac{kT}{\hbar\omega}, \tag{14.4}$$

for $kT \gg \hbar\omega$, and we see that not only does the assumed constancy of $\rho$ yield the correct functional form, but we also learn that the appropriate factor to use in the classical partition function is $h_0 = h$: Planck's constant. In classical calculations it does not matter what value for $h_0$ is taken, as most of the thermal properties are obtained by taking derivatives of the logarithm of $Z$, and $h_0$ therefore disappears. Only in the absolute values of the entropy and chemical potential does it remain, but even here it cancels out when taking differences in these quantities.

In general, the classical canonical partition function of a system of $N$ particles in a 3-d space may be written as

$$Z = \frac{1}{h^{3N}N!}\int \prod_{i=1}^{N} \mathrm{d}^3\mathbf{p}_i \mathrm{d}^3\mathbf{x}_i\ \exp\left(-\frac{H}{kT}\right), \tag{14.5}$$

where the Hamiltonian $H$ is a function that corresponds to the energy, dependent on the positions and momenta of the particles, and where a factor of $1/N!$ has been inserted to account for particle indistinguishability, this being the correct factor for a system with a particle density well below the quantum concentration. We would not impose such a correction if the particles were distinguishable.

Equation (14.5) goes beyond the treatment of gases or independently tethered oscillators, because the Hamiltonian can include particle–particle interaction terms. It is a starting point for calculating the statistical properties of general classical systems, and hence a powerful tool in the broad field of theoretical condensed matter physics.

## 14.2 Virial Expansion

Let us now study a system of two particles of equal mass that interact through a potential $\phi(r_{12})$, where $r_{12}$ is their separation. The Hamiltonian is $H(\mathbf{p}_1, \mathbf{p}_2, \mathbf{x}_1, \mathbf{x}_2) = |\mathbf{p}_1|^2/(2m) + |\mathbf{p}_2|^2/(2m) + \phi(|\mathbf{x}_1 - \mathbf{x}_2|)$ and the canonical partition function is

$$Z_2 = \frac{1}{h^6 2!} \int \prod_{i=1}^{2} \mathrm{d}^3\mathbf{p}_i \mathrm{d}^3\mathbf{x}_i \, \exp\left(-\frac{H}{kT}\right). \tag{14.6}$$

The integration is performed over all values of momenta, and all particle positions within a box of volume $V$.

The momentum integrals are of the form $\int_{-\infty}^{\infty} \mathrm{d}p \, \exp(-p^2/(2mkT)) = (2\pi mkT)^{1/2}$ and we can transform to spatial coordinates $\mathbf{R} = (\mathbf{x}_1 + \mathbf{x}_2)/2$ and $\mathbf{r}_{12} = \mathbf{x}_1 - \mathbf{x}_2$, with a Jacobian of unity, such that

$$Z_2 = \frac{1}{2}\left(\frac{2\pi mkT}{h^2}\right)^3 \int \mathrm{d}^3\mathbf{R}\mathrm{d}^3\mathbf{r}_{12} \, \exp\left(-\frac{\phi(r_{12})}{kT}\right) = \frac{1}{2}V^2 n_q^2 \alpha, \tag{14.7}$$

where $n_q$ is the quantum concentration. Assuming the integral converges, we have defined

$$\alpha = \frac{1}{V}\int_0^{\infty} 4\pi r_{12}^2 \exp\left(-\frac{\phi(r_{12})}{kT}\right) \mathrm{d}r_{12}, \tag{14.8}$$

which would be unity if the interaction potential were zero, and otherwise expresses the deviation from the behaviour of two noninteracting classical particles, which would be represented by a partition function $Z_2^{\mathrm{ig}} = (1/2)(n_q V)^2 = (1/2)(Z_1^{\mathrm{ig}})^2$ following the pattern set by (9.19) and (9.13) with spin $s = 0$.

It is convenient to write

$$\alpha = 1 + \frac{1}{V}\int_0^{\infty} 4\pi r^2 f(r,T)\mathrm{d}r = 1 + \frac{\mathcal{B}(T)}{V}, \tag{14.9}$$

in terms of a temperature-dependent quantity $\mathcal{B}(T)$, where

$$f(r,T) = \exp\left(-\frac{\phi(r)}{kT}\right) - 1, \tag{14.10}$$

is called the Mayer function, such that

$$Z_2 = \frac{1}{2}V^2 n_q^2(T)\left(1 + \frac{\mathcal{B}(T)}{V}\right), \tag{14.11}$$

from which thermodynamic properties follow. For example, the pressure is

$$p = kT\left(\frac{\partial \ln Z_2}{\partial V}\right)_T = \frac{2kT}{V} - kT\left(1+\frac{\mathcal{B}}{V}\right)^{-1}\frac{\mathcal{B}}{V^2}, \tag{14.12}$$

and we can regard the second term as a nonideal correction to the ideal gas pressure represented by the first term.

The procedure can be extended to a gas of $N$ particles in a 3-d box. The partition function is now

$$\begin{aligned}
Z_N &= \frac{1}{N!}n_q^N \int \prod_{i=1}^{N} \mathrm{d}^3\mathbf{x}_i \, \exp\left(-\frac{\sum_{j>i}\phi(r_{ij})}{kT}\right) \\
&= \frac{1}{N!}n_q^N \int \prod_{i=1}^{N} \mathrm{d}^3\mathbf{x}_i \prod_{j>i} \exp\left(-\frac{\phi(r_{ij})}{kT}\right) \\
&= \frac{1}{N!}n_q^N \int \prod_{i=1}^{N} \mathrm{d}^3\mathbf{x}_i \prod_{j>i}(1 + f(r_{ij},T)) \\
&= \frac{1}{N!}n_q^N \int \prod_{i=1}^{N} \mathrm{d}^3\mathbf{x}_i \, (1 + f(r_{12},T))(1 + f(r_{13},T))\cdots \\
&\approx \frac{1}{N!}n_q^N \left[V^N + \frac{N(N-1)}{2}\int \prod_{i=3}^{N} \mathrm{d}^3\mathbf{x}_i \mathrm{d}^3\mathbf{x}_1 \mathrm{d}^3\mathbf{x}_2 \, f(r_{12},T)\right] \\
&= \frac{1}{N!}(Vn_q)^N \left[1 + \frac{N(N-1)}{2V^2}\int \mathrm{d}^3\mathbf{x}_1 \mathrm{d}^3\mathbf{x}_2 \left(\exp\left(-\frac{\phi(r_{12})}{kT}\right) - 1\right)\right],
\end{aligned} \tag{14.13}$$

where $r_{ij}$ is the distance between particles $i$ and $j$. On the second to last line, only contributions to the integrand involving a single $f$ function are retained, and it is recognised that there are $N(N-1)/2$ of them corresponding to the number of particle pairs.

We therefore write

$$Z_N \approx Z_N^{\mathrm{ig}}\left[1 + \frac{N(N-1)}{2V}\mathcal{B}\right], \tag{14.14}$$

where $Z_N^{\mathrm{ig}} = (Z_1^{\mathrm{ig}})^N/N!$ and we obtain the pressure

$$\begin{aligned}
p = kT\left(\frac{\partial \ln Z_N}{\partial V}\right)_{T,N} &\approx \frac{NkT}{V} - kT\left[1 + \frac{N(N-1)}{2V}\mathcal{B}\right]^{-1}\frac{N(N-1)\mathcal{B}}{2V^2} \\
&\approx nkT - \frac{1}{2}kT\mathcal{B}n^2.
\end{aligned} \tag{14.15}$$

We have inserted $n = N/V$ and assumed $N \gg 1$. We take $N^2\mathcal{B}/V$ to be small and have neglected terms proportional to $n^3$ and beyond.

If we compare (14.15) with the virial expansion of the equation of state of a nonideal gas in (3.42), it is apparent that we have derived a microscopic form for the second virial coefficient:

$$B_2(T) = -\frac{1}{2}\mathcal{B} = \int_0^\infty 2\pi r^2 \left(1 - \exp\left(-\frac{\phi(r)}{kT}\right)\right) \mathrm{d}r, \tag{14.16}$$

that appears in the equation of state

$$\frac{p}{kT} = \sum_{i=1}^{\infty} B_i(T) n^i. \tag{14.17}$$

Our interest only in the quadratic correction to the ideal gas law in (14.15) is the reason why we neglected terms involving $f(r_{12})f(r_{13})$ in the integrand of the partition function (14.13).

For an attractive hard sphere interaction potential where $\phi(r)$ is equal to infinity for $r < r_\mathrm{p}$, corresponding to a strong repulsion at that particle separation, and small in magnitude in comparison with $kT$ for $r \geq r_\mathrm{p}$, as illustrated in Figure 14.1, we can write

$$\begin{aligned} B_2(T) &= \frac{2\pi}{3} r_\mathrm{p}^3 + \int_{r_\mathrm{p}}^\infty 2\pi r^2 \left[1 - \exp\left(-\frac{\phi(r)}{kT}\right)\right] \mathrm{d}r \\ &\approx \frac{2\pi}{3} r_\mathrm{p}^3 + \int_{r_\mathrm{p}}^\infty 2\pi r^2 \frac{\phi(r)}{kT} \mathrm{d}r = b - \frac{a}{kT}, \end{aligned} \tag{14.18}$$

which then takes the form argued in (3.61) on the basis of classical thermodynamics. Thus the $b$ parameter in the van der Waals equation of state is of the order of the particle volume and $a$ is a measure of the mutual interaction energy of a particle pair. If $\phi$ took a different form, such as the Lennard–Jones expression also shown in Figure 14.1, or if $\phi(r) \sim kT$, the second virial coefficient would have a more elaborate

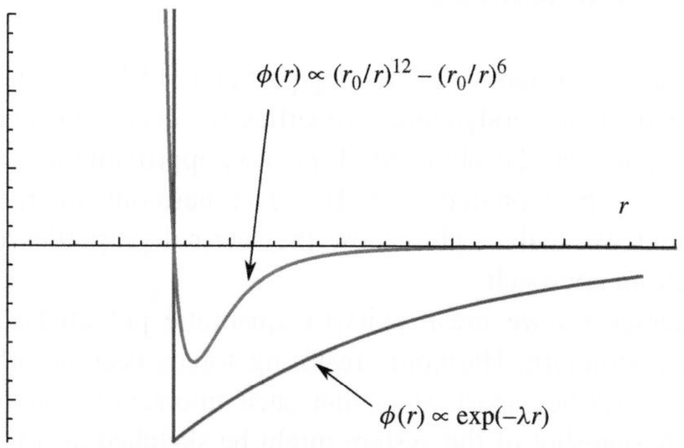


**Figure 14.1** Two commonly chosen interaction pair potentials: in red the so-called Lennard–Jones 6–12 potential with range parameter $r_0$, and in blue an exponentially decreasing attractive part, specified by parameter $\lambda$, with a hard repulsion.

temperature dependence and bear a more complicated relationship to the strength and range of the interaction.

But for the case of weakly attractive hard spheres, we have

$$Z_N \approx Z_N^{\rm ig}(1 - nNB_2) = Z_N^{\rm ig}\left[1 - nN\left(b - \frac{a}{kT}\right)\right]. \tag{14.19}$$

The mean energy of the gas per unit volume $e = \frac{kT^2}{V}\frac{{\rm d}\ln Z^N}{{\rm d}T}$ follows the same form as (3.60):

$$e(n,T) \approx \frac{3}{2}nkT - n^2kT^2\frac{{\rm d}B_2(T)}{{\rm d}T} \approx \frac{3}{2}nkT + n^2\int_{r_{\rm p}}^{\infty} 2\pi r^2\phi(r){\rm d}r, \tag{14.20}$$

where the second, nonideal term is clearly a potential energy to supplement the kinetic energy of the first term. The dependence on $n^2$ arises because we are considering a pairwise interaction and therefore the energy should be proportional to the number of particle pairs in the system.

The entropy of the gas follows (3.57), namely

$$S_{\rm G} = \frac{\langle E\rangle}{T} + k\ln Z_N = S_{\rm G}^{\rm ig} - \frac{N^2k}{V}\frac{{\rm d}(TB_2(T))}{{\rm d}T} = S_{\rm G}^{\rm ig} - nkNb, \tag{14.21}$$

suggesting that the interactions reduce the entropy of the gas, per particle, by an amount proportional to the volume $Nb$ that each particle is unable to explore owing to their mutual repulsion.

By including more terms in the partition function, further contributions to the pressure, energy and entropy of the gas can be obtained, corresponding to higher virial coefficients, but they rapidly become complicated to compute. Nevertheless, they are in principle calculable from information about microscopic interactions.

## 14.3 Harmonic Structures

The partition function of a system of interacting particles (14.5) can, in principle, provide us with all the statistical thermodynamic properties we need, but there are only a few cases where exact results can be obtained. Typically, approximations have to be made, such as making a virial expansion in density. However, harmonic interparticle interactions are an exception, and we shall explore how the thermal properties of a solid can be modelled using such an approach.

By harmonic interactions we mean pairwise quadratic potentials, usually involving near neighbours in a structure. Harmonic restoring forces become arbitrarily strong as the participants move further apart, such that each interaction can be regarded as an unbreakable bond. A snapshot of the system might be sketched as a network of connections as shown in Figure 14.2, and this has an important effect on the way we construct the partition function. Each particle is labelled by its position within the network. The dynamics are incapable of swapping particle positions in a manner that would create a configuration that is indistinguishable from another. The factor of $1/N!$ therefore need not appear in the partition function.

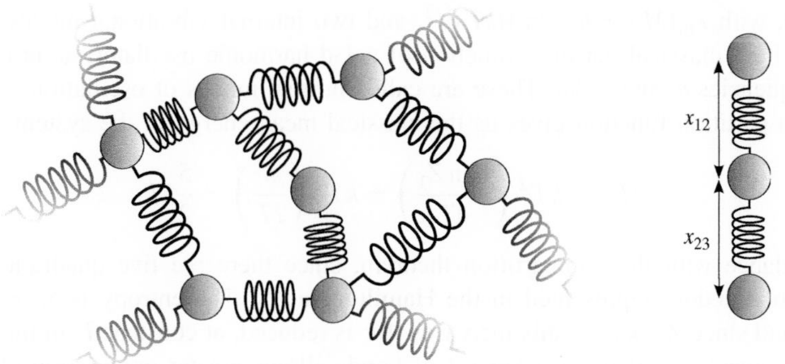


**Figure 14.2** A network of unbreakable harmonic bonds linking atoms in a general structure, and in a triatomic linear molecule.

### 14.3.1 Triatomic Molecule

As an example, let us consider a molecule consisting of three atoms of equal mass, as illustrated on the right in Figure 14.2. Harmonic interactions with spring constant $\kappa$ exist between nearest neighbours and the Hamiltonian is $H_3 = (p_1^2 + p_2^2 + p_3^2)/(2m) + (1/2)\kappa(x_1 - x_2 - r_0)^2 + (1/2)\kappa(x_2 - x_3 - r_0)^2$ where $r_0$ is the particle separation that minimises each bond energy. The atoms are constrained to lie on a line of length $L \gg r_0$.

The partition function is

$$Z_3 = \frac{1}{h^3}\int \prod_{i=1}^{3} \mathrm{d}p_i \mathrm{d}x_i \, \exp\left(-\frac{H_3}{kT}\right)$$
$$= \frac{(2\pi mkT)^{\frac{3}{2}}}{h^3}\int \prod_{i=1}^{3} \mathrm{d}x_i \, \exp\left(-\frac{(\kappa(x_1 - x_2 - r_0)^2 + \kappa(x_2 - x_3 - r_0)^2)}{2kT}\right). \quad (14.22)$$

Such an integral can be evaluated by a suitable transformation of the $x_i$ to so-called normal coordinates in terms of which the Hamiltonian may be written as separate quadratic terms. By inspection, these are

$$x_{12} = x_1 - x_2, \quad x_{23} = x_2 - x_3, \quad X = \tfrac{1}{3}(x_1 + x_2 + x_3), \quad (14.23)$$

for which the Jacobian is unity. The partition function becomes

$$Z_3 = \frac{(2\pi mkT)^{\frac{3}{2}}}{h^3}\int \mathrm{d}x_{12}\mathrm{d}x_{23}\mathrm{d}X \, \exp\left(-\frac{(\kappa(x_{12} - r_0)^2 + \kappa(x_{23} - r_0)^2)}{2kT}\right)$$
$$= \frac{[2\pi(3m)kT]^{\frac{1}{2}}}{h}\frac{L}{3^{\frac{1}{2}}}\left[\frac{2\pi mkT}{h^2}\right]\left[\frac{2\pi kT}{\kappa}\right] = \frac{L}{\lambda_{\mathrm{th}}(3m)}\left(\frac{kT}{\hbar 3^{\frac{1}{2}}\omega}\right)\left(\frac{kT}{\hbar\omega}\right), \quad (14.24)$$

where $\omega = (\kappa/m)^{1/2}$. The separability of the Hamiltonian gives rise to factors in the partition function that represent the 1-d translation of the centre of mass $X$ of the molecule, of form $[L/\lambda_{\mathrm{th}}(3m)]$ analogous to the 3-d version for a particle of mass $m$

in (9.13), with $\lambda_{\text{th}}(M) = h/(2\pi MkT)^{1/2}$, and two internal vibrational modes each represented by a classical partition function of a 1-d harmonic oscillator, as in (14.3), but with frequencies $\omega$ and $\sqrt{3}\omega$. These are called normal modes of oscillation.

Such a partition function gives us the classical mean energy of the system

$$\langle E\rangle = kT^2\left(\frac{\partial \ln Z_3}{\partial T}\right) = kT^2\left(\frac{5}{2T}\right) = \frac{5}{2}kT, \qquad (14.25)$$

in accordance with the equipartition theorem, since there are five quadratic terms or degrees of freedom represented in the Hamiltonian $H_3$. The entropy is $S_{\text{G}} = \langle E\rangle/T + k\ln Z_3$, and since $Z_3 \propto \kappa^{-1}$, this increases if $\kappa$ is reduced, at constant $T$, in line with our intuitive expectation that a weaker set of bonds allows greater exploration of different microscopic configurations.

### 14.3.2 Einstein Solid

It is an intuitive leap to claim now that the partition function of any harmonically bound set of particles in 3-d space may be written as a product of expressions describing the translation of, and rigid rotation about, the centre of mass, that we shall immediately neglect, together with factors corresponding to all the normal modes of oscillation of the structure. We write $Z_{\text{vib}} = \prod_i Z_{\omega_i}$ and recognise that there are $3N-6$ vibrational modes. There are six coordinates per particle in three dimensions, and therefore $6N-6$ degrees of freedom of relevance to the equipartition theorem as there are no quadratic potential energy terms in the Hamiltonian associated with the three spatial coordinates of the centre of mass position or the three coordinates that describe the orientation of the structure with respect to specific axes. Six of the degrees of freedom correspond to the kinetic energy of bulk translation and rotation, leaving $6N-12$ for vibration, two for the potential and kinetic energy of each of the $3N-6$ modes.

The partition function $Z_\omega$ corresponding to each mode will take the form (6.24) appropriate to a quantum oscillator. We can then write

$$\ln Z_{\text{vib}} = \sum_i \ln Z_{\omega_i} \approx \int_0^\infty g(\omega)\ln Z_\omega \mathrm{d}\omega, \qquad (14.26)$$

in terms of a density of states of vibrational modes $g(\omega)$, in a manner familiar now from our treatments of particles or electromagnetic modes in a box. However, the evaluation of the spectrum of vibrations, or equivalently the density of states $g(\omega)$, is rather complicated. Einstein made the very simple approximation that all vibrational modes had the same frequency $\omega_{\text{E}}$, as explored in Section 6.6. Such a step, together with the use of the partition function of a quantum oscillator at frequency $\omega_{\text{E}}$, leads to the Einstein model of the thermal heat capacity of a solid

$$C_V = \frac{3N\hbar^2\omega_{\text{E}}^2}{4kT^2\sinh^2\left(\frac{\hbar\omega_{\text{E}}}{2kT}\right)}, \qquad (14.27)$$

given earlier in (6.29), in which the difference between $3N-6$ and $3N$ is neglected. However, such an assumption is too bold and the model does not match experimental data. The Debye model, that we discuss next, is much more realistic.

### 14.3.3 Debye Solid

Peter Debye (1884–1966) developed a model of the heat capacity of solids that is superior to Einstein's. He proposed that the spectrum of vibrational modes of a harmonically bound structure should take the same form as the spectrum of sound vibrations in the solid, a natural choice and quite accurate, at least for the lower frequencies or larger wavelengths of the corresponding disturbances.

In a continuum treatment of elastic solids, longitudinal and transverse sound waves may be transmitted at all frequencies at respective speeds $v_s^l$ and $v_s^t$, related to the elastic constants of the material. Debye proposed that the density of states of oscillatory modes took the same form as the density of modes of electromagnetic waves in a box, but with the speed of light replaced by an average speed of sound $v_s = (1/3)v_s^l + (2/3)v_s^t$, where we recognise that there are two transverse modes and only one longitudinal mode. Thus from (13.5) we have

$$g(\omega) = \frac{3V\omega^2}{2v_s^3\pi^2}, \tag{14.28}$$

for a solid of volume $V$, where the factor of three accounts for the three modes. In order that the total number of modes should equal $3N - 6 \approx 3N$, this spectrum has an upper limit at the so-called Debye frequency $\omega_D$ given by

$$3N = \int_0^{\omega_D} g(\omega)\mathrm{d}\omega = \frac{3V}{2v_s^3\pi^2}\int_0^{\omega_D} \omega^2 \mathrm{d}\omega = \frac{V\omega_D^3}{2v_s^3\pi^2}, \tag{14.29}$$

and so the Debye frequency may be written in terms of particle density $n$ as $\omega_D = v_s(6\pi^2 n)^{1/3}$.

The partition function is therefore given by

$$\ln Z_{\mathrm{vib}} = \frac{3V}{2v_s^3\pi^2}\int_0^{\omega_D} \omega^2 \ln Z_\omega \mathrm{d}\omega, \tag{14.30}$$

with $Z_\omega = [2\sinh(\hbar\omega\beta/2)]^{-1}$ from (6.24), and thermodynamic properties can then be determined. From (6.17), the vibrational heat capacity is

$$\begin{aligned}
C_V &= \frac{\mathrm{d}\langle E_{\mathrm{vib}}\rangle}{\mathrm{d}T} = \frac{1}{kT^2}\frac{\mathrm{d}^2 \ln Z_{\mathrm{vib}}}{\mathrm{d}\beta^2} = \frac{3V}{2v_s^3\pi^2 kT^2}\int_0^{\omega_D} \omega^2 \frac{\mathrm{d}^2 \ln Z_\omega}{\mathrm{d}\beta^2}\mathrm{d}\omega \\
&= \frac{3V}{2v_s^3\pi^2 kT^2}\int_0^{\omega_D} \omega^2 \frac{\mathrm{d}}{\mathrm{d}\beta}\left[-\frac{2\cosh\left(\frac{\hbar\omega\beta}{2}\right)}{2\sinh\left(\frac{\hbar\omega\beta}{2}\right)}\frac{\hbar\omega}{2}\right]\mathrm{d}\omega \\
&= \frac{3V\hbar^2}{8v_s^3\pi^2 kT^2}\int_0^{\omega_D} \frac{\omega^4}{\sinh^2\left(\frac{\hbar\omega\beta}{2}\right)}\mathrm{d}\omega.
\end{aligned} \tag{14.31}$$

At high temperature, such that $\hbar\omega_D/kT \ll 1$, the approximation $\sinh(\hbar\omega\beta/2) \approx \hbar\omega\beta/2$ can be used and we obtain

$$C_V \approx \frac{3Vk}{2v_s^3\pi^2}\int_0^{\omega_D} \omega^2 \mathrm{d}\omega = \frac{3Vk}{2v_s^3\pi^2}\frac{\omega_D^3}{3} = \frac{Vk}{2v_s^3\pi^2}v_s^3\left(\frac{6\pi^2 N}{V}\right) = 3Nk, \tag{14.32}$$

as would be expected in the classical limit, while for low temperature we write

$$C_V = \frac{3V\hbar^2}{8v_s^3\pi^2 kT^2}\left(\frac{kT}{\hbar}\right)^5 \int_0^{\hbar\omega_D\beta} \frac{x^4}{\sinh^2\left(\frac{x}{2}\right)}\mathrm{d}x \approx \frac{3Vk(kT)^3}{8v_s^3\pi^2\hbar^3}\int_0^{\infty}\frac{x^4}{\sinh^2\left(\frac{x}{2}\right)}\mathrm{d}x. \quad (14.33)$$

The integral is equal to $16\pi^4/15$, and so the low temperature heat capacity is

$$C_V \approx \frac{2\pi^2}{5}Vk\left(\frac{kT}{\hbar v_s}\right)^3 = \frac{12\pi^4}{5}Nk\left(\frac{kT}{\hbar\omega_D}\right)^3 = \frac{12\pi^4}{5}Nk\left(\frac{T}{T_D}\right)^3, \quad (14.34)$$

where we define the Debye temperature $T_D = \hbar\omega_D/k$. This is in excellent agreement with the temperature dependence of solids at low temperature. The Debye heat capacity is sketched in Figure 14.3, showing the approach towards the classical result $3Nk$ at high temperatures. It might superficially resemble the Einstein heat capacity shown in Figure 6.5 but the dependence $C_V \propto T^3$ at low temperature shown in the inset is distinctive and matches the data much better.

Finally, we determine the vibrational entropy of the Debye solid using

$$S(T) = \int_0^T \frac{C_V(T')}{T'}\,\mathrm{d}T', \quad (14.35)$$

together with (14.33), and plot it in Figure 14.3. We find that $S \sim \ln T$ at high temperatures when $C_V \to 3Nk$ while at low temperatures $S = (4/5)\pi^4 Nk(T/T_D)^3$, as indicated in the inset of Figure 14.3. This resembles the temperature dependence of the entropy of the photon gas in (13.25), for reasons that will be clear when considering the similarities in derivation. Quanta of vibrational energy are called phonons, and consequently this may be regarded as the entropy of a low temperature *phonon* gas.

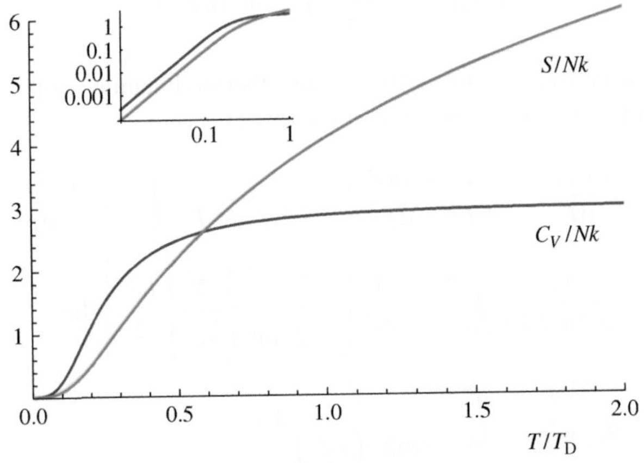


**Figure 14.3** Temperature dependence of the vibrational heat capacity (blue) and entropy per particle (green) according to the Debye model. The $T^3$ behaviour at low temperature is demonstrated in the inset using logarithmic axes.

## Exercises

**14.1** Evaluate the second virial coefficient that corresponds to an attractive square well potential between pairs of particles, such that $\phi(r) = -\phi_0$ for $r \leq r_0$ and $\phi(r) = 0$ for $r > r_0$. Determine the entropy of the gas in the classical regime.

**14.2** Evaluate the classical partition function of two particles moving in two dimensions and interacting through a harmonic potential, and identify the factor of the partition function that corresponds to rotation about the centre of mass.

**14.3** Evaluate the classical partition function $Z_4$ of a linear molecule of four particles interacting through harmonic nearest neighbour forces and factorise it in a manner similar to (14.24).

**14.4** Evaluate the entropy per particle of the Einstein solid at low temperature.

# 15

# Thermodynamics away from Equilibrium

The next few chapters are devoted to the topic of entropy generation in systems that are out of equilibrium. We examine the traditional modelling of this phenomenon, involving an extension of classical thermodynamics, and then discuss how the same process might be viewed within statistical thermodynamics.

## 15.1 Nonequilibrium Classical Thermodynamics

The rate of internal generation of entropy in a system exposed to a reservoir in classical thermodynamics was given in (2.67) as

$$\frac{\mathrm{d}S_i}{\mathrm{d}t} = \left(\frac{1}{T(t)} - \frac{1}{T_\mathrm{r}(t)}\right)\frac{\mathrm{d}E}{\mathrm{d}t} + \left(\frac{p(t)}{T(t)} - \frac{p_\mathrm{r}(t)}{T_\mathrm{r}(t)}\right)\frac{\mathrm{d}V}{\mathrm{d}t} - \left(\frac{\mu(t)}{T(t)} - \frac{\mu_\mathrm{r}(t)}{T_\mathrm{r}(t)}\right)\frac{\mathrm{d}N}{\mathrm{d}t}, \quad (15.1)$$

which takes the form of rates of change of extensive system properties $E$, $V$ and $N$ multiplied by differences in certain intensive properties of the system and reservoir. This expression is intuitively valuable, but is based on several extensions to the meaning of thermodynamics. The system is assumed to possess a temperature and chemical potential even though it is not in equilibrium with the reservoir. The meaning placed on (15.1) requires us to employ a rather more general thermodynamics, one where the state variables serve as characteristics of nonequilibrium systems, while retaining their mathematical relationship to one another, and where equations such as (15.1) relate their evolution in time.

### 15.1.1 Energy and Particle Currents and their Conjugate Thermodynamic Driving Forces

We now develop a general framework for these ideas. Figure 15.1 illustrates a network of systems between which energy and particles are exchanged. For simplicity we do not

*Statistical Physics: An Entropic Approach*, First Edition. Ian Ford.

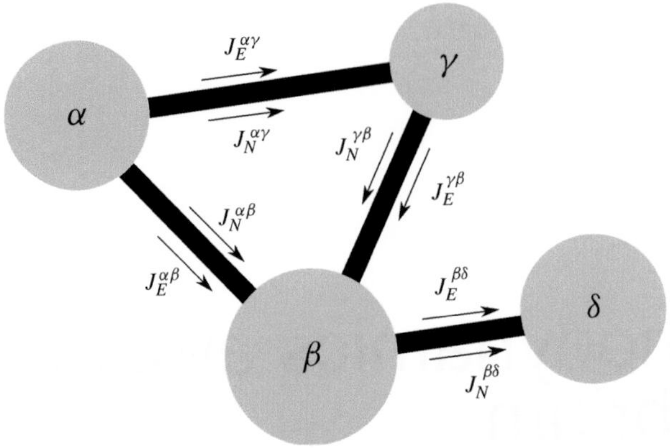


**Figure 15.1** In nonequilibrium thermodynamics we consider exchanges of energy and particles between systems that are considered to be large enough to be in a state of local quasi-equilibrium, but unlike reservoirs, their temperatures and chemical potentials respond to the flows and become time dependent. Currents flow between the systems, and entropy is generated.

consider volume exchanges. The rates of transfer of energy and particles from system $\alpha$ to system $\beta$ are to be called currents and are denoted $J_E^{\alpha\beta}$ and $J_N^{\alpha\beta}$, respectively. The notation implies that $J_E^{\alpha\beta} = -J_E^{\beta\alpha}$. Such transfers between systems allow us to develop evolution equations for the energy and particle number associated with each system:

$$\frac{dE_\alpha}{dt} = \sum_\beta J_E^{\beta\alpha} \quad , \quad \frac{dN_\alpha}{dt} = \sum_\beta J_N^{\beta\alpha} . \tag{15.2}$$

The first of these is equivalent to familiar expressions used to model the equalisation of temperature between systems. The second relation is analogous to the conservation of charge in electrical circuits, and bears a resemblance to Kirchhoff's current law.

Following (15.1), the transfers give rise to entropy production, namely

$$\frac{dS_i^{\alpha\beta}}{dt} = \left(\frac{1}{T_\beta} - \frac{1}{T_\alpha}\right) J_E^{\alpha\beta} - \left(\frac{\mu_\beta}{T_\beta} - \frac{\mu_\alpha}{T_\alpha}\right) J_N^{\alpha\beta} . \tag{15.3}$$

Each system in the network is presumed to be in a state of quasi-equilibrium, and hence can act as a reservoir for the transfer of energy and particles to a linked neighbour. The systems are finite in size, and so their temperatures and chemical potentials evolve with time, but they act like reservoirs in that we consider the entropy production to be taking place in the links between the systems, and not in the systems themselves. It is through these links that the currents pass and across which the differences in intensive thermodynamic variables apply.

Next we consider a very particular network of systems of equal volume, arranged in a line along the $x$-axis and coupled by nearest neighbours as shown in Figure 15.2. For simplicity, we ignore particle exchange for now and consider only the effect of

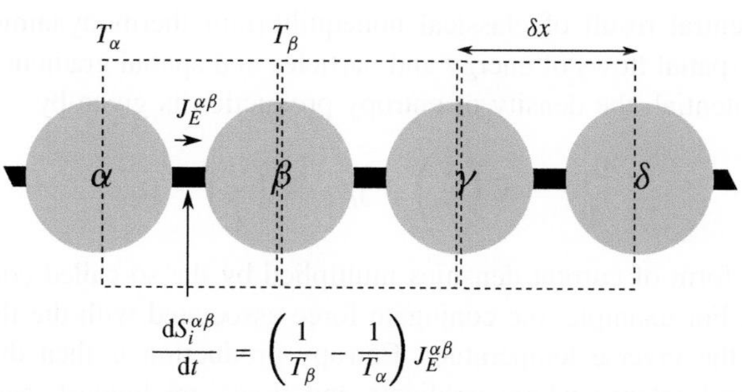


**Figure 15.2** A set of systems characterised by temperatures $T_\alpha, T_\beta, \cdots$ arranged along the $x$-axis and exchanging energy according to currents $J_E^{\alpha\beta}$ etc. Entropy is generated in the links between systems, but we can also define a density of entropy production over the volume elements shown as dashed boxes, as well as current densities $j_E$ through the interfaces between boxes.

energy transfers on the total entropy production of the network, which is

$$\frac{dS_{\text{tot}}}{dt} = \sum_{\beta>\alpha} \frac{dS_i^{\alpha\beta}}{dt} = \sum_{\beta>\alpha} \left(\frac{1}{T_\beta} - \frac{1}{T_\alpha}\right) J_E^{\alpha\beta}, \tag{15.4}$$

taking the form of a sum over contributions from each link. The notation $\beta > \alpha$ is to ensure that links are not counted twice. We now convert this sum into an integral by drawing space-filling cuboidal boxes of volume $\delta V = \mathcal{A}\delta x$ around each link, where $\mathcal{A}$ is the cross-sectional area and $\delta x$ the thickness. We define a density of entropy production $ds_i^{\alpha\beta}/dt = d(S_i^{\alpha\beta}/\delta V)/dt$ in the box containing link $\alpha\beta$, and a current per unit area or current density $j_E^{\alpha\beta} = J_E^{\alpha\beta}/\mathcal{A}$, such that

$$\frac{dS_{\text{tot}}}{dt} = \sum_{\beta>\alpha} \frac{ds_i^{\alpha\beta}}{dt}\delta V = \sum_{\beta>\alpha} \delta V \frac{1}{\delta x}\left(\frac{1}{T_\beta} - \frac{1}{T_\alpha}\right) j_E^{\alpha\beta}. \tag{15.5}$$

Next we consider small $\delta x$ while insisting that each system is large enough to be characterised by thermodynamic variables, such that we can write

$$\frac{dS_{\text{tot}}}{dt} = \int dV \, \frac{d}{dx}\left(\frac{1}{T}\right) j_{Ex}(x), \tag{15.6}$$

where we have now introduced a spatial gradient of a continuous variable $T^{-1}$, and $j_{Ex}$ is the current density in the $x$ direction.

Extending this argument to three dimensions, and including particle exchange as well, we can write the total entropy production as the integral

$$\frac{dS_{\text{tot}}}{dt} = \int \frac{ds_{\text{tot}}}{dt} dV = \int dV \left[\nabla\left(\frac{1}{T}\right) \cdot \mathbf{j}_E - \nabla\left(\frac{\mu}{T}\right) \cdot \mathbf{j}_N\right], \tag{15.7}$$

where we have now introduced a particle current density $\mathbf{j}_N$.

This is the central result of classical nonequilibrium thermodynamics. In a system where there are spatial flows of energy and particles and spatial gradients of temperature and chemical potential, the density of entropy production is given by

$$\frac{\mathrm{d}s_{\mathrm{tot}}}{\mathrm{d}t} = \nabla\left(\frac{1}{T}\right) \cdot \mathbf{j}_E - \nabla\left(\frac{\mu}{T}\right) \cdot \mathbf{j}_N, \tag{15.8}$$

which takes the form of current densities multiplied by the so-called *conjugate* thermodynamic forces. For example, the conjugate force associated with the flow of energy is the gradient of the inverse temperature. Entropy production is then distributed across space, but is most intense where gradients and flows are highest, namely where the system deviates most from equilibrium.

Further development follows if we express the gradient of $\mu/T$ in terms of those of inverse temperature and particle density $n$. We write

$$\mathrm{d}(\mu/T) = \mu\mathrm{d}(T^{-1}) + T^{-1}\mathrm{d}\mu = \mu\mathrm{d}(T^{-1}) + T^{-1}\left(\frac{\partial\mu}{\partial T}\right)_n \mathrm{d}T + T^{-1}\left(\frac{\partial\mu}{\partial n}\right)_T \mathrm{d}n, \tag{15.9}$$

and so the spatial gradients are related by

$$\nabla\left(\frac{\mu}{T}\right) = \left[\mu - T\left(\frac{\partial\mu}{\partial T}\right)_n\right]\nabla\left(\frac{1}{T}\right) + \frac{1}{T}\left(\frac{\partial\mu}{\partial n}\right)_T \nabla n. \tag{15.10}$$

Furthermore

$$T\left(\frac{\partial\mu}{\partial T}\right)_n = T\left(\frac{\partial\mu}{\partial T}\right)_{N,V} = -T\left(\frac{\partial S}{\partial N}\right)_{V,T}, \tag{15.11}$$

using the Maxwell relation (3.39), and so

$$\mu - T\left(\frac{\partial\mu}{\partial T}\right)_n = \mu + T\left(\frac{\partial S}{\partial N}\right)_{T,V} = \left(\frac{\partial F}{\partial N}\right)_{V,T} + T\left(\frac{\partial S}{\partial N}\right)_{V,T} = \left(\frac{\partial E}{\partial N}\right)_{V,T} = \hat{e}, \tag{15.12}$$

using (3.35) and $F = E - TS$, where $\hat{e}$ is the energy change associated with adding one particle to a system without a change in temperature or volume. If we recognise that the energy current density is a sum of components corresponding to heat flow and particle flow, or $\mathbf{j}_E = \mathbf{j}_Q + \hat{e}\mathbf{j}_N$, then we can write

$$\begin{aligned}\frac{\mathrm{d}s_{\mathrm{tot}}}{\mathrm{d}t} &= \nabla\left(\frac{1}{T}\right) \cdot [\mathbf{j}_Q + \hat{e}\mathbf{j}_N] - \left[\hat{e}\nabla\left(\frac{1}{T}\right) + \frac{1}{T}\left(\frac{\partial\mu}{\partial n}\right)_T \nabla n\right] \cdot \mathbf{j}_N \\ &= -\frac{1}{T^2}\nabla T \cdot \mathbf{j}_Q - \frac{1}{T}\left(\frac{\partial\mu}{\partial n}\right)_T \nabla n \cdot \mathbf{j}_N,\end{aligned} \tag{15.13}$$

where we can regard the conjugate thermodynamic forces associated with the heat and particle currents to be proportional to the temperature and particle density gradients, respectively, which is what we might have expected on the grounds of intuition. If we allowed system volume changes to take place as well, this would lead to an additional term on the right hand side to match the relevant term in (15.1).

A system where heat flows as a consequence of a gradient in temperature is shown in Figure 15.3. Hotter regions are shown in darker colour, and the magnitude and direction

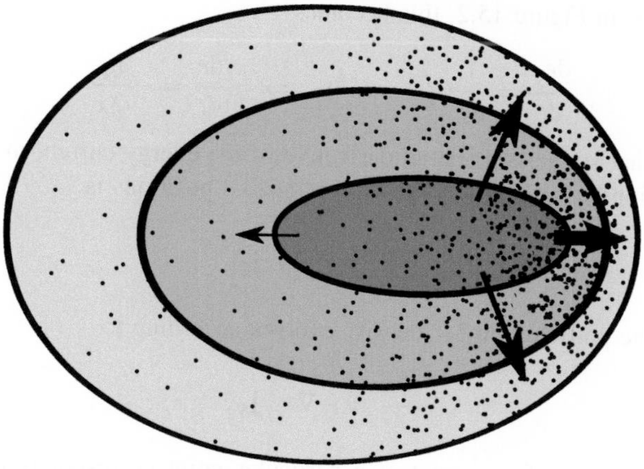

Figure 15.3 Contours of temperature and vectors of heat current density, together with the density of entropy production shown as a speckle pattern.

of the heat current densities are indicated with arrows. The density of entropy production is represented by the speckles and is strongest where the gradients and flows are largest.

The heat current density is commonly taken to be proportional to the temperature gradient:

$$\mathbf{j}_Q = -\kappa \nabla T, \tag{15.14}$$

where $\kappa$ is the thermal conductivity, and similarly

$$\mathbf{j}_N = -D \nabla n, \tag{15.15}$$

is often used, where $D$ is the diffusion coefficient. These are empirical expressions known as Fourier's law and Fick's law, respectively. More complicated transport laws might be imagined, with a particle current responding to a temperature gradient, for example, but here we ignore this for simplicity. Thus

$$\frac{\mathrm{d}s_{\mathrm{tot}}}{\mathrm{d}t} = \frac{\kappa}{T^2}(\nabla T)^2 + \frac{D}{T}\left(\frac{\partial \mu}{\partial n}\right)_T (\nabla n)^2, \tag{15.16}$$

and this expression makes it quite apparent that the local rate of entropy production in the flow is never negative, in accordance with the second law. This follows since $\kappa$ and $D$ are both positive, and the chemical potential at constant $T$ increases with particle density: using the Gibbs–Duhem equation (3.66),

$$\left(\frac{\partial \mu}{\partial n}\right)_T = \left(\frac{\partial \mu}{\partial p}\right)_T \left(\frac{\partial p}{\partial n}\right)_T = v\left(\frac{\partial p}{\partial n}\right)_T \geq 0, \tag{15.17}$$

due to the fact that the volume per particle $v$ is positive, and empirically the pressure of a system never decreases with an increase in particle density at constant temperature.

The transport laws (15.14) and (15.15) also allow us to cast the energy and particle conservation equations (15.2) into differential form. We write $E_\alpha = e_\alpha \delta V$ and

$$\frac{\mathrm{d}e_\alpha}{\mathrm{d}t}\delta V = \sum_\beta \delta V \frac{1}{\delta x} j_E^{\beta\alpha}, \tag{15.18}$$

and for system $\gamma$ in Figure 15.2, this becomes

$$\frac{de_\gamma}{dt} = \frac{1}{\delta x}(j_E^{\beta\gamma} - j_E^{\gamma\delta}) \quad \Rightarrow \quad \frac{de}{dt} = -\frac{\delta j_{Ex}}{\delta x}, \tag{15.19}$$

using notation that suggests a spatial derivative of the energy current in the $x$ direction. The generalisation to 3-d and a continuum of spatial positions is

$$\frac{\partial e}{\partial t} = -\nabla \cdot \mathbf{j}_E, \tag{15.20}$$

and similarly, the particle number density evolves according to

$$\frac{\partial n}{\partial t} = -\nabla \cdot \mathbf{j}_N. \tag{15.21}$$

A positive divergence of a current at a particular point corresponds intuitively to an overall flow away from that point, and hence a fall in the density of the appropriate conserved quantity, so the sign on the right hand side of these so-called continuity equations makes intuitive sense. Inserting (15.15), we deduce that

$$\frac{\partial n}{\partial t} = \nabla \cdot (D\nabla n) = D\nabla^2 n, \tag{15.22}$$

with the last form applying as long as $D$ is the same at all spatial points. This is known as the diffusion equation. The counterpart for heat flow is obtained by combining (2.56), (3.44) and (15.2) and writing $de = c_V dT + \hat{e}dn$ where $c_V$ is the heat capacity per unit volume, as well as $\mathbf{j}_E = \mathbf{j}_Q + \hat{e}\mathbf{j}_N$ such that (15.20) becomes

$$c_V \frac{\partial T}{\partial t} = -\nabla \cdot \mathbf{j}_Q = \kappa \nabla^2 T, \tag{15.23}$$

assuming that $\hat{e}$, $c_V$ and $\kappa$ are also constants. This final form is known as the heat equation.

### 15.1.2 Entropy Production in Constrained and Evolving Systems

Entropy is produced when thermodynamic state variables change nonquasistatically as a consequence of the flows of energy and particles. It is therefore associated with the *evolution* or *relaxation* of a system. But entropy is also generated if state variables are time independent but there are nonzero currents. This is a constrained system since it must be maintained away from equilibrium by external boundary conditions. It is known as a nonequilibrium steady state.

We briefly describe an example of this mode of entropy production using three coupled systems $\alpha$, $\beta$ and $\gamma$ illustrated in Figure 15.4. We consider energy transfers only, and employ an equation for the evolution of the energy of system $\beta$ of the form

$$\frac{dE_\beta}{dt} = J_E^{\alpha\beta} - J_E^{\beta\gamma}, \tag{15.24}$$

with energy currents given by a version of Fourier's law (Newton's law of cooling): $J_E^{\alpha\beta} = -K(T_\beta - T_\alpha)$ where $K$ is a constant. In a case of relaxation, systems $\alpha$ and $\gamma$

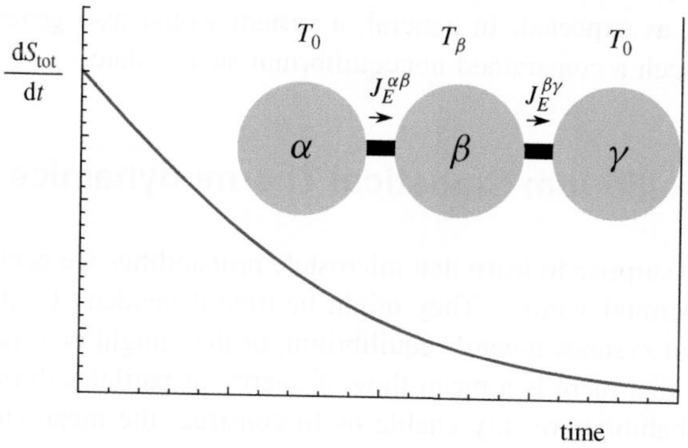


**Figure 15.4** The decline in the rate of production of entropy as a system relaxes towards equilibrium. In contrast, if $T_\alpha = T_1 > T_\gamma = T_2$, then a nonequilibrium steady state would be possible with a time independent rate of entropy production.

are held at a temperature $T_0$ and by introducing a heat capacity $C_\beta = dE_\beta/dT_\beta$ we find that the time dependence of the temperature $T_\beta$ of system $\beta$ is given by

$$\frac{dT_\beta}{dt} = \frac{K}{C_\beta}[T_0 - T_\beta - (T_\beta - T_0)] = -\frac{2K}{C_\beta}(T_\beta - T_0), \tag{15.25}$$

and so $T_\beta - T_0 = \Delta T \exp(-2Kt/C_\beta)$, where $\Delta T$ is the value of $T_\beta - T_0$ at $t = 0$. From (15.4) the rate of entropy production is

$$\begin{aligned}\frac{dS_{tot}}{dt} &= \left(\frac{1}{T_\beta} - \frac{1}{T_0}\right) J_E^{\alpha\beta} + \left(\frac{1}{T_0} - \frac{1}{T_\beta}\right) J_E^{\beta\gamma} = 2\left(\frac{1}{T_\beta} - \frac{1}{T_0}\right) K(T_0 - T_\beta) \\ &= \frac{2K(\Delta T)^2 \exp\left(-\frac{4Kt}{C_\beta}\right)}{T_0\left[T_0 + \Delta T \exp\left(-\frac{2Kt}{C_\beta}\right)\right]},\end{aligned} \tag{15.26}$$

and this is plotted in Figure 15.4: it steadily declines as equilibrium is approached.

In contrast, consider a case of a nonequilibrium steady state brought about by fixed boundary conditions $T_\alpha = T_1$ and $T_\gamma = T_2 < T_1$. By inspection, the steady state temperature of system $\beta$ is given by $(T_1 + T_2)/2$, and the steady state rate of entropy production is

$$\begin{aligned}\frac{dS_{tot}}{dt} &= \left(\frac{1}{T_\beta} - \frac{1}{T_1}\right) J_E^{\alpha\beta} + \left(\frac{1}{T_2} - \frac{1}{T_\beta}\right) J_E^{\beta\gamma} \\ &= \left(\frac{2}{T_1 + T_2} - \frac{1}{T_1}\right) K\frac{1}{2}(T_1 - T_2) + \left(\frac{1}{T_2} - \frac{2}{T_1 + T_2}\right) K\frac{1}{2}(T_1 - T_2) \\ &= \left[\frac{T_1 - T_2}{T_1(T_1 + T_2)} + \frac{T_1 - T_2}{T_2(T_1 + T_2)}\right]\frac{K}{2}[T_1 - T_2] = \frac{K}{2}\frac{(T_1 - T_2)^2}{T_1 T_2},\end{aligned} \tag{15.27}$$

which is positive, as expected. In general, a system would also generate entropy as it evolves towards such a constrained nonequilibrium steady state.

## 15.2 Nonequilibrium Statistical Thermodynamics

It will come as no surprise to learn that microstate probabilities are central to nonequilibrium statistical thermodynamics. They might be time dependent, to allow a description of the relaxation of systems towards equilibrium, or they might be time independent but constrained, such that there is a mean flow of energy or particles through the system.

Microstate probabilities readily enable us to construct the mean energy and particle number, but just as in classical nonequilibrium thermodynamics, concepts such as temperature, chemical potential and entropy do not necessarily extend to nonequilibrium situations. For example, both the Boltzmann and Gibbs entropies are defined for equilibrium conditions and can be represented in terms of equilibrium probabilities. But the Shannon entropy $S_I = -k\sum_i P_i(t)\ln P_i(t)$ is constructed to be more general and might conceivably be employed for situations where the probabilities $P_i$ evolve in time. This does not guarantee that $S_I$ matches the properties of the nonequilibrium entropy that appears in classical thermodynamics, of course. First we must consider how probabilities evolve, and what this might mean.

### 15.2.1 Probability Flow and the Principle of Equal a Priori Probabilities

If probabilities are a distillation of our best judgement of the likelihood that a system might be found in a given microscopic state, then they might naturally be time dependent in certain situations, for example, in the period just after the acquisition of information. The time dependence should then reflect, at least in some way, the microscopic dynamics of the system, but it is easy to show that this introduces a problem.

Consider a system of three oscillators with a familiar triangular phase space as shown in Figure 15.5, evolving according to a known set of microscopic dynamical laws, and starting from a known initial probability distribution over the phase space. For example, let us take the system at $t = 0$ to have a probability 1/3 of being in each of the microstates at the vertices of the triangle, indicated by the columns. The initial Shannon entropy is $-k\sum P_i \ln P_i = (-k(1/3)\ln(1/3)) \times 3 = k\ln 3$. But since we know the dynamical

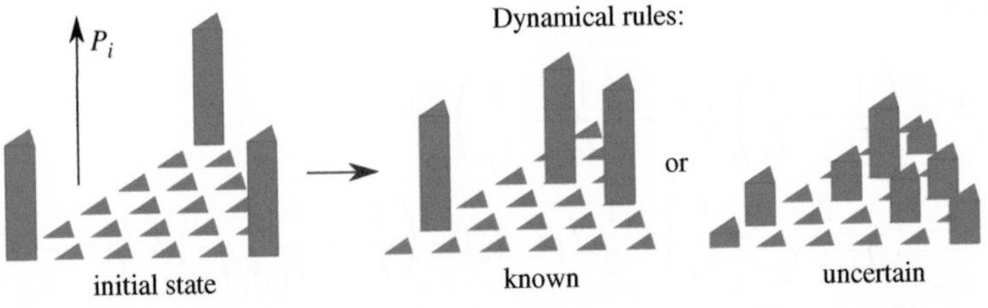


**Figure 15.5** The transport of probability as time progresses: either rigidly and deterministically according to Liouville's theorem if the rules of dynamics are known, or stochastically, with some spreading out, if they are uncertain.

rules, at a later time we remain just as uncertain as to the identity of the occupied microstate: it could be the end point of a trajectory starting from any of the three possible initial states. This situation is shown as the first choice of outcome in Figure 15.5: the probabilities have been shifted to three new microstates, and the Shannon entropy remains equal to $k \ln 3$.

The point is that such a model cannot provide us with a working representation of nonequilibrium thermodynamics: the system is not initially in equilibrium (since the initial $P_i$ do not satisfy the principle of equal a priori probabilities) and yet its entropy remains constant as time progresses; it does not seem to be able to relax such that the microstate probabilities equalise. In systems that evolve according to known laws of motion, the microstate probabilities are transported along trajectories in phase space, a result known as Liouville's theorem, and the Shannon entropy is conserved. This difficulty was known to Gibbs and various interpretations have been sought.

If we are to employ Shannon entropy in general discussions of entropy change, and by implication the Gibbs and Boltzmann forms for equilibrium situations, then the concession we must make is the insistence that the dynamical rules determining trajectories in the system phase space are known. If we are not sure where a trajectory starting at a particular microstate will end up after a certain time has elapsed, then Liouville's theorem is invalid and the Shannon entropy can change. The rigid transport of probability shown in Figure 15.5 would be replaced by the second option: a spreading out of the probability amongst further microstates.

What can such a situation mean? It could be that we actually do not know the dynamical rules exactly, and the residual uncertainty in parameters or even in the form of the equations of motion can give rise to uncertainty in destination. But leaving aside these problems, it might instead be the case that the microstates illustrated in Figure 15.5 are not sufficiently fine in scale to allow us to state with precision where the trajectories emanating from them might lead. They might actually represent a collection of finer-scale microstates, each of which would give rise to a different trajectory, and we cannot predict which one is actually followed. This would require the trajectories to diverge from one another, which does not always happen, but arguably is a natural state of affairs in a complex system: it is an aspect of *chaotic dynamics*.

As an extension to this, if a system with fine grained microstates is coupled to a reservoir that is by definition coarse grained, then system trajectories are not exactly predictable because the initial microstate of the reservoir is not known, and the dynamical rules describing the system–reservoir interactions would reflect this. A distinguishing feature of models that can lead to the growth of uncertainty is that the dynamical rules involve macrostate as well as microstate variables.

We can therefore make progress by accepting that phase space trajectory dynamics might be uncertain rather than deterministic, owing to deficiencies in our microscopic perception. We shall develop models of uncertain, or *stochastic* dynamics in Chapter 16, but for now, we surmise that such a model might give rise to equal a priori probabilities, or something similar to them, in an isolated system, as it would have the effect of spreading probabilities out across phase space.

As we saw in Section 8.3, equal microstate probabilities give rise to a global maximum in Shannon entropy. Shannon entropy is a measure of the uncertainty of a probability distribution, and an appealing intuitive view is that the dynamics naturally increase this measure because of inherent dynamical uncertainties, until it can be increased no longer. The principle of equal a priori probabilities for isolated systems would then be a consequence of the dynamics. We take this further in the next section.

### 15.2.2 The Dynamical Basis of the Principle of Entropy Maximisation

In this section we consider in general terms how the evolution of probabilities and a natural increase in uncertainty might account for the procedure where we perform a constrained maximisation of the Shannon entropy in order to identify a state of equilibrium.

We imagine that the dynamics are such that the total Shannon entropy of a world described in part by macroscopic variables should increase with time. We then divide the world into a system and its environment and suppose that the total entropy is a sum of a system entropy that we represent using the Shannon expression in terms of time-dependent microstate probabilities $P_i(t)$, and a reservoir entropy $S_\mathrm{r}$ that also depends on time as a result of mean energy or particle exchanges with the system.

Incremental changes in the reservoir entropy satisfy the fundamental relation (2.49) or (8.21), which we write in the form

$$T_\mathrm{r}\mathrm{d}S_\mathrm{r} = \mathrm{d}\langle E_\mathrm{r}\rangle - \mu_\mathrm{r}\mathrm{d}\langle N_\mathrm{r}\rangle + \sum_j f_{x_j}\mathrm{d}x_j, \tag{15.28}$$

where the brackets represent averages over the state of the reservoir, which is presumed to remain in quasi-equilibrium throughout. The last contribution is related to changes in parameters $x_j$ such as volume. For now, we ignore these, and we also ignore contributions due to particle exchange. Then the rate of change of total nonequilibrium entropy is

$$\frac{\mathrm{d}S_\mathrm{tot}}{\mathrm{d}t} = \frac{\mathrm{d}S_I}{\mathrm{d}t} + \frac{\mathrm{d}S_\mathrm{r}}{\mathrm{d}t} = \frac{\mathrm{d}S_I}{\mathrm{d}t} + \frac{1}{T_\mathrm{r}}\frac{\mathrm{d}\langle E_\mathrm{r}\rangle}{\mathrm{d}t} = \frac{\mathrm{d}S_I}{\mathrm{d}t} - \frac{1}{T_\mathrm{r}}\frac{\mathrm{d}\langle E\rangle}{\mathrm{d}t} = \frac{\mathrm{d}}{\mathrm{d}t}(S_I - T_\mathrm{r}^{-1}\langle E\rangle), \tag{15.29}$$

where $\langle E\rangle = \sum_i E_i P_i$, noting that the total energy $E + E_\mathrm{r}$ is fixed, such that the rates of change of the average system and reservoir energies are equal and opposite.

The evolution of $S_\mathrm{tot}$ or $S_I - T_\mathrm{r}^{-1}\langle E\rangle$ towards a maximum as a result of the dynamics of the $P_i$ is equivalent to the maximisation of $-k\sum_i P_i \ln P_i - \lambda'\sum_i E_i P_i$ over the $P_i$ subject to the normalisation condition $\sum_i P_i = 1$, where $\lambda' = T_\mathrm{r}^{-1} = k\beta$; in other words, it is identical to the axiomatic procedure for identifying the equilibrium probabilities for the canonical ensemble discussed in Section 8.3. If we were to retain the term in (15.28) corresponding to particle exchange, we would recover the Shannon procedure for deriving equilibrium probabilities for the grand canonical ensemble as well.

The constrained maximisation of Shannon entropy, and the equilibrium microstate probabilities used in the various ensembles of statistical thermodynamics, would then simply be a reflection of the increase in total *uncertainty* embodied in $S_\mathrm{tot}$, which would be a consequence of the dynamics of probabilities, a topic that we address next.

## Exercises

**15.1** Calculate the total entropy produced in the relaxation discussed in Section 15.1.2.

**15.2** Solve the heat equation to obtain the steady state temperature profile for a copper rod of length 1 m and cross-sectional area 1 $cm^2$ with its ends maintained at temperatures of 0 °C and 40 °C, and determine the rate of production of entropy if the thermal conductivity of copper is 385 $Wm^{-1} K^{-1}$.

**15.3** Consider a classical oscillator with known initial position and velocity. Sketch the probability distribution over its $(x, v)$ phase space at a later time if left isolated. Does its entropy increase?

# 16

# The Dynamics of Probability

In this chapter we discuss mathematical models that attempt to represent the evolution of uncertainty, that is to say the time dependence of the probabilities of occupation of microstates of a system. These matters are not often covered in an undergraduate text on statistical physics, but are central to an understanding of the topic. The material can seem forbidding, hence a hazard warning sign!

Caution: Entropy

## 16.1 The Discrete Random Walk

We start with a simple case: the symmetric discrete random walk in 1-d. A particle is initially at the origin, and every timestep of length $\tau$ it makes a move through a distance $a$, but to the left or right with equal probability. We presume that there are underlying *deterministic* dynamical rules that produce such steps, but that we lack the details of their operation, and so instead we employ a *stochastic* scheme that captures the same events through the explicit incorporation of randomness. If the probabilities of a step to left or right were different, the scheme would be called an asymmetric random walk.

Notice the 'Markov' property of this rule for updating the configuration: the position after the $(n+1)$th timestep is determined only from information about the situation after the $n$th timestep. We could make the update rule depend on information further back in time, and such a scheme with an extended memory is called 'non-Markovian'. However, we shall limit ourselves to Markovian dynamics.

We can readily generate a set of realisations of a symmetric random walk by the selection of steps through repeatedly tossing a coin, as illustrated in Figure 16.1. After $n$ steps (we have chosen $n = 5$), the particle will lie somewhere in the range from $-na$ to $+na$. The probability $P_n(x_m)$ of ending up at position $x_m = ma$ after time $n\tau$ is the probability of taking $(n+m)/2$ steps to the right and $(n-m)/2$ steps to the left, summed over all possible sequences. One of several such paths that reach the point $m = -1$ is illustrated, and an exhaustive search yields nine others.

*Statistical Physics: An Entropic Approach*, First Edition. Ian Ford.

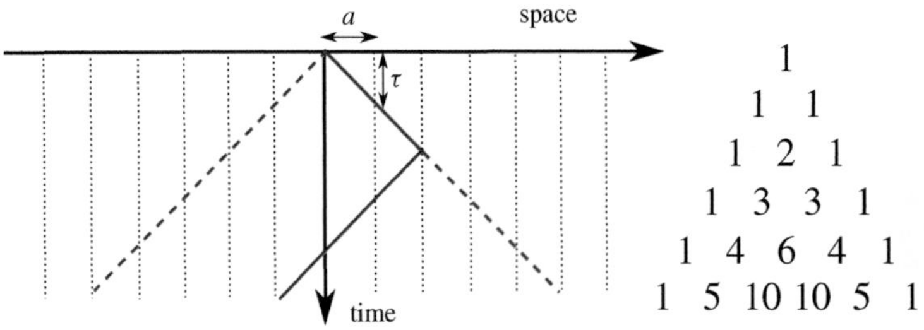


**Figure 16.1** Three realisations of the symmetric random walk in 1-d, showing the unique ways to reach $x = \pm 5a$ after five steps starting from the origin, and one of the ten paths that end up at $x = -a$. Pascal's triangle on the right illustrates the computation of the numbers of such paths.

In fact we can determine the number of paths leading to each destination by constructing Pascal's triangle, where numbers in each row are obtained by summing the two numbers positioned diagonally above, as demonstrated in Figure 16.1. This reflects the process of generating a path by making $n$ appropriate 50 : 50 decisions, or equivalently it follows from the fact that each point can be reached from one of two points at the preceding timestep. The probability of reaching a destination is the relevant number in Pascal's triangle multiplied by the probability of each individual path, which is $\left(\frac{1}{2}\right)^n$, and this determines the shape of the probability distribution over the phase space of particle positions, and hence the various statistical properties of the walk. In general, the $\ell$th moment of the probability distribution after step $n$ is defined as

$$\langle x^\ell \rangle_n = \sum_{m=-\infty}^{+\infty} x_m^\ell P_n(x_m). \tag{16.1}$$

The $\ell = 1$ moment is the mean displacement while the $\ell = 2$ moment is the mean square displacement.

If we cannot list every possible realisation of a random walk, we can instead numerically generate as many as might be necessary to give us sufficiently accurate frequencies of various outcomes, and then use these as estimates of the probability distribution resulting from the dynamics. Such simulation is called 'Monte Carlo': each realisation is generated randomly, just as the motion of a ball on the roulette wheel is supposed to be controlled by chance.

However, there are ways to calculate evolving probability distributions such as $P_n(x_m)$ for the symmetric random walk in 1-d, and in the next section, we develop a powerful technique.

## 16.2 Master Equations

We solved the symmetric random walk problem by explicitly counting trajectories emanating from an initial point into a phase space. There is an equivalent point of view where we imagine the propagation of *probability* into the phase space, such that the probability distribution evolves as time progresses. The rules of propagation are based on the multiplicative and additive laws of probability, or intuitively on the redistribution of probability brought about by the available decision processes.

For example, in the case of a symmetric random walk, probability is split into two whenever a choice of step is made. The probability that the particle is at position $x_m$ at timestep $n+1$ is therefore one half the probability that it was located at $x_{m-1}$ at timestep $n$, plus one half the probability that it was at $x_{m+1}$. In short, the probability evolves according to the rule

$$P_{n+1}(x_m) = \tfrac{1}{2}P_n(x_{m-1}) + \tfrac{1}{2}P_n(x_{m+1}), \tag{16.2}$$

which conveys an idea of probability as a property that is divided as decisions are made, but accumulated if the outcomes are the same. On reflection, this is precisely the way Pascal's triangle works.

More generally, if there are many possible transitions in the previous timestep that have the effect of bringing a particle to the position $x_m$, we would write

$$P_{n+1}(x_m) = \sum_{m'=-\infty}^{\infty} T(x_m - x_{m'}|x_{m'})P_n(x_{m'}), \tag{16.3}$$

where $T(\Delta x|x)$ is the transition probability for making a step of size $\Delta x$ given a starting position of $x$. In principle, this could be time dependent. Note that the scheme is Markovian: the probability distribution at timestep $n+1$ depends only on the probability distribution at timestep $n$, and on transition probabilities that have no memory of previous times. Equation (16.3) is called a *master equation*. The origin of the name is a little obscure, but it is appropriate because it provides the basic framework for the dynamics of probabilities.

For the symmetric random walk, there are only two nonzero transition probabilities, describing steps of length $a$ to left or right, and they are both equal to $1/2$. The transition probability is normalised such that

$$\sum_{m=-\infty}^{\infty} T(x_m - x_{m'}|x_{m'}) = 1, \tag{16.4}$$

since there is a probability of unity that *some* transition is made starting from position $x_{m'}$.

The probability of reaching position $m$ at time $n+1$ is a sum of probabilities of all possible paths that lead to this point. In the last section we stated that the probability of a particular path of $n$ steps is equal to $(1/2)^n$. More formally, this path probability is a product of transition probabilities for each step taken along the path, multiplied by the probability of starting at the appropriate initial position. We shall consider this representation of path probabilities again in Chapter 17.

For the random walk, there are only two possible transitions starting at $x_{m'}$ and we write

$$T(x_m - x_{m'}|x_{m'}) = \tfrac{1}{2}(\delta_{m-1\ m'} + \delta_{m+1\ m'}), \tag{16.5}$$

where we employ the Kronecker delta $\delta_{ij}$, which is unity if the two indices are equal, but zero otherwise. The two terms in the brackets represent, respectively, steps to the

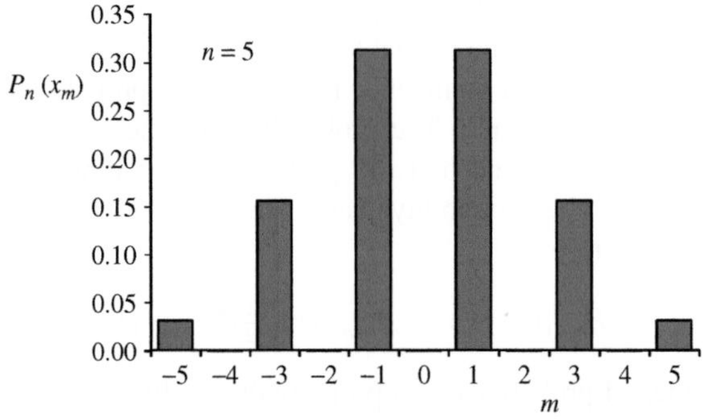


**Figure 16.2** Probability distribution for the symmetric random walk in 1-d after five timesteps, starting from $m = 0$.

right ($m = m' + 1$) and to the left ($m = m' - 1$). Hence, as proposed earlier on intuitive grounds, the master equation for this system is

$$P_{n+1}(x_m) = \sum_{m'=-\infty}^{\infty} \frac{1}{2}(\delta_{m-1\ m'} + \delta_{m+1\ m'})P_n(x_{m'}) = \frac{1}{2}P_n(x_{m-1}) + \frac{1}{2}P_n(x_{m+1}), \quad (16.6)$$

since the properties of the Kronecker delta are such that $\sum_j \delta_{ij} F_j = F_i$.

### 16.2.1 Solution to the Random Walk

We can often employ a discrete Fourier transform to solve master equations where the probability distribution is defined for positions $x_m$ with positive and negative integer $m$. We define the *characteristic function*

$$G_n(k) = \sum_{m=-\infty}^{\infty} e^{ikx_m} P_n(x_m), \quad (16.7)$$

such that by multiplying by $e^{ikx_m}$ and summing, the master equation (16.6) can be turned into

$$\sum_{m=-\infty}^{\infty} e^{ikx_m} P_{n+1}(x_m) = \frac{1}{2} \sum_{m=-\infty}^{\infty} e^{ik(x_{m-1}+a)} P_n(x_{m-1}) + \frac{1}{2} \sum_{m=-\infty}^{\infty} e^{ik(x_{m+1}-a)} P_n(x_{m+1}), \quad (16.8)$$

where we note that $x_{m-1} + a = x_m = x_{m+1} - a$. In the first sum on the right hand side, we define a new summation index $M = m - 1$ and in the second we define $M' = m + 1$. The bounds on the sums are unchanged, and we get

$$\sum_{m=-\infty}^{\infty} e^{ikx_m} P_{n+1}(x_m) = \frac{1}{2} \sum_{M=-\infty}^{\infty} e^{ik(x_M+a)} P_n(x_M) + \frac{1}{2} \sum_{M'=-\infty}^{\infty} e^{ik(x_{M'}-a)} P_n(x_{M'}), \quad (16.9)$$

which clearly corresponds to

$$G_{n+1}(k) = \tfrac{1}{2}e^{ika} G_n(k) + \tfrac{1}{2}e^{-ika} G_n(k) = \cos(ka) G_n(k). \quad (16.10)$$

Since the particle is at the origin at $t = 0$, we have $P_0(x_m) = \delta_{m0}$ such that $G_0(k) = 1$ from (16.7). By iteration, therefore, we have

$$G_n(k) = \cos^n(ka), \tag{16.11}$$

and all we have to do now is invert the discrete Fourier transform to obtain $P_n(x_m)$. The simplest procedure is to use the binomial expansion to write

$$G_n(k) = (p+q)^n = \sum_{\ell=0}^{n} p^{n-\ell} q^{\ell} \frac{n!}{\ell!(n-\ell)!}, \tag{16.12}$$

with $p = \frac{1}{2}\mathrm{e}^{ika}$ and $q = \frac{1}{2}\mathrm{e}^{-ika}$. Writing $m = n - 2\ell$, this becomes

$$G_n(k) = \frac{1}{2^n} \sum_{m=n}^{-n} \frac{n!}{\left(\frac{n-m}{2}\right)!\left(\frac{n+m}{2}\right)!} \mathrm{e}^{ika[(n-\ell)-\ell]}, \tag{16.13}$$

noting that $m$ is spaced in increments of two. Note also that $(n-\ell)-\ell$ is equal to $m$ and that $ma = x_m$. Hence

$$G_n(k) = \frac{1}{2^n} \sum_{m=-n}^{n} \frac{n!}{\left(\frac{n-m}{2}\right)!\left(\frac{n+m}{2}\right)!} \mathrm{e}^{ikx_m}, \tag{16.14}$$

and by referring to the definition of the characteristic function (16.7), we read off the coefficients of $\mathrm{e}^{ikx_m}$ to extract the probabilities:

$$P_n(x_m) = \frac{1}{2^n} \frac{n!}{\left(\frac{n-m}{2}\right)!\left(\frac{n+m}{2}\right)!} \tag{16.15}$$

for $|m| \le n$ and even $(n-m)$, and $P_n(x_m) = 0$ otherwise. This distribution is shown in Figure 16.2 for $n = 5$. This is entirely consistent with the numbers in Pascal's triangle that were used earlier to generate the probability distribution by explicitly counting paths.

### 16.2.2 Entropy Production during a Random Walk

The probabilities of walking to positions $x_m$ after time $n\tau$ allow us to determine the time dependence of the Shannon entropy for this process. It is clearly given by

$$S_I = -k \sum_m P_n(x_m) \ln P_n(x_m) = -k \sum_m \frac{1}{2^n} \frac{n!}{\left(\frac{n-m}{2}\right)!\left(\frac{n+m}{2}\right)!} \ln\left[\frac{1}{2^n} \frac{n!}{\left(\frac{n-m}{2}\right)!\left(\frac{n+m}{2}\right)!}\right], \tag{16.16}$$

and it is equal to zero at $n = 0$, corresponding to complete certainty in location, and increases with time as uncertainty develops, as shown in Figure 16.3.

If the space on which the walk takes place is infinite, the entropy continues to grow, but if the space is a ring, such that a move to the right from $x = Ma$ takes the particle to $x = -Ma$, and vice versa, then intuitively all the probabilities will evolve towards a constant equal to $(2M+1)^{-1}$. The Shannon entropy then increases towards an asymptotic value of $k\ln(2M+1)$. What we find in such a case is an analogue of the free

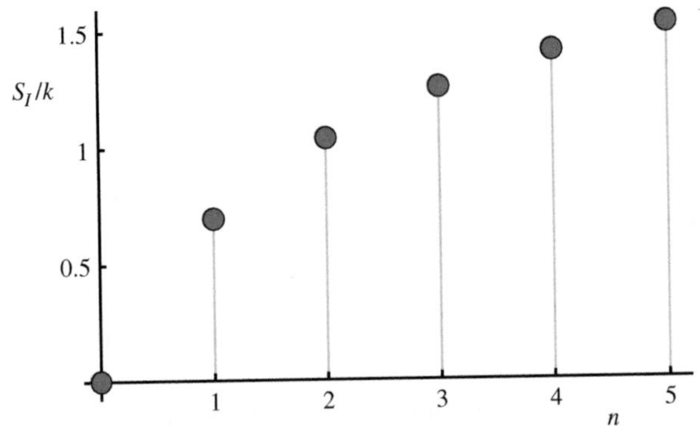


**Figure 16.3** Shannon entropy after $n$ steps of a 1-d random walk.

expansion of an isolated, initially constrained system into a larger phase space, with dynamics that generate a final equilibrium state with equal microstate probabilities. The Shannon entropy increases throughout the expansion until it reaches a value given by Boltzmann's expression. Dynamics such as these can clearly form a basis for developing a treatment of nonequilibrium processes.

## 16.3 The Continuous Random Walk and the Fokker–Planck Equation

We have discussed the evolving probabilities on a discrete phase space for a particle undergoing a random walk, and it is also of interest to consider the evolution of probability density functions in continuous time over a continuum of positions. This would give us insight into the change in statistical properties of the classical kinetic energy of a particle as it is heated, for example. We shall approach this by deriving the appropriate equations that describe a continuous random walk, where the spatial step and the time step both become infinitesimal.

It is convenient to define the probability density function $p(x,t)$ such that $p(x,t)\mathrm{d}x$ is the probability that the particle lies in the region $x-(1/2)\mathrm{d}x \to x+(1/2)\mathrm{d}x$. We also define a Markovian transition probability *density* $\mathcal{T}(x-x'|x')$ such that $\mathcal{T}(x-x'|x')\mathrm{d}x'$ is the probability that a transition through displacement $\Delta x = x-x'$ is made from a point lying in the region $x' \pm \mathrm{d}x'/2$ in a period $\tau$ starting from time $t$. An integral form of the master equation (16.3) for the discrete case may then be written as

$$p(x,t+\tau) = \int_{-\infty}^{\infty} p(x',t)\mathcal{T}(x-x'|x',t)\mathrm{d}x' = \int_{-\infty}^{\infty} p(x-\Delta x,t)\mathcal{T}(\Delta x|x-\Delta x,t)\mathrm{d}\Delta x, \tag{16.17}$$

which is often called the Chapman–Kolmogorov equation. Note that $\mathcal{T}$ has dimensions of inverse length and is normalised according to $\int \mathcal{T}(x-x'|x',t)\mathrm{d}x' = \int \mathcal{T}(\Delta x|x,t)\mathrm{d}\Delta x = 1$, the analogue of (16.4).

We can turn this integral equation into a differential equation by performing a Taylor expansion of the integrand:

$$p(x-\Delta x,t)\mathcal{T}(\Delta x|x-\Delta x,t) = p(x,t)\mathcal{T}(\Delta x|x,t) + \sum_{n=1}^{\infty}\frac{1}{n!}(-\Delta x)^n\frac{\partial^n(p(x,t)\mathcal{T}(\Delta x|x,t))}{\partial x^n}, \quad (16.18)$$

such that (16.17) becomes

$$p(x,t+\tau)-p(x,t) = \int \mathrm{d}\Delta x\sum_{n=1}^{\infty}\frac{1}{n!}(-\Delta x)^n\frac{\partial^n(p(x,t)\mathcal{T}(\Delta x|x,t))}{\partial x^n}. \quad (16.19)$$

Now we define

$$M_n(x,t) = \frac{1}{\tau}\int_{-\infty}^{\infty}\mathrm{d}\Delta x(\Delta x)^n\mathcal{T}(\Delta x|x,t), \quad (16.20)$$

in which case we can write

$$\frac{1}{\tau}(p(x,t+\delta t)-p(x,t)) = \sum_{n=1}^{\infty}\frac{(-1)^n}{n!}\frac{\partial^n(M_n(x,t)p(x,t))}{\partial x^n}. \quad (16.21)$$

This is an infinite order differential-difference equation describing the evolution of the probability density function $p$, given an underlying transition probability density $\mathcal{T}$. It is often called the Kramers–Moyal equation.

We now consider a random walk in the limit of continuous space and time. At each point in the walk, defined by position $x$ and time $t$, there is a choice of a step to the right of length $a+u$, or a step to the left of length $a-u$, (with $a>u$), both with probability 1/2. The transition probability density is taken to be independent of time and position and we can write

$$\mathcal{T}(\Delta x|x,t) = \mathcal{T}(\Delta x) = \tfrac{1}{2}[\delta(\Delta x-(a+u))+\delta(\Delta x+(a-u))], \quad (16.22)$$

using the Dirac delta function, defined to be infinite where its argument vanishes, and zero everywhere else, and satisfying the important normalisation condition $\int_{-\infty}^{\infty}\delta(y)\mathrm{d}y=1$ and sifting condition $\int_{-\infty}^{\infty}f(y)\delta(y-y_0)\mathrm{d}y=f(y_0)$. The transition probability density (16.22) specifies that transitions of precisely $\Delta x=(a+u)$ and $\Delta x=-(a-u)$ are allowed, and no others. We then evaluate

$$M_1 = \frac{1}{\tau}\int\mathrm{d}\Delta x\;\Delta x\frac{1}{2}[\delta(\Delta x-a-u)+\delta(\Delta x+a-u)] = \frac{1}{2\tau}(a+u-a+u) = \frac{u}{\tau}$$

$$M_2 = \frac{1}{\tau}\int\mathrm{d}\Delta x\;(\Delta x)^2\frac{1}{2}[\delta(\Delta x-a-u)+\delta(\Delta x+a-u)] = \frac{a^2+u^2}{\tau}$$

$$M_3 = \frac{1}{\tau}\int\mathrm{d}\Delta x\;(\Delta x)^3\frac{1}{2}[\delta(\Delta x-a-u)+\delta(\Delta x+a-u)] = \frac{3a^2u+u^3}{\tau} \quad (16.23)$$

$$M_4 = \frac{1}{\tau}\int \mathrm{d}\Delta x\ (\Delta x)^4 \frac{1}{2}[\delta(\Delta x - a - u) + \delta(\Delta x + a - u)] = \frac{a^4 + 6a^2u^2 + u^4}{\tau}$$

and so on.

In order to model the continuous random walk, we allow $a$, $u$ and $\tau$ to go to zero such that $\lim(u/\tau)$, and $\lim(a^2/\tau)$ are both finite, in which case only $M_1$ and $M_2$ are retained while the other $M_n$ vanish. Furthermore, the difference on the left hand side of (16.21) becomes a time derivative and we arrive at

$$\frac{\partial p(x,t)}{\partial t} = -\frac{\partial (M_1 p(x,t))}{\partial x} + \frac{1}{2}\frac{\partial^2 (M_2 p(x,t))}{\partial x^2}. \tag{16.24}$$

If we generalise to a walk where the parameters $a$ and $u$ describing the step lengths depend on $x$ and $t$, the $M_1$ and $M_2$ also become functions of position and time. This general result is known as the Fokker–Planck equation: a second order partial differential equation describing the evolution of a probability density, in this case for a continuous version of the 1-d random walk.

### 16.3.1 Wiener Process

The limit of the discrete 1-d random walk with equal step lengths to left and right, that is with $u = 0$, and with an $a$ that is independent of $x$ and $t$, is called the Wiener process. The probability density $p(x,t)$ evolves according to (16.24) with $M_1 = 0$ and $M_2 = 2D$, where $D$ is a constant. The Fokker–Planck equation then takes the form of the diffusion equation previously used to describe particle transport (15.22):

$$\frac{\partial p(x,t)}{\partial t} = D\frac{\partial^2 p(x,t)}{\partial x^2}. \tag{16.25}$$

We can solve the diffusion equation using a continuum version of the method employed in the discrete random walk case. We define the characteristic function

$$G(k,t) = \int_{-\infty}^{\infty} p(x,t)\mathrm{e}^{\mathrm{i}kx}\,\mathrm{d}x, \tag{16.26}$$

such that

$$p(x,t) = \frac{1}{2\pi}\int_{-\infty}^{\infty} G(k,t)\mathrm{e}^{-\mathrm{i}kx}\,\mathrm{d}k, \tag{16.27}$$

and we Fourier transform both sides of the diffusion equation to get

$$\int_{-\infty}^{\infty} \frac{\partial p}{\partial t}\mathrm{e}^{\mathrm{i}kx}\,\mathrm{d}x = D\int_{-\infty}^{\infty} \frac{\partial^2 p(x,t)}{\partial x^2}\mathrm{e}^{\mathrm{i}kx}\,\mathrm{d}x. \tag{16.28}$$

Now we perform two integrations by parts, and assume that $p$ and $\partial p/\partial x$ go to zero as $x \to \pm\infty$, such that

$$\int_{-\infty}^{\infty} \frac{\partial^2 p(x,t)}{\partial x^2}\mathrm{e}^{\mathrm{i}kx}\,\mathrm{d}x = \left[\frac{\partial p(x,t)}{\partial x}\mathrm{e}^{\mathrm{i}kx}\right]_{-\infty}^{\infty} - \mathrm{i}k\int_{-\infty}^{\infty} \frac{\partial p(x,t)}{\partial x}\mathrm{e}^{\mathrm{i}kx}\,\mathrm{d}x$$

$$= -\mathrm{i}k\left(\left[p(x,t)\mathrm{e}^{\mathrm{i}kx}\right]_{-\infty}^{\infty} - \mathrm{i}k\int_{-\infty}^{\infty} p(x,t)\mathrm{e}^{\mathrm{i}kx}\,\mathrm{d}x\right) = -k^2 G(k,t), \tag{16.29}$$

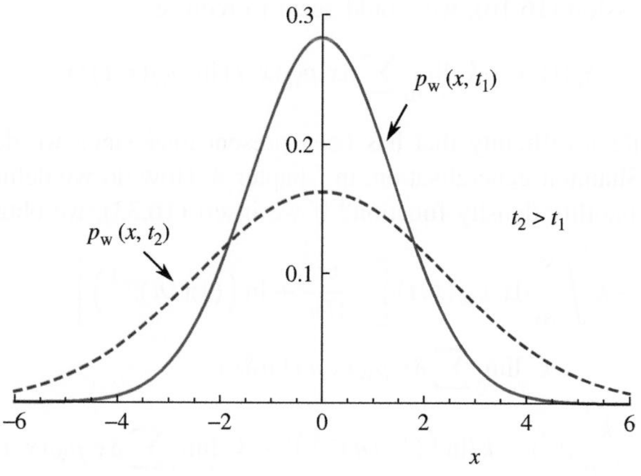


**Figure 16.4** Evolving Gaussian probability density function characterising a Wiener process.

and hence (16.28) becomes

$$\frac{\partial G}{\partial t} = -k^2 DG. \tag{16.30}$$

Since the particle starts at the origin the initial condition is $p(x,0) = \delta(x)$, such that $G(k,0) = 1$, and hence we can integrate (16.30) to obtain

$$G(k,t) = \exp(-k^2 Dt). \tag{16.31}$$

The inverse Fourier transform can now be performed:

$$\begin{aligned} p(x,t) &= \frac{1}{2\pi}\int_{-\infty}^{\infty} \exp(-k^2 Dt - \mathrm{i}kx)\mathrm{d}k \\ &= \frac{1}{2\pi}\int_{-\infty}^{\infty} \exp\left[-\left(k + \frac{\mathrm{i}x}{2Dt}\right)^2 Dt - \frac{x^2}{4Dt}\right]\mathrm{d}k, \end{aligned} \tag{16.32}$$

and by relabelling the integration variable, shifting the integration contour in the complex plane, and using the Gaussian integral $\int_{-\infty}^{\infty} \exp(-\alpha x^2)\mathrm{d}x = (\pi/\alpha)^{1/2}$ we arrive at the solution

$$p_W(x,t) = \frac{1}{(4\pi Dt)^{\frac{1}{2}}} \exp\left(-\frac{x^2}{4Dt}\right). \tag{16.33}$$

The result is a Gaussian with a mean of zero and a variance $2Dt$, and the suffix W reminds us that it is a description of the Wiener process. This is the counterpart of the probability distribution of the discrete random walk we considered in (16.15). An illustration of the evolution of the distribution is given in Figure 16.4.

### 16.3.2 Entropy Production in the Wiener Process

We studied the evolution of Shannon entropy in the discrete random walk and we might expect to find similar behaviour in the case of the Wiener process. Taking the continuum limit of the expression (16.16), we would seem to require

$$S_I(t) = -k \lim_{\delta x \to 0} \sum \delta x \; p_W(x,t) \ln(p_W(x,t)\delta x). \qquad (16.34)$$

We here encounter a difficulty that has been present ever since we defined the Gibbs entropy, and its Shannon generalisation, in Chapter 8. How do we define the entropy of a continuous probability density function? If we insert (16.33), we obtain

$$\begin{aligned} S_I(t) &= -k \int_{-\infty}^{\infty} \mathrm{d}x \; p_W(x,t) \left[ -\frac{x^2}{4Dt} + \ln\left( (4\pi Dt)^{-\frac{1}{2}} \right) \right] \\ &\quad -k \lim_{\delta x \to 0} \sum \delta x \; p_W(x,t) \ln(\delta x) \\ &= \frac{k}{4Dt} \langle x^2 \rangle - k \langle \ln\left( (4\pi Dt)^{-\frac{1}{2}} \right) \rangle - k \lim_{\delta x \to 0} \sum \delta x \; p_W(x,t) \ln(\delta x) \\ &= \frac{k}{2} + \frac{k}{2} \ln(4\pi Dt) - k \langle \ln \delta x \rangle. \end{aligned} \qquad (16.35)$$

The problem with this result is that the last term diverges to $\infty$ when we approach a continuum limit. This will not do.

The difficulty is somewhat avoided if we consider instead the difference in Shannon entropies at two times $t_1$ and $t_2 > t_1$: using a compact notation, we write for a general pdf

$$\begin{aligned} \Delta S_I &= -k \int_{-\infty}^{\infty} \mathrm{d}x \; p(x,t_2) \ln(p(x,t_2)\delta x) - \left( -k \int_{-\infty}^{\infty} \mathrm{d}x \; p(x,t_1) \ln(p(x,t_1)\delta x) \right) \\ &= -k \int_{-\infty}^{\infty} \mathrm{d}x \; p(x,t_2) \ln p(x,t_2) + k \int_{-\infty}^{\infty} \mathrm{d}x \; p(x,t_1) \ln p(x,t_1), \end{aligned} \qquad (16.36)$$

reducing for the Wiener process to

$$\Delta S_I = \frac{k}{2} \ln(4\pi D t_2) - \frac{k}{2} \ln(4\pi D t_1) = \frac{k}{2} \ln\left( \frac{t_2}{t_1} \right), \qquad (16.37)$$

which steadily increases with $t_2 - t_1$. This resembles the behaviour of $S_I$ for the discrete walk shown in Figure 16.3.

It makes sense to define the entropy of a system described by a pdf over a continuous phase space using the expression

$$S_{\mathrm{sys}}(t) = -k \int_{-\infty}^{\infty} \mathrm{d}x \; p(x,t) \ln\left( \frac{p(x,t)}{p_{\mathrm{ref}}} \right), \qquad (16.38)$$

having inserted a constant probability density $p_{\mathrm{ref}}$ to ensure that the argument of the logarithm is dimensionless. The difference in values of $S_{\mathrm{sys}}$ for two pdfs $p_1(x,t)$ and $p_2(x,t)$ is identical to the difference in $S_I$ for the same cases, but $S_{\mathrm{sys}}$ is mathematically better behaved than $S_I$.

We can test these ideas for the case of a single spin zero particle in a 3-d box of volume $V$. We expect that $p(\mathbf{x},t) = 1/V$ in equilibrium, in which case $S_{\mathrm{sys}} = k \ln(p_{\mathrm{ref}} V)$. We know that the correct entropy for such a system in canonical equilibrium at temperature $T$, calculated from quantum mechanics in Section 9.3, is $S_G = T^{-1}\langle E \rangle + k \ln Z_1 = \frac{3}{2}k +$

$k\ln(n_q V) = k\ln(e^{3/2} n_q V)$ and so for consistency we should choose $p_{\text{ref}} = e^{3/2} n_q$, where $n_q(T)$ is the quantum concentration. Just as we determined the parameter $h_0$ in the partition function over a continuous phase space in Section 14.1, by making reference to a quantum mechanical version of the same problem where the phase space is discrete, we can similarly determine the correct form of the entropy for continuous systems.

## 16.4 Brownian Motion

We shall use the model of a continuous random walk to understand some aspects of the phenomenon of Brownian motion. In 1827, the botanist Robert Brown (1773–1858) described the jiggling motion, when suspended in water, of tiny particles of a few tens of microns in diameter found inside pollen grains. Later in the century, with the emergence of kinetic gas theory, attempts were made to ascribe this motion to the fluctuating bombardment of the particles by the surrounding molecules. However, a predictive model based on this assumption was not available until Einstein in 1905 adopted a statistical approach. He abandoned any attempt to describe the microscopic bombardment in detail and just introduced an unknown transition probability density $\mathcal{T}$ describing shifts in particle position $\Delta x$ over a time interval $\tau$. The first and second moments of the transition probability density (divided by $\tau$) correspond to $M_1$ and $M_2$ in the Fokker–Planck equation.

We can interpret $M_1 = \langle \Delta x \rangle / \tau$ as a mean or *drift* velocity $v_D$, and $(1/2)M_2 = \langle (\Delta x)^2 \rangle / (2\tau)$ may again be written as a diffusion coefficient $D$, both assumed to be constants. For Brownian motion in 1-d, the pdf should therefore satisfy

$$\frac{\partial p(x,t)}{\partial t} = -v_D \frac{\partial p(x,t)}{\partial x} + D \frac{\partial^2 p(x,t)}{\partial x^2}, \tag{16.39}$$

and if a particle is released from a position $x_0$, the initial condition is $p(x,0) = \delta(x - x_0)$.

The remaining task is to relate this behaviour to other phenomena. Einstein assumed that the steady state probability density function for the particle position in a gravitational field was the canonical equilibrium pdf previously introduced by Gibbs and Boltzmann, and derived in Section 6.1.3, namely $p(x,\infty) \propto \exp(-mgx/kT)$, where $g$ is gravitational acceleration. Quite clearly, the particle is in thermal contact with its environment through the jiggling it receives from the surrounding molecules, and $T$ is the temperature of that environment. Inserting this into (16.39), Einstein found that he could relate $D$ to the drift velocity $v_D$ and the prevailing temperature:

$$0 = v_D \left(\frac{mg}{kT}\right) p(x,\infty) + D \left(\frac{mg}{kT}\right)^2 p(x,\infty) \quad \Rightarrow \quad D = \frac{kT}{mg} v_D. \tag{16.40}$$

Einstein then chose the drift velocity to be equal to the settling velocity of a small sphere of mass $m$ and radius $r$ under gravity, which is calculable from classical mechanics. For micron size spherical particles and low velocities such that the fluid flow pattern is laminar, the downward force $mg$ is balanced by a force $\alpha v_D$, where $\alpha$ is the drag coefficient given by Stokes' law, $\alpha = 6\pi r \mu_v$, and $\mu_v$ is the viscosity of the host fluid.

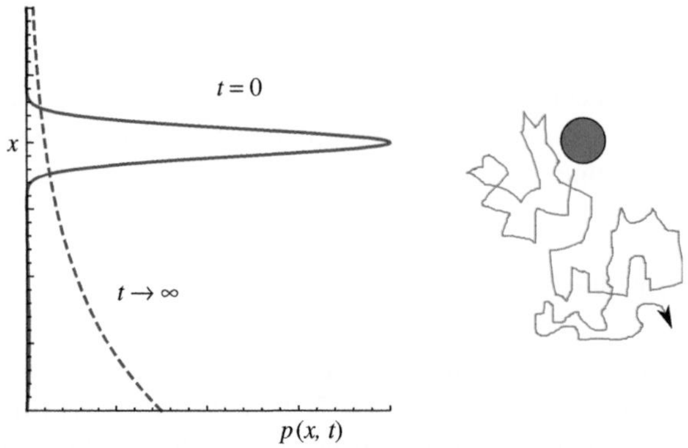


**Figure 16.5** Illustration of Brownian motion in a gravitational field. A particle follows a complicated path that is difficult to predict in detail and is best represented stochastically. The outcome is an evolving pdf of particle height $x$ from an arbitrary initial distribution towards a final Boltzmann distribution. Downward drift under gravity is eventually balanced by upward diffusion brought about by the density gradient.

Hence we obtain an expression for the diffusion coefficient for Brownian motion:

$$D = \frac{kT}{\alpha} = \frac{kT}{6\pi r \mu_v}, \tag{16.41}$$

which is known as the Einstein relation. Such a relationship between the particle diffusion rate, its size and the properties of the host medium was confirmed by Perrin in 1909.

It is sometimes claimed that Einstein's work confirmed that molecules exist. In fact it did this only indirectly. The hypothesis that molecular bombardment jiggled the particles had been around for some years, but quantifying the effect was too difficult. The main thrust of Einstein's work was to reduce the complex dynamics down to something very straightforward: stochastic particle motion characterised by a diffusion equation with drift. This captured the essence of the dynamics and then by making the behaviour consistent with other effects, the diffusion rate could be deduced. This is a key strategy in statistical physics.

Thus, Einstein seemed to confirm that a cloud of suspended particles large enough to be visible under a microscope will relax towards a canonical probability distribution function that is exponential with height, as sketched in Figure 16.5. The visible motion of the particles could be viewed as a scaled up, and slowed down, version of the kinetic turmoil operating at the truly molecular level. This work provided strong support for Boltzmann's then not universally accepted view that matter was composed of particles and that their dynamics could be described statistically. Unfortunately, these developments came too late for Boltzmann, who took his own life in 1906.

## 16.5 Transition Probability Density for a Harmonic Oscillator

We shall discuss one further case of probability dynamics, describing the Brownian motion of a particle tethered to a point by a harmonic force with a spring constant

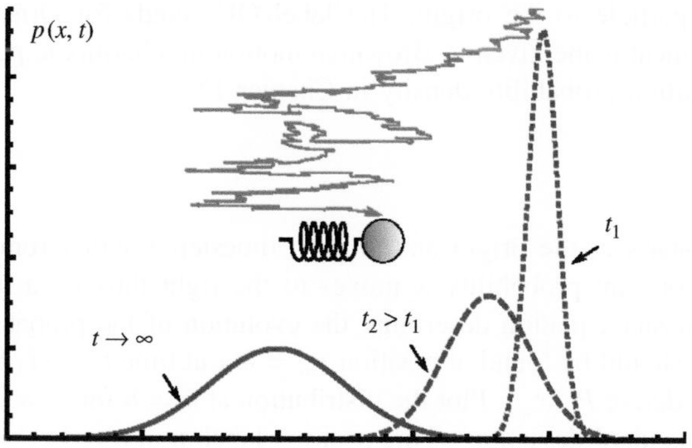


**Figure 16.6** 1-d Brownian motion in a harmonic well. The pdf is given by the Ornstein–Uhlenbeck expression (16.44) at three different times. The mean position drifts from the release point towards the tether point, and the variance of position evolves towards that of the canonical distribution as $t \to \infty$. An example of a realisation of the motion is shown.

$\kappa$, as illustrated in Figure 16.6. We proceed heuristically as we did in the last section by identifying the $x$-dependent drift velocity through a balance between the harmonic force and a drag force. Therefore we write $\alpha v_D(x) = -\kappa x$ and the Fokker–Planck equation as

$$\frac{\partial p(x,t)}{\partial t} = \frac{\kappa}{\alpha}\frac{\partial (xp(x,t))}{\partial x} + D\frac{\partial^2 p(x,t)}{\partial x^2} = \frac{\partial}{\partial x}\left(\frac{\kappa}{\alpha}xp(x,t) + D\frac{\partial p(x,t)}{\partial x}\right). \quad (16.42)$$

We expect the canonical equilibrium pdf of particle position to be $p(x,\infty) \propto \exp\left(-\kappa x^2/(2kT)\right)$, as derived in Section 6.1.2, and since this satisfies

$$\frac{\kappa}{\alpha}xp(x,\infty) + D\frac{\partial p(x,\infty)}{\partial x} = \frac{\kappa}{\alpha}xp(x,\infty) - D\frac{\kappa x}{kT}p(x,\infty) = 0, \quad (16.43)$$

using (16.41), it also satisfies (16.42).

If the particle is released at $x = x_0$ at $t = 0$, we would need to solve (16.42) subject to the initial condition $p(x,0) = \delta(x - x_0)$ in order to understand the time dependence of the tethered Brownian motion. Unfortunately, this is a rather lengthy derivation, and so we instead simply quote the solution:

$$p_{\mathrm{OU}}(x,t) = \sqrt{\frac{\kappa}{2\pi kT(1-\mathrm{e}^{-\frac{2\kappa t}{\alpha}})}}\exp\left(-\frac{\kappa\left(x - x_0\mathrm{e}^{-\frac{\kappa t}{\alpha}}\right)^2}{2kT\left(1-\mathrm{e}^{-\frac{2\kappa t}{\alpha}}\right)}\right), \quad (16.44)$$

which can be checked by explicit insertion into (16.42).

This should be viewed as a transition probability density, as it is a solution conditional on a definite position $x_0$ at the initial time. It is a Gaussian distribution with time-dependent mean and variance, and its evolution is sketched in Figure 16.6. It satisfies our intuitive expectation that the distribution spreads out from the release point, but in contrast to the Wiener process, it does not continue to spread forever, because of the tethering of the particle to the origin. The label OU stands for Ornstein–Uhlenbeck, which is the technical name given to Brownian motion in a harmonic potential. We shall employ this transition probability density in Chapter 17.

## Exercises

**16.1** A particle starts at the origin and at each timestep $\tau$ either remains in position, or with a constant probability $\lambda$ moves to the right through a distance $a$. Write down the master equation describing the evolution of the probability $P_n(x_m)$ that the particle should be found at position $x_m = ma$ at time $t = n\tau$. Solve this master equation to derive $P_n(x_m)$. Plot the distribution at $n = 8$ for $\lambda = 0.2$. Calculate the mean and standard deviation of the position label $m$ at $n = 8$.

**16.2** In the 'Ehrenfest Urn' problem, a particle moves randomly on a grid of positions $x = ma$ with $m$ an integer in the range $-L \leq m \leq L$, and with timestep $\tau$. The probability, when at position $m'$, of a step to the right $m' \to m' + 1$ is $T_+ = \frac{1}{2}\left(1 - \frac{m'}{L}\right)$ and the probability of a step to the left $m' \to m' - 1$ is $T_- = \frac{1}{2}\left(1 + \frac{m'}{L}\right)$. Evaluate the coefficients $M_{1-4}$ of the Kramers–Moyal equation for this process. Take the continuum limits $a \to 0$, $\tau \to 0$, $L \to \infty$ such that $a^2/\tau \to 2D$ and $La^2 \to 2\sigma^2$, where $D$ and $\sigma$ are constants, to show that the Fokker–Planck equation describing the evolution of the pdf $p(x,t)$ is

$$\frac{\partial p}{\partial t} = \frac{D}{\sigma^2}\frac{\partial (xp)}{\partial x} + D\frac{\partial^2 p}{\partial x^2}.$$

Verify by substitution that $p(x) = (2\pi\sigma^2)^{-1/2} \exp\left(-x^2/2\sigma^2\right)$ is the time-independent solution to this equation.

**16.3** Mr and Mrs Ehrenfest keep $N$ rabbits, and house them in two rabbit hutches, one blue and one pink. Every morning they select one of the rabbits at random and move it to the other hutch. The probability that on day $n$ there are $m$ rabbits in the pink hutch is $P_n(m)$, with $0 \leq m \leq N$. Show that the master equations describing the process are

$$P_{n+1}(m) = \frac{m+1}{N}P_n(m+1) + \frac{N-m+1}{N}P_n(m-1),$$

except for $m = 0$ and $m = N$, for which $P_{n+1}(0) = (1/N)P_n(1)$ and $P_{n+1}(N) = (1/N)P_n(N-1)$. Verify that the following time-independent probability distribution satisfies the master equations

$$P_n(m) = \hat{P}(m) = \frac{1}{2^N}\frac{N!}{m!(N-m)!}.$$

After some time, the Ehrenfests find that the number of rabbits has increased. Express $\hat{P}(m)$ in terms of $\mu = m - \frac{1}{2}N$ and show that if $N \gg 1$ and $|\mu| \ll N/2$ then the distribution may be approximated by

$$\hat{P}(\mu) \propto \exp\left(-\frac{\mu^2}{2\sigma^2}\right),$$

and show that the variance $\sigma^2$ is $N/4$. You might need to use the following approximations: $\ln k! \approx k \ln k - k$ for $k \gg 1$ and $\ln(1+x) \approx x - \frac{1}{2}x^2$ for $|x| \ll 1$.

**16.4** A set of radioactive atoms decays with time. The process may be modelled approximately using the Fokker–Planck equation

$$\frac{\partial p(n,t)}{\partial t} = \lambda \frac{\partial (np(n,t))}{\partial n} + \frac{\lambda}{2}\frac{\partial^2 (np(n,t))}{\partial n^2},$$

where $p(n,t)$ is the probability that $n$ atoms remain after time $t$ and where $\lambda$ is a constant. Using an integration by parts, derive an expression for $\mathrm{d}\langle n \rangle/\mathrm{d}t$, where the mean population is given by $\langle n \rangle = \int_0^\infty np(n,t)\, \mathrm{d}n$. You may make the approximation that both $p(n,t)$ and $\partial p(n,t)/\partial n$ vanish at $n = 0$ and $\infty$. Also derive an expression for $\mathrm{d}\langle n^2 \rangle/\mathrm{d}t$.

**16.5** Estimate the root mean square displacement of a particle of radius 1 μm after one minute if it is suspended in air at room temperature. The viscosity of air is $18 \times 10^{-6}\,\mathrm{Nsm^{-2}}$.

**16.6** Sketch the time evolution of the system entropy of a harmonically tethered Brownian particle arising from the instantaneous halving of the spring constant, assuming it had previously been in equilibrium.

# 17

# Fluctuation Relations

In recent years, the field of nonequilibrium statistical thermodynamics has developed in ways that cast new light on the concept of entropy and its production, and it is the aim of this chapter to provide a brief introduction to ideas that are summed up under the generic name of fluctuation relations. This is fairly challenging stuff, hence another warning sign!

Caution: Entropy

## 17.1 Forward and Backward Path Probabilities: a Criterion for Equilibrium

It is crucial to appreciate what is meant by equilibrium in statistical thermodynamics. On a macroscopic scale, it is quite clear that this means a state of a system where nothing changes or flows, but on the microscopic scale, such a concept cannot apply as particles are in motion, and there are fluctuations in the amount of energy and material in a system due to exchanges with the environment. We have stated that equilibrium on the microscale means that the statistical properties of the system are independent of time, and furthermore that there are no mean flows through the system between different parts of the environment. We shall extend this concept now in the following way.

As a system evolves, sequences of events will be observed, and we might wish to ascribe a probability to each sequence. For example, in the one dimensional Brownian motion of a particle in a harmonic potential $\phi = (1/2)\kappa x^2$ with a drag coefficient $\alpha$, considered in Section 16.5, we can imagine a sequence of events consisting of an observation of the particle at time $t = 0$ in the position range $x_0 \pm (1/2)\mathrm{d}x_0$, followed by its observation at a later time $t$ in the range $x \pm (1/2)\mathrm{d}x$. The sequence, described here as a *path*, is sketched in Figure 17.1 in terms of a trajectory between two points in the $x - t$ plane, although it should be borne in mind that we are not specifying exactly how the particle reaches its destination, so the trajectory shown is only one possibility.

*Statistical Physics: An Entropic Approach*, First Edition. Ian Ford.

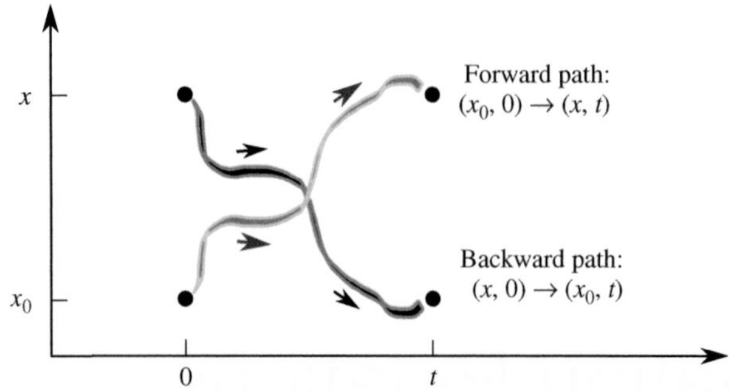


**Figure 17.1** Sketch of a forward sequence of events, or path, consisting of the observation of a Brownian particle at coordinates $(x_0, 0)$ and then $(x, t)$, and the backward sequence corresponding to the opposite order.

The probability of such a sequence of events, with its implicit inclusion of all intervening trajectories, is given by

$$\mathcal{P}(x,t;x_0,0) = \mathcal{T}(x - x_0|x_0)p(x_0,0)\mathrm{d}x_0\mathrm{d}x, \tag{17.1}$$

which is written in such a way as to resemble the notation used in the discussion of the random walk in Sections 16.2 and 16.3. It is a product of the probability of the initial event, the observation in the region of $x_0$, multiplied by the probability of observation in the region of $x$ *given* that motion began at $x_0$. In fact for the system in question we can use the transition probability density for the Ornstein–Uhlenbeck process introduced in (16.44), and write

$$\mathcal{P}(x,t;x_0,0) = p_{\mathrm{OU}}(x - x_0,t)p(x_0,0)\mathrm{d}x_0\mathrm{d}x. \tag{17.2}$$

The key concept is that a state of equilibrium is a situation where the probabilities of realising a sequence of events and the exact opposite sequence are precisely the same. This is so important as to bear repeating. Equilibrium means that the likelihood that the system goes backward is the same as its likelihood of going forward. We can demonstrate that this is so by writing the probability of the reverse of the sequence in question as

$$\mathcal{P}(x_0,t;x,0) = p_{\mathrm{OU}}(x_0 - x,t)p(x,0)\mathrm{d}x\mathrm{d}x_0, \tag{17.3}$$

and examining the ratio of these so-called forward and backward path probabilities:

$$\begin{aligned}
\frac{\mathcal{P}(x,t;x_0,0)}{\mathcal{P}(x_0,t;x,0)} &= \exp\left(-\frac{\kappa\left(x - x_0\mathrm{e}^{-\frac{\kappa t}{\alpha}}\right)^2}{2kT\left(1-\mathrm{e}^{-\frac{2\kappa t}{\alpha}}\right)} + \frac{\kappa\left(x_0 - x\mathrm{e}^{-\frac{\kappa t}{\alpha}}\right)^2}{2kT\left(1-\mathrm{e}^{-\frac{2\kappa t}{\alpha}}\right)}\right)\exp\left(-\frac{\kappa x_0^2}{2kT} + \frac{\kappa x^2}{2kT}\right) \\
&= \exp\left(-\frac{\kappa\left(x^2 + x_0^2\mathrm{e}^{-\frac{2\kappa t}{\alpha}}\right)}{2kT\left(1-\mathrm{e}^{-\frac{2\kappa t}{\alpha}}\right)} + \frac{\kappa\left(x_0^2 + x^2\mathrm{e}^{-\frac{2\kappa t}{\alpha}}\right)}{2kT\left(1-\mathrm{e}^{-\frac{2\kappa t}{\alpha}}\right)}\right)\exp\left(-\frac{\kappa x_0^2}{2kT} + \frac{\kappa x^2}{2kT}\right) \\
&= 1,
\end{aligned} \tag{17.4}$$

having inserted (16.44) as well as the equilibrium canonical pdf $p(x,0) \propto \exp(-\kappa x^2/2kT)$.

What this tells us is that not only does the equilibrium state have time independent statistical properties, such as the mean and variance of particle position, but also that any sequence of events, at least of the kind we have considered, is equally likely to be seen going forward or going backward. The ratio of path probabilities is a powerful statistical expression of what is meant by *reversibility* in a randomly evolving system. Everything we observe is just as likely to be seen in reverse, if the system is in an equilibrium state. We have illustrated this for a simple tethered Brownian particle, but the same definition of equilibrium can be extended to more general cases. For a system in equilibrium, we cannot tell statistically whether we are viewing a movie of its behaviour running forward or backward in time.

## 17.2 Time Asymmetry of Behaviour and a Definition of Entropy Production

Having established what is meant by equilibrium, it is straightforward to define what we mean when we say a system is out of equilibrium. For such a system, the probabilities of observing some sequences and their reversals are not equal. This is a deeper statistical meaning to the concept of the irreversibility of a process. We shall use the ratio of path probabilities as we did in the previous section to obtain a measure of how far the system might be away from equilibrium.

The manner in which we do so requires some careful explanation. We shall consider a 1-d harmonic oscillator that is relaxing towards equilibrium while the parameters $\kappa$, $\alpha$ and $T$ remain constant. The statistical state of the system is represented by a time-dependent probability density function $p(x,t)$. We observe the system at $t = 0$, at $t = \Delta t$ and again at $t = 2\Delta t$, defining two observational intervals of length $\Delta t$. We consider the probability of observing a path from $x_0$ to $x$ in the first of these intervals, and quite independently, the probability of a path from $x$ to $x_0$ during the second, as illustrated in Figure 17.2.

If the system were in an equilibrium state with time-independent statistics, then according to the argument presented in the last section, the two probabilities would be equal, for all possible paths, but in general they will differ. So we define

$$\exp\left(\frac{\Delta s_i(x,\Delta t;x_0,0)}{k}\right) = \frac{\mathcal{P}(x,\Delta t;x_0,0)}{\mathcal{P}(x_0,2\Delta t;x,\Delta t)} = \frac{p_{OU}(x-x_0,\Delta t)\,p(x_0,0)}{p_{OU}(x_0-x,\Delta t)\,p(x,\Delta t)}, \quad (17.5)$$

where $\Delta s_i(x,\Delta t;x_0,0)$ is a property that we assign to the path from $x_0$ to $x$. In contrast to (17.4), we consider $p(x,t)$ to be noncanonical in form. We expect to find that the ratio is not unity for all sequences, or equivalently that $\Delta s_i(x,\Delta t;x_0,0)$ differs from zero depending on the start and end positions of the path. The key task now is to look at the statistics of $\Delta s_i$, and specifically to examine how it fluctuates away from zero.

We start by evaluating the mean of the quantity $\exp(-\Delta s_i(x,\Delta t;x_0,0)/k)$ over all possible paths $x_0 \to x$, each of which occurs with probability $\mathcal{P}(x,\Delta t;x_0,0)$. The mean

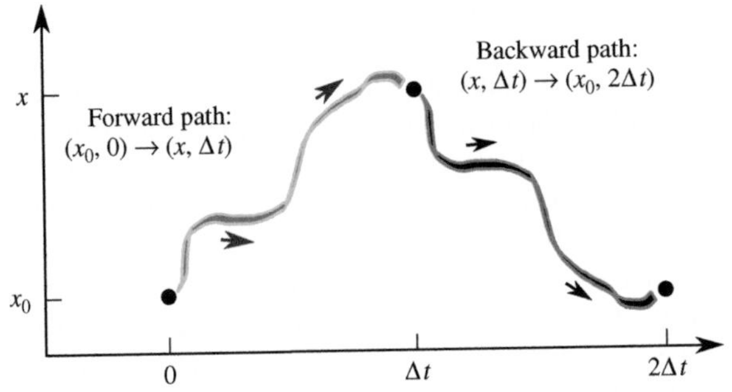


**Figure 17.2** Forward and backward paths defined over successive periods $\Delta t$ when a system is relaxing towards equilibrium.

is written as

$$\left\langle \exp\left(-\frac{\Delta s_i}{k}\right)\right\rangle = \int_{-\infty}^{\infty}\int_{-\infty}^{\infty} \exp\left(-\frac{\Delta s_i(x, \Delta t; x_0, 0)}{k}\right) \times p_{\mathrm{OU}}(x - x_0, \Delta t)\, p(x_0, 0)\, \mathrm{d}x\mathrm{d}x_0, \tag{17.6}$$

but by inserting the definition (17.5), this can be written

$$\left\langle \exp\left(-\frac{\Delta s_i}{k}\right)\right\rangle = \int_{-\infty}^{\infty}\int_{-\infty}^{\infty} p_{\mathrm{OU}}(x_0 - x, \Delta t)\, p(x, \Delta t)\, \mathrm{d}x\mathrm{d}x_0 = 1. \tag{17.7}$$

This is unity since the final integral is simply a normalisation of the probability that a backward path is observed, starting and finishing anywhere at all. This straightforwardly derived but powerful result is called the *integral fluctuation relation*.

We now make use of the inequality $\exp(z) \geq 1 + z$ that holds for any real value of $z$. This implies that the average of $\exp(z)$ over any distribution of the variable $z$ should satisfy $\langle \exp(z)\rangle \geq 1 + \langle z\rangle$ and if we insert $z = -\Delta s_i/k$ the consequence is

$$\frac{\langle \Delta s_i\rangle}{k} \geq 1 - \left\langle \exp\left(-\frac{\Delta s_i}{k}\right)\right\rangle = 0, \tag{17.8}$$

using (17.7).

We conclude that $\Delta s_i$, a property associated with the evolution of a tethered Brownian particle as it evolves over the period $0 \leq t \leq \Delta t$, (and implicitly a reflection of projected behaviour over a further period $\Delta t$), is positive when averaged over all possible paths, unless the system initially takes a canonical equilibrium pdf over its phase space. If the system does take the equilibrium distribution, $\Delta s_i$ is zero for all observed behaviours, and so its average is zero too. This is starting to sound rather familiar!

Let us therefore consider whether the value of $\Delta s_i$ for a given period, averaged over all possible system behaviour under the prescribed stochastic or random dynamics, and starting from an initial situation specified by a given pdf over position, might correspond to the associated internal production of entropy, namely $\Delta S_i = \langle \Delta s_i\rangle$. We need to check this assertion explicitly.

## 17.3 The Relaxing Harmonic Oscillator

A relaxation process involving a 1-d classical harmonic oscillator in thermal contact with a heat bath is governed by the Fokker–Planck equation

$$\frac{\partial p(x,t)}{\partial t} = \frac{\kappa}{\alpha}\frac{\partial (xp(x,t))}{\partial x} + \frac{kT}{\alpha}\frac{\partial^2 p(x,t)}{\partial x^2}, \tag{17.9}$$

as in (16.42), where $\kappa$ is the spring constant and $\alpha$ the drag coefficient. This equation applies for 'overdamped' conditions, where the drag force on the particle is strong, and where we need not concern ourselves with deviations from the Maxwell–Boltzmann distribution of particle velocities during the relaxation, but can focus instead on changes in the distribution over position.

We consider the evolution of $p(x,t)$ starting from the initial condition

$$p(x,0) = \left(\frac{\kappa_0}{2\pi kT}\right)^{\frac{1}{2}} \exp\left(-\frac{\kappa_0 x_0^2}{2kT}\right), \tag{17.10}$$

with $\kappa_0 \neq \kappa$. We are interested in the statistical behaviour of the quantity $\Delta s_i$ associated with the relaxation of the system pdf towards equilibrium, as illustrated in Figure 17.3.

It may be shown that the solution to the Fokker–Planck equation (17.9) with this initial condition takes the Gaussian form

$$p(x,t) = \left(\frac{\tilde{\kappa}(t)}{2\pi kT}\right)^{\frac{1}{2}} \exp\left(-\frac{\tilde{\kappa}(t)x^2}{2kT}\right), \tag{17.11}$$

and by inserting this into (17.9), the function $\tilde{\kappa}(t)$ can be shown to satisfy

$$\frac{d\tilde{\kappa}}{dt} = -\frac{2}{\alpha}\tilde{\kappa}(\tilde{\kappa} - \kappa), \tag{17.12}$$

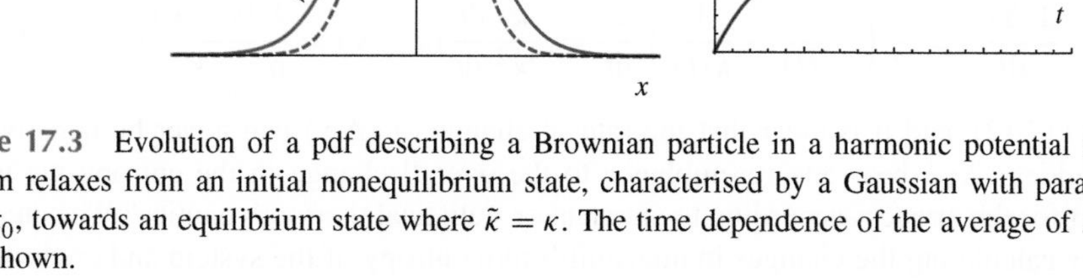

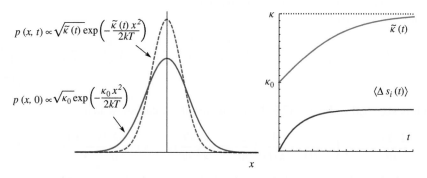


**Figure 17.3** Evolution of a pdf describing a Brownian particle in a harmonic potential as the system relaxes from an initial nonequilibrium state, characterised by a Gaussian with parameter $\tilde{\kappa} = \kappa_0$, towards an equilibrium state where $\tilde{\kappa} = \kappa$. The time dependence of the average of $\Delta s_i$ is also shown.

with $\tilde{\kappa}(0) = \kappa_0$. This is equivalent to $\mathrm{d}z/\mathrm{d}t = 2\alpha^{-1}(1 - \kappa z)$, where $z = \tilde{\kappa}^{-1}$, which may be solved for $z(t)$ to give

$$\tilde{\kappa}(t) = \frac{\kappa_0 \kappa}{\kappa_0 + \mathrm{e}^{-\frac{2\kappa t}{\alpha}}(\kappa - \kappa_0)}, \tag{17.13}$$

which is also illustrated in Figure 17.3.

We can then write

$$\begin{aligned}\Delta s_i(x, t; x_0, 0) &= k \ln\left(\frac{p_{\mathrm{OU}}(x - x_0, t)\, p(x_0, 0)}{p_{\mathrm{OU}}(x_0 - x, t)\, p(x, t)}\right) = -\frac{\kappa}{2T}(x^2 - x_0^2) + k \ln\left(\frac{p(x_0, 0)}{p(x, t)}\right) \\ &= -\frac{\kappa}{2T}(x^2 - x_0^2) + \frac{k}{2}\ln\frac{\kappa_0}{\tilde{\kappa}(t)} - \frac{\kappa_0 x_0^2}{2T} + \frac{\tilde{\kappa}(t) x^2}{2T},\end{aligned} \tag{17.14}$$

and we can average this over the probability distribution of paths to get

$$\langle \Delta s_i \rangle = \int p_{\mathrm{OU}}(x - x_0, t) p(x_0, 0) \Delta s_i(x, t; x_0, 0)\, \mathrm{d}x \mathrm{d}x_0. \tag{17.15}$$

In order to proceed, we note that

$$p(x, t) = \int p_{\mathrm{OU}}(x - x_0, t) p(x_0, 0)\, \mathrm{d}x_0, \tag{17.16}$$

which is the Chapman–Kolmogorov equation (16.17) for this situation, representing the transfer of probability to the final position $x$ by all possible paths, so that we can write

$$\langle x^2 \rangle = \int p_{\mathrm{OU}}(x - x_0, t) p(x_0, 0) x^2\, \mathrm{d}x \mathrm{d}x_0 = \int p(x, t) x^2 \mathrm{d}x = \frac{kT}{\tilde{\kappa}}, \tag{17.17}$$

by insertion of (17.11), and similarly

$$\langle x_0^2 \rangle = \int p_{\mathrm{OU}}(x - x_0, t) p(x_0, 0) x_0^2\, \mathrm{d}x \mathrm{d}x_0 = \int p(x_0, t) x_0^2 \mathrm{d}x_0 = \frac{kT}{\kappa_0}, \tag{17.18}$$

since $\int p_{\mathrm{OU}}(x - x_0, t)\, \mathrm{d}x = 1$ by normalisation.

Hence we can write

$$\langle \Delta s_i \rangle = k\left[-\frac{\kappa}{2}\left(\frac{1}{\tilde{\kappa}(t)} - \frac{1}{\kappa_0}\right) + \frac{1}{2}\ln\frac{\kappa_0}{\tilde{\kappa}(t)} - \frac{1}{2} + \frac{1}{2}\right] = \frac{k}{2}\left(\frac{\kappa}{\kappa_0} - \frac{\kappa}{\tilde{\kappa}(t)} + \ln\frac{\kappa_0}{\tilde{\kappa}(t)}\right), \tag{17.19}$$

which is sketched in Figure 17.3. From this result we can show that

$$\frac{\mathrm{d}\langle \Delta s_i \rangle}{\mathrm{d}t} = \frac{k}{2}\left(\frac{\kappa}{\tilde{\kappa}^2(t)} - \frac{1}{\tilde{\kappa}(t)}\right)\frac{\mathrm{d}\tilde{\kappa}}{\mathrm{d}t} = \frac{k}{2\tilde{\kappa}^2}\frac{\mathrm{d}\tilde{\kappa}}{\mathrm{d}t}(\kappa - \tilde{\kappa}) = \frac{k}{\alpha}\frac{(\tilde{\kappa} - \kappa)^2}{\tilde{\kappa}}, \tag{17.20}$$

using (17.12), and it is clear that the rate of change of $\langle \Delta s_i \rangle$ can never be negative.

So does $\langle \Delta s_i \rangle$ bear any resemblance to the overall change in the entropy we would expect for this process according to classical or statistical thermodynamics? We can check this by calculating the changes in nonequilibrium entropy of the system and environment resulting from the process. We calculate the change in system entropy defined according to (16.38), namely

$$\Delta S_{\rm sys} = -k\int_{-\infty}^{\infty} {\rm d}x\; p(x,t)\ln p(x,t) + k\int_{-\infty}^{\infty} {\rm d}x_0\; p(x_0,0)\ln p(x_0,0)$$
$$= -k\left[\frac{1}{2}\ln\left(\frac{\tilde{\kappa}}{2\pi kT}\right) - \frac{\tilde{\kappa}}{2kT}\frac{kT}{\tilde{\kappa}}\right] + k\left[\frac{1}{2}\ln\left(\frac{\kappa_0}{2\pi kT}\right) - \frac{\kappa_0}{2kT}\frac{kT}{\kappa_0}\right]$$
$$= \frac{k}{2}\ln\frac{\kappa_0}{\tilde{\kappa}}, \qquad (17.21)$$

and as the work done during the process is zero, the change in reservoir entropy is related to the energy transfer as follows:

$$\Delta S_{\rm r} = \frac{\langle \Delta E_{\rm r}\rangle}{T} = -\frac{\langle \Delta E\rangle}{T} = -\frac{1}{T}\left(\int_{-\infty}^{\infty} {\rm d}x\; p(x,t)\frac{\kappa x^2}{2} - \int_{-\infty}^{\infty} {\rm d}x_0\; p(x_0,0)\frac{\kappa x_0^2}{2}\right)$$
$$= -\frac{\kappa}{2T}\left(\frac{kT}{\tilde{\kappa}} - \frac{kT}{\kappa_0}\right), \qquad (17.22)$$

such that

$$\Delta S_{\rm tot} = \Delta S_{\rm r} + \Delta S_{\rm sys} = \frac{k}{2}\left(\frac{\kappa}{\kappa_0} - \frac{\kappa}{\tilde{\kappa}(t)} + \ln\frac{\kappa_0}{\tilde{\kappa}(t)}\right), \qquad (17.23)$$

and just as desired, this matches $\langle \Delta s_i\rangle$ in (17.19).

We now draw an analogy with the example of entropy production due to the transfer of heat between a reservoir and an ideal gas considered in Section 2.7. In Figure 17.3 we illustrate the case $\kappa > \kappa_0$, and it should be recognised that this is equivalent to a process of cooling the Brownian particle from an initial state at a temperature $T_0 = T\kappa/\kappa_0 > T$ towards the reservoir temperature $T$. This is made apparent by writing the initial pdf as $p(x_0,0) \propto \exp(-\kappa_0 x_0^2/2kT) = \exp(-\kappa x_0^2/2kT_0)$, and noting that as $t \to \infty$ the pdf evolves towards the form $p(x,\infty) \propto \exp(-\kappa x^2/2kT)$. The limit of $\langle \Delta s_i\rangle$ for $t \to \infty$ when equilibrium is restored is then

$$\langle \Delta s_i\rangle_{t\to\infty} = \frac{k}{2}\left(\frac{\kappa}{\kappa_0} - 1 + \ln\frac{\kappa_0}{\kappa}\right) = \frac{k}{2}\left(\frac{T_0 - T}{T} + \ln\frac{T}{T_0}\right), \qquad (17.24)$$

since $\tilde{\kappa}(t\to\infty) = \kappa$, and this depends on the initial and final temperatures in exactly the same way as we found for the cooling of the ideal gas in (2.30). The cycle is complete.

## 17.4 Entropy Production Arising from a Single Random Walk

There is one major conceptual leap that remains to be made. We have introduced a quantity $\Delta s_i$ that is associated with a specific path taken by the oscillator. When averaged over all possible paths generated by the dynamics, its value appears to be equal to the thermodynamic entropy produced in the nonquasistatic relaxation process initiated by the release of a constraint. So $\Delta s_i$ is sampled from a distribution whose mean is $\Delta S_i$.

The conceptual leap is to consider that $\Delta s_i$ is also an entropy production, but one that is associated with a particular outcome of the dynamics, unlike $\Delta S_i$, which is associated

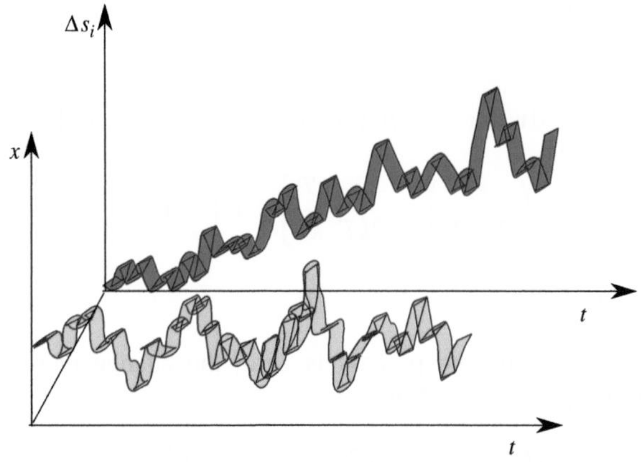


**Figure 17.4** A single trajectory $x(t)$ taken by a Brownian particle has an associated entropy production $\Delta s_i(t)$ that can rise and fall as time progresses. When averaged over all trajectories it increases monotonically in correspondence with the second law.

with the whole set of possibilities. Except this is an entropy production that does not satisfy the second law. We know this because its distribution is not restricted to non-negative values. If it were, then $\exp(-\Delta s_i/k)$ would always be less than one, and yet the integral fluctuation relation (17.7) requires the average of this quantity to be unity. The entropy production of individual paths fluctuates, and can occasionally be negative. This is illustrated in Figure 17.4.

There are advantages in taking this view because it removes some of the conceptual difficulties that surround entropy production. With this interpretation, the entropy production is just a quantity associated with a particular path along which a system might develop in time. It is not necessary to imagine repeating an observation and taking an average in order to determine the entropy production. A single random walk of a particle would do, along with a description of the dynamics of probability that accompany the process. The latter is not necessarily a representation of the behaviour of many repeated paths, but is a judgement about the probability of a single realisation of the process. As the walk proceeds, it clocks up a tally of $\Delta s_i$ entropy production, and on the whole it goes up, but sometimes it goes down.

Clearly $\Delta s_i$ is defined in terms of probabilities of various events, and therefore relies on employing a stochastic rather than a deterministic model for the dynamics of a system. Entropy production is related to the increase in uncertainty, and arguably it will occur only if there is randomness in the dynamics. It is natural to take this view because it ties in with the original statistical interpretation of entropy proposed by Gibbs.

An entropy production that can be associated with the path taken by a system through its phase space implies that there is a system entropy that depends on the microstate. From (16.38), we have

$$S_{\rm sys} = -k \int_{-\infty}^{\infty} dx\, p(x,t) \ln\left(\frac{p(x,t)}{p_{\rm ref}}\right), \tag{17.25}$$

where $p_{ref}$ is a constant, suggesting that the system entropy of a microstate is $s_{sys} = -k\ln(p(x,t)/p_{ref})$ such that $S_{sys} = \langle s_{sys}\rangle$. We shall not pursue this any further, but again it is conceptually useful to recognise that we can conceive of an entropy of a system that does not require averaging over different arrangements; that is a property even of a *microstate*.

The entropy production $\Delta s_i$ is related to other path-dependent quantities. We have already seen this explicitly in (17.14) in terms of particle positions, but there is an identity, which we shall not prove, such that

$$T\Delta s_i^{ee} = \Delta w - \Delta F, \tag{17.26}$$

where $\Delta s_i^{ee}$ is the entropy production for a path undertaken between microstates when the system evolves from initial to final *equilibrium* states, which naturally requires that the duration of the process is long enough for relaxation to be completed. $\Delta w$ is the work done on the system over the course of such a path, and $\Delta F$ is the associated difference in system free energy. This is a counterpart of the result $T\Delta S_i = \Delta W - \Delta F$ for an isothermal nonquasistatic work process in classical thermodynamics derived in (3.22). The average of $\Delta w$ over all possible behaviours of the system, each of which draws different amounts of work from the environment, is equal to $\Delta W$. This connection will be explored further in the next section.

## 17.5 Further Fluctuation Relations

From the integral fluctuation relation (17.7) and expression (17.26), we can deduce the result

$$\left\langle \exp\left(-\frac{(\Delta w - \Delta F)}{kT}\right)\right\rangle = 1, \tag{17.27}$$

for a process where the system starts and finishes in equilibrium. Let us focus on a process specified by a time-dependent change in the confining volume while the system remains in contact with a reservoir at constant temperature. The change in free energy $\Delta F$ associated with the process will depend on the initial and final volumes. The work done will depend on the history of compressions and expansions, as well as the detailed path followed by the system over the course of the process.

We now imagine starting with the system in equilibrium, imposing such a sequence of volume changes, and then continuing for an extended period of relaxation at constant volume until equilibrium is restored. Equation (17.27) holds for such a process, but we recognise that $\Delta w$ and $\Delta F$ no longer evolve once the volume stops changing. This has the implication that the period of relaxation does not change the value of $\langle\exp(-(\Delta w - \Delta F)/kT)\rangle$ and hence (17.27) holds at an arbitrary point in the process, not just at a final equilibrium.

The expression (17.27) is called the *Jarzynski equality*. It has more practical implications than the integral fluctuation relation (17.7) because it involves readily measured quantities, namely work and free energy, instead of the more ethereal entropy production.

We can see the Jarzynski equality operating in practice for the harmonic oscillator example. We take the initial spring constant of the oscillator to be $\kappa_0$, such that the

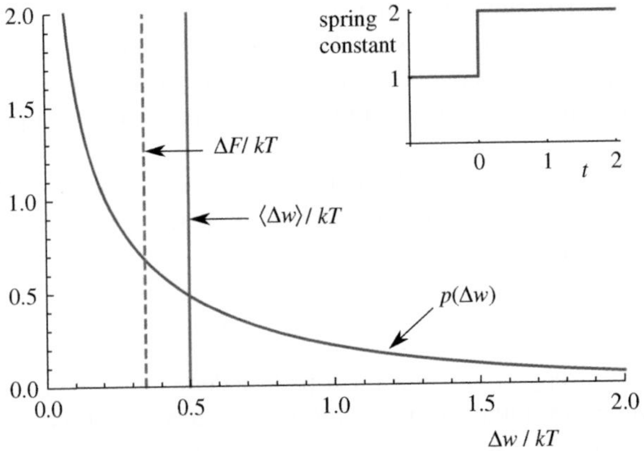


**Figure 17.5** The pdf $p(\Delta w)$ of work done on a system arising from a step change in spring constant from $\kappa_0 = 1$ to $\kappa = 2$ at $t = 0$, starting in equilibrium. The average of $\Delta w$ is greater than the change in free energy, as implied by the Jarzynski equality (17.27).

Gaussian $p(x_0, 0) \propto \exp(-\kappa_0 x_0^2/2kT)$ specified in (17.10) is the initial canonical equilibrium pdf. Then we consider a process consisting of a step change in spring constant from $\kappa_0$ to $\kappa$ at $t = 0$, as illustrated in Figure 17.5.

The work done on the system is the input of potential energy over the course of the process, and this is entirely performed at the shift in spring constant at $t = 0$. We have

$$\Delta w = \tfrac{1}{2}(\kappa - \kappa_0)x_0^2, \tag{17.28}$$

and using (17.18), the mean work is

$$\Delta W = \langle \Delta w \rangle = \frac{1}{2}(\kappa - \kappa_0)\langle x_0^2 \rangle = \frac{1}{2}\frac{(\kappa - \kappa_0)kT}{\kappa_0}. \tag{17.29}$$

The difference in the free energy corresponding to the change in spring constant from $\kappa_0$ to $\kappa$ is $\Delta F = -kT\ln(Z(\kappa)/Z(\kappa_0)) = (1/2)kT\ln(\kappa/\kappa_0)$, where $Z(\kappa) = kT/\hbar\omega$ is the canonical partition function of the 1-d classical harmonic oscillator given in (14.3), using $\omega = (\kappa/m)^{1/2}$. We can therefore establish that the quantity that appears in the Jarzynski equality takes the form

$$\Delta w - \Delta F = \frac{1}{2}kT\left[(\kappa - \kappa_0)\frac{x_0^2}{kT} - \ln\frac{\kappa}{\kappa_0}\right], \tag{17.30}$$

and this is equal to $T\Delta s_i(x, \infty; x_0, 0)$ from (17.14), bearing in mind that $\tilde{\kappa} \to \kappa$ as $t \to \infty$. It is then clear that the averaged result $\Delta W - \Delta F = T\langle \Delta s_i(x, \infty; x_0, 0)\rangle$ also holds. Furthermore,

$$\Delta W - \Delta F = \langle \Delta w \rangle - \Delta F = \frac{1}{2}kT\left(\frac{\kappa - \kappa_0}{\kappa_0} - \ln\frac{\kappa}{\kappa_0}\right) \geq 0, \tag{17.31}$$

in explicit agreement with the classical second law (3.22).

We can now verify the Jarzynski equality:

$$\begin{aligned}\langle \exp(-\Delta w/kT)\rangle &= \int \mathrm{d}x_0 p(x_0,0)\exp\left(-\frac{(\kappa-\kappa_0)\,x_0^2}{2kT}\right)\\ &= \left(\frac{\kappa_0}{2\pi kT}\right)^{\frac{1}{2}}\int \mathrm{d}x_0 \exp\left(-\frac{\kappa_0 x_0^2}{2kT}\right)\exp\left(-\frac{(\kappa-\kappa_0)\,x_0^2}{2kT}\right)\\ &= \left(\frac{\kappa_0}{\kappa}\right)^{\frac{1}{2}} = \exp\left(-\frac{\Delta F}{kT}\right),\end{aligned} \tag{17.32}$$

as required. The averaging is performed over the initial position $x_0$ alone since the work performed in this example takes place at the instant when the spring constant changes.

Since we know how $x_0$ is distributed, we can go further and derive the pdf of $\Delta w$ for this case. First, we identify the initial position $x_0$ that gives rise to work $\Delta w$ using (17.28):

$$x_0 = \pm\left(\frac{2\Delta w}{\kappa-\kappa_0}\right)^{\frac{1}{2}}, \tag{17.33}$$

and it should be noticed that if $\kappa > \kappa_0$ we expect only positive values of $\Delta w$, and only negative values if $\kappa < \kappa_0$. In the example illustrated in Figure 17.3, therefore, we shall find only positive values of work. Next we obtain the pdf of $x_0^2$ from the pdf of $x_0$ using $p(x_0^2)\mathrm{d}x_0^2 = p(x_0)\mathrm{d}x_0$ so $p(x_0^2) = (1/2)p(x_0)/|x_0|$, taking care to ensure that $p(x_0^2)$ is positive, and hence we write

$$\begin{aligned}p(\Delta w-\Delta F) &\propto \left(\frac{\kappa-\kappa_0}{2\Delta w}\right)^{\frac{1}{2}}\exp\left[-\frac{\kappa_0}{2kT}\left(\frac{2\Delta w}{\kappa-\kappa_0}\right)\right]\\ &\propto \frac{1}{(\Delta w)^{\frac{1}{2}}}\exp\left(-\frac{\kappa_0}{(\kappa-\kappa_0)}\frac{\Delta w}{kT}\right),\end{aligned} \tag{17.34}$$

which we plot in Figure 17.5. The implication of (17.31) is that $\langle \Delta w\rangle > \Delta F$ and this is reflected in the distribution, but just as clearly, it is easily feasible that the particle can follow a path such that $\Delta w < \Delta F$.

Further results may be proved that add to our understanding of the fluctuations in work and entropy production in the course of various processes. However, this is not the place to prove them or even to illustrate them with explicit examples; see Further Reading.

The *Crooks relation* states that the outcome of a sequence of mechanical actions on a system in a heat bath is related to the outcome of the reversed sequence of actions. It is a connection between the probability density functions of the work $\Delta w$ performed on the system in the course of such *forward* and *backward* processes, $p_{\mathrm{F}}(\Delta w)$ and $p_{\mathrm{B}}(\Delta w)$, respectively. A forward process might consist of the movement of a piston to compress a gas in a cylinder, while the backward process would be the opposite movement to expand the gas. The validity of the Crooks relation requires that the system should start in canonical equilibrium at the same temperature $T$ for both processes. The relation reads

$$p_{\mathrm{B}}(-\Delta w) = p_{\mathrm{F}}(\Delta w)\exp\left(-\frac{(\Delta w-\Delta F)}{kT}\right), \tag{17.35}$$

where $\Delta F$ is the change in free energy of the system associated with the forward process, evaluated for example on the basis of an isothermal change in volume of the expanded gas.

According to (17.35), the probability $p_{\mathrm{F}}(\Delta w)$ that the forward process should require work $\Delta w$, and the probability $p_{\mathrm{B}}(-\Delta w)$ of receiving the same amount of work *from* the system during the backward process are related to each other. If the work done in the forward process is greater than $\Delta F$, which is overwhelmingly likely for the nonquasistatic processing of macroscopic systems, where we expect classical thermodynamics to prevail, then the probability of getting the work back again is exponentially small, according to (17.35). The Crooks relation is an expression of the vanishingly small likelihood of mechanical reversibility in the nonquasistatic thermodynamic limit. On the other hand, in a quasistatic process, we expect $\Delta w$ to equal $\Delta F$ for any path, in which case, the recovery of the work in the reversal of the process is guaranteed.

A more general result in a similar vein is the *detailed fluctuation relation*, or the equivalent *Evans–Searles fluctuation theorem*. In the context developed here, this concerns the entropy production brought about by forward and reverse processes, and reads

$$p_{\mathrm{B}}(-\Delta s_i) = p_{\mathrm{F}}(\Delta s_i) \exp\left(-\frac{\Delta s_i}{k}\right), \tag{17.36}$$

with the stipulation that the forward process followed by the backward process should return the pdf of the system to its initial form. While the Crooks relation determines the extent to which work put in can be taken out again, (17.36) quantifies the likelihood that a reverse operation on a system might negate the entropy generated in a forward process. Sketches of entropy production for a forward process and its backward version are shown in Figure 17.6. Clearly, the likelihood of observing events that leave the overall entropy unchanged is exponentially suppressed, unless the changes in $\Delta s_i$ for each operation are small compared with $k$, and this is itself very unlikely except for microscopic systems, or processes that are exceedingly slow.

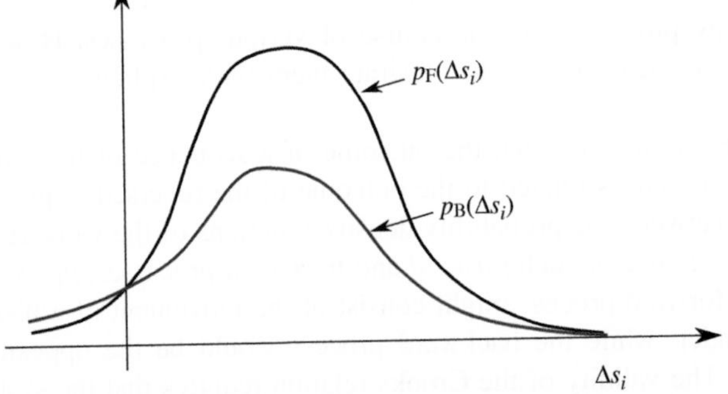


**Figure 17.6** Entropy production has a positive mean in a forward process and its backward counterpart. The probability that the backward process should negate the entropy generated in the forward process is exponentially small, according to the detailed fluctuation relation (17.36).

## 17.6 The Fundamental Basis of the Second Law

The fluctuation relations emerged from studies of the thermodynamic behaviour of small systems. Their principal value is that they tell us something about the statistics of work done or heat transferred in circumstances when there are significant fluctuations in system behaviour. But equally importantly they can provide us with an interpretation of entropy production in terms of the statistical reversibility of sequences of events. We have explored how this definition can be linked to established thermodynamic concepts using a simple example of an oscillator, but the concept can be applied more generally.

The probability of a particular sequence of events in systems modelled by random dynamics can be compared with the probability of the reverse sequence, under the reversal of external driving forces, if there are any. We define a quantity $\Delta s_i$ that is positive for the more likely of the two, and negative but of equal magnitude for the other. We average the quantity over the two sequences according to their relative likelihoods, and naturally the result is positive. We consider all possible sequences and their reversals, and the average remains positive. This is thermodynamic entropy production, at least within the framework of stochastic or random dynamics. It is an indication that the system has a preference to evolve in a certain manner rather than in the opposite direction.

But hold on. It is believed that real dynamics are fundamentally deterministic and *reversible* in the following sense. If all the atoms in the world had their velocities reversed, then the equations of motion would dictate that they retraced their steps: this would then guarantee the reversal of all the 'irreversible' phenomena observed up to that point. This observation was made by Josef Loschmidt (1821–1895) in response to Boltzmann's attempts to relate the second law to Newtonian mechanics. Matters are different in a model with stochastic dynamics: we typically find that forward and backward sequences of events have different likelihoods, and reversing the velocities does not imply that previous behaviour is retraced. An interpretation of entropy production based on ratios of path probabilities within a stochastic framework has given us some understanding of the second law. But we ought to accept that probabilistic dynamics is a model, where our lack of knowledge about, or lack of interest in the complete dynamics affecting a system are represented through the coarse graining and randomisation of some aspect of the interactions. In a sense, therefore, entropy and its production seem not to be as objectively real as some other system variables, but will depend on the way we choose to represent the behaviour. With this in mind, in Chapter 18 we once again address the question 'what is entropy?'.

### Exercises

**17.1** Show that (17.11) is a solution of (17.9) and that the solution of (17.12) is (17.13).

**17.2** Using the distribution of work $p(\Delta w)$, starting in equilibrium, for a step up in spring constant from $\kappa_0$ to $\kappa$ given in (17.34), determine the distribution for a step down from $\kappa$ to $\kappa_0$ and hence verify the Crooks relation (17.35).

# 18 Final Remarks

The main purpose of this book was to describe statistical models of the behaviour of physical systems that can account for the laws of classical thermodynamics. In particular, we needed to develop an interpretation of entropy that leaves as few questions unanswered as possible. The tools that emerged from this allow us to enquire into the likely microscopic or macroscopic behaviour of a system, principally when it is coupled to environments of various kinds. We have considered gases of particles at high and low temperatures and densities, with or without pairwise interactions; magnets; vacancies in crystals; harmonic oscillators; solids and electromagnetic radiation. Much more can be done with these tools, and there are many studies in the literature describing further applications, including the vitally important confrontation between models and experimental data.

The basis of statistical physics is the principle of equal a priori probabilities, which states that all microscopic configurations of an isolated system are equally likely to be realised when the system is in equilibrium. The system might be likened to a multiple-faced die that is thrown repeatedly: each face, or each microstate, is imagined to be equally likely to come up. The time-averaged properties of an isolated system at equilibrium are then equal to ensemble averages over all possible microstates.

The principle is simple and appealing, but it is hard to find an absolutely satisfactory justification for it. Perhaps the best argument is that it is sometimes appropriate to model parts of the world using stochastic dynamical rules, and that these must convey a system into its microstates with equal likelihood if there is no discernible reason why any should be favoured over the rest. While this view might be questioned, the conclusions that follow seem to be consistent with the equilibrium behaviour of complex physical systems.

The centrepiece of statistical and classical thermodynamics is the concept of entropy, a quantity that has intrigued generations of scientists, and caused no small amount of confusion. When equilibrium is disturbed by the lifting of a constraint, such as when a partition between boxes containing different gases is removed, an isolated system will undergo an increase in entropy: the famous second law of thermodynamics. In all processes, except for those that proceed extremely slowly, entropy is generated. The entropy of the universe tends towards a maximum. But what *is* entropy?

---

*Statistical Physics: An Entropic Approach*, First Edition. Ian Ford.

We have emphasised repeatedly that entropy need not be enveloped in mystery. It is a thermodynamic property of a physical system, obtainable from measurements of the heat capacity or from the equation of state. We have plotted its value against parameters such as temperature for a variety of cases. It would be quite ordinary were it not for the second law and the insistence that for an isolated system it can go up but never go down. This suggests that entropy is not a quantity that can be described in the same manner as particle positions and momenta, which are allowed to reverse their evolution. Determining an appropriate interpretation of entropy has been the most puzzling issue in the development of statistical physics.

The meaning of entropy that makes the most sense to me personally is that it expresses the uncertainty in the microscopic state of a system. Such uncertainty is an inevitable consequence of a failure to include all the details of the dynamics that describe the evolution of the system. This neglect might arise because the universe is very complicated, and we only wish to focus on certain features. So when studying the motion of a pendulum, we happily apply Newton's laws of motion for the bob, but are not inclined to consider the motion of the gas molecules that collide with the bob as it swings. Ignoring the molecules will neglect the fact that energy will leak away from the pendulum; so we have to do something or we cannot account for the fact that it will eventually appear to stop moving. One way is to represent the dissipation of energy in a random, uncertain way. It is at this level of description that a quantity can arise that does not retrace its steps. It is microscopic uncertainty itself, and entropy is its measure.

In short, if we know the present microscopic state of only parts of the universe, or equivalently know only some of the dynamical rules that control its evolution, then our certainty about the state of the universe will naturally decline in the future, and this is the second law.

Is entropy fundamental? Not in the same way as energy or particle number. It is an *emergent property* of a system, meaning that it shows up when we have to deal with a complex system in a coarse-grained way. A single particle or set of particles, with known coordinates, does not possess entropy until it is coupled to a coarsely specified environment for a period of time, whereby the uncertainty in the initial state of the environment leads to uncertainty in the state of the system. A system does not possess entropy if we know precisely where all the particles are, and how fast they are going. But typically we do not possess all the microscopic information about a complex system, and the missing information is essentially its entropy. It exists because we neglect details in our models: it is emergent. The most remarkable thing is that microscopic uncertainty can be measured with thermometers and pressure gauges.

So entropy *production* is a reflection of the loss of information (an increase in *missing* information) about the microscopic state of a system as time progresses, when we are able to follow the evolution only on a macroscopic, coarse-grained scale. The definition of entropy production given in Chapter 17 in terms of the probabilities of observing a certain sequence of events and its reverse has a particular resonance, since it connects directly with *reversibility*. The law of increasing entropy may perhaps be regarded as simply a shorthand, or a slogan, to describe the increase in uncertainty arising from the roughly modelled dynamics of complex systems. Because it is expressed in the form of a law, it can offer an intuitive understanding of many kinds of behaviour.

A line of enquiry that reveals the connection between entropy and information involves a character known as Maxwell's Demon and it would be remiss to leave him out of the discussion. In trying to understand the meaning of the second law, Maxwell imagined that a tiny creature could observe the motion of particles within a container and, by judicious use of a trapdoor in a partition placed across the middle, use this information to organise the particles between the two subvolumes just as he desired. For example, faster particles could be allowed to propagate into the left hand side and slower ones directed towards the right, producing a separation of a gas into hot and cold parts without doing work (assuming the trapdoor and the observation require none) which would be in violation of the second law!

The implications of these actions have been discussed at length. Maxwell's motivation was to show that the second law was statistical in nature: even if no intelligence was at work, a trapdoor flapping open and shut at random could conceivably produce such a separation, except it would be incredibly unlikely and extremely short-lived. But from the standpoint of equating entropy with uncertainty, the Demon is nothing more than an experimenter who makes a microscopic measurement and is thereby able to reduce his uncertainty about the world. He is acquiring data, or reducing the missing information, and is thereby able to narrow down the pdf of a few microscopic variables of the system. Even if he took no action, simply making the observation reduces the uncertainty in the system microstate, albeit temporarily, since the dynamics would presumably then proceed unobserved in a manner that is difficult to predict. I do a similar thing when I open a door briefly to find out what is happening on the other side. Opening the trapdoor to allow a fast particle to pass through preserves the reduction in missing information, and the result is a fall in the thermodynamic entropy of the gas.

The debate about the Demon revolves around how such events are consistent with the second law. One resolution suggests that the microscopic state of the Demon before the process is known, but after he has completed his task his state is less certain. He has sorted the gas into a state of reduced microscopic uncertainty, but from the point of view of an observer who does not have access to the microscopic state of the Demon, he has acquired some uncertainty himself. He is the repository of an unspecified stream of data about the sorting process; in short he has a memory. An analysis of how he might be returned to his initial state reveals that external work has to be dissipated as heat to the environment. An alternative resolution is that obtaining the initial microscopic information fundamentally requires the Demon to perform work, which is also to be dissipated as heat, with consequent entropy production. The debate is still ongoing.

Entropy has been related to uncertainty since the time of Gibbs, who represented it in terms of a probability distribution, which is a specification of uncertainty. The natural question is 'whose uncertainty?' My uncertainty might be different from yours. I might have made more measurements of system parameters. It is natural to be suspicious of a quantity when its value doesn't seem to be objective, but rather depends on how closely an individual, in the same manner as the Demon, decides to investigate and to measure. After all, entropy appears in thermodynamics alongside quantities that do have an objective reality, such as energy. No wonder entropy causes so much confusion!

But the suspicion is unfounded. We develop thermodynamics within a framework of a chosen set of macroscopic system variables, and the entropy that comes into the discussion expresses microscopic uncertainty within that framework. If we decide to make further measurements and identify, for example, a macroscopic parameter such as magnetisation, then we are required to define entropy within such a broadened framework making it a function of the new state variable as well as the old ones. As long as everyone is agreed on the level or the coarseness of the description, entropy is a well-defined measure of the remaining uncertainty, since it is to be calculated on the basis of measurements, such as heat capacities, made within that framework.

Of course, the framework might be insufficient; for example, if we neglect magnetisation, there will be physical effects that we cannot explain, and indeed phenomena that might even suggest that the entropy of an isolated system goes down. Such is our confidence in the second law that this usually means that the neglected macroscopic parameter needs to be incorporated into the thermodynamic framework to make everything work out satisfactorily.

The novelist and scientist C.P. Snow considered that entropy and the second law of thermodynamics deserve to be more widely appreciated by society. In his book *The Two Cultures and the Scientific Revolution*, he famously wrote

> 'A good many times I have been present at gatherings of people who, by the standards of the traditional culture, are thought highly educated and who have with considerable gusto been expressing their incredulity at the illiteracy of scientists. Once or twice I have been provoked and have asked the company how many of them could describe the second law of Thermodynamics. The response was cold: it was also negative. Yet I was asking something that is the scientific equivalent of: Have you read a work of Shakespeare's?'

Entropy and its increase are indeed culturally important because they present us with a deep perspective on events in the world. Every macroscopic thermodynamic process, indeed any macroscopic event we can conceive of generates entropy as a result of the natural tendency for energy to be shared equitably between interacting particles. This is equivalent to saying that every process is macroscopically irreversible: the sharing is unlikely to be undone. Such a view leads us to conclude that the universe is heading towards a state of complete entropy maximisation, known as the heat death, where matter and energy are uniformly spread out, where no change is perceptible on the macroscopic scale, and where 'time', or at least evolution, appears to have ended.

The future might be bleak, but at least it is straightforward to comprehend. It is rather more puzzling to understand why we presently exist in a relatively low entropy universe, such that the driving forces for changes on the macroscopic scale are not yet exhausted. The most compelling interpretation is that the low entropy is a remnant feature of the Big Bang. Since that event, the universe has undergone an expansion and cooling analogous to the free expansion of a gas, but luckily for us there is still more energy conversion and sharing to be done.

It will be clear by now that I prefer to view entropy as microscopic uncertainty, and not to rely too heavily on the traditional interpretation in terms of disorder. I find it hard to consider the universe to be disordered in a way that can easily be defined, but it certainly *is* disorderly, and evolving in a tremendously complex way on various spatial and temporal scales[1]. The proper understanding of such behaviour is to be found in the analysis of the effective equations of motion that apply at each scale, but common to

[1] One of the advantages of being disorderly is that one is constantly making exciting discoveries. – A.A. Milne

most of them is the property that the overall thermodynamic entropy increases with time. The intrinsic disorderliness can mix and share the constituents of the universe, exploring new modes of behaviour, with the consequence that solving the equations of motion is a difficult task. We can associate entropy with the uncertainty in our current perception of a physical system, and the second law with the inevitable increase of this uncertainty into the future if we are obliged to employ such effective equations of motion.

And finally, we should note that the universe is not simply following a programme of decay and decline. Organised structures, including life itself, have arisen from the original impetus of the Big Bang and from the rules of dynamics that have been in play. These have developed or are maintained in a manner that can be rationalised in terms of overall entropy increase. The impetus ought to run out eventually, of course, and it would seem that a gloomy future associated with the heat death does lie ahead, but the universe has nevertheless burned very, very brightly, and the second law is, in some sense, a celebration of this behaviour, and not just a sign warning us of impending doom.

# Further Reading

The following are undergraduate level textbooks that I find offer valuable guidance on various points.


Adkins C J 1983 *Equilibrium Thermodynamics*, 3rd edn., Cambridge.

Blundell S J and Blundell K M 2006 *Concepts in Thermal Physics*, Oxford University Press.

Bowley R and Sanchez M 1999 *Introductory Statistical Mechanics*, Oxford University Press.

Callen H B 1985 *Thermodynamics and an Introduction to Thermostatistics*, Wiley, New York.

Cowan B 2005 *Topics in Statistical Mechanics*, Imperial College Press.

de Podesta M 1996 *Understanding the Properties of Matter*, UCL Press.

Glazer M and Wark J 2001 *Statistical Mechanics: A Survival Guide*, Oxford University Press.

Guenault T 2007 *Statistical Physics*, Springer.

Hook J R and Hall H E 1991 *Solid State Physics*, Wiley.

Houghton J 2009 *Global Warming: The Complete Briefing*, 4th edn., Cambridge.

Kondepudi K 2008 *Introduction to Modern Thermodynamics*, Wiley.

Mandl F 1988 *Statistical Physics*, 2nd edn., Wiley.

Reif F 1965 *Fundamentals of Statistical and Thermal Physics*, McGraw-Hill.


The following make very interesting reading for historical context.


Cercignani C 1998 *Ludwig Boltzmann: The Man Who Trusted Atoms*, Oxford University Press.

Gibbs J W 1960 *Elementary Principles in Statistical Mechanics*, Dover Publications, New York, (originally published in 1902 by C. Scribner).

Jaynes E 2003 *Probability Theory: The Logic of Science*, Cambridge.


The following are at a more advanced level but relevant to Chapter 16.


Lemons D S 2002 *An Introduction to Stochastic Processes in Physics*, Johns Hopkins University Press.

Risken H 1992 *The Fokker–Planck Equation: Methods of Solutions and Applications*, Series in Synergetics, Springer.



*Statistical Physics: An Entropic Approach*, First Edition. Ian Ford.

The following are frontier level references for Chapter 17.

Evans DJ and Searles DJ 1994 Equilibrium microstates which generate second law violating steady states. *Phys. Rev. E* **50**, 1645–1648.

Crooks GE 1999 Entropy production fluctuation theorem and the nonequilibrium work relation for free energy differences. *Phys. Rev. E* **60**, 2721–2726.

Jarzynski C 1997 Nonequilibrium equality for free energy differences. *Phys. Rev. Lett.* **78**, 2690–2693.

Seifert U 2005 Entropy production along a stochastic trajectory and an integral fluctuation theorem. *Phys. Rev. Lett.* **95**, 040602.

Spinney RE and Ford IJ 2013 Fluctuation relations: a pedagogical overview. In *Nonequilibrium Statistical Physics of Small Systems: Fluctuation Relations and Beyond*, Wiley-VCH, Weinheim, ISBN 978-3-527-41094-1.

# Index



*Statistical Physics: An Entropic Approach*, First Edition. Ian Ford.

Printed and bound by CPI Group (UK) Ltd, Croydon, CR0 4YY

27/10/2024

14580303-0002